Food Chemistry: Sensory Analysis and Mechanisms

Food Chemistry: Sensory Analysis and Mechanisms

Edited by **Logan Bowman**

SYRAWOOD
PUBLISHING HOUSE

New York

Published by Syrawood Publishing House,
750 Third Avenue, 9th Floor,
New York, NY 10017, USA
www.syrawoodpublishinghouse.com

Food Chemistry: Sensory Analysis and Mechanisms
Edited by Logan Bowman

International Standard Book Number: 978-1-68286-087-8 (Hardback)

Contents

Preface

Taste analysis is a complex sensory procedure and many sensory mechanisms are involved in its perception. This book presents all the processes associated with sensory analysis of food in the most comprehensible manner. This book brings together detailed explanations of the various concepts like mechanisms of taste and flavor, evolution of taste organs, multi-sensory perception of flavor, influence of genetics in flavor chemistry and perception, etc. This book presents researches that have transformed this discipline and aided its advancement. Scientists and students actively engaged in this field will find this book full of crucial and unexplored concepts.

This book unites the global concepts and researches in an organized manner for a comprehensive understanding of the subject. It is a ripe text for all researchers, students, scientists or anyone else who is interested in acquiring a better knowledge of this dynamic field.

I extend my sincere thanks to the contributors for such eloquent research chapters. Finally, I thank my family for being a source of support and help.

Editor

Food aroma affects bite size

René A de Wijk[1,2*], Ilse A Polet[1,2], Wilbert Boek[3], Saskia Coenraad[3] and Johannes HF Bult[1,4]

Abstract

Background: To evaluate the effect of food aroma on bite size, a semisolid vanilla custard dessert was delivered repeatedly into the mouth of test subjects using a pump while various concentrations of cream aroma were presented retronasally to the nose. Termination of the pump, which determined bite size, was controlled by the subject via a push button. Over 30 trials with 10 subjects, the custard was presented randomly either without an aroma, or with aromas presented below or near the detection threshold.

Results: Results for ten subjects (four females and six males), aged between 26 and 50 years, indicated that aroma intensity affected the size of the corresponding bite as well as that of subsequent bites. Higher aroma intensities resulted in significantly smaller sizes.

Conclusions: These results suggest that bite size control during eating is a highly dynamic process affected by the sensations experienced during the current and previous bites.

Background

Eating and drinking serve to transfer food and drink from the mouth towards the throat before they enter the stomach and intestines. Before they enter these parts of the digestive system, foods are 'predigested'; that is, broken down in the mouth via mechanical and enzymatic degradation. The resulting fragments are mixed with saliva into a consistent bolus that is safe to swallow. The amount of food that is processed each time in the mouth (the bite size) is highly variable between consumers [1], between foods, and even within the same food when a single property such as viscosity is varied [2]. Solid foods that require more breaking down in the mouth typically result in smaller bite sizes than semisolid foods and liquids. Smaller bite sizes are known to elicit weaker food sensations [3], lower flavor release [4] and more satiation [5,6]. Furthermore, bite sizes tend to be smaller for unfamiliar foods and foods that are liked less [7]. Finally, bite sizes become smaller as the consumer becomes satiated [2]. These results suggest that bite size is actively regulated during eating in response to sensory and/or digestive factors.

Insight into the dynamics of bite size regulation may not only be relevant from a theoretical point of view, but may also assist in the development of foods that are

more satiating and are therefore consumed in smaller quantities. Given this objective, the use of aromas to affect bite size is particularly interesting since these can be manipulated without raising the caloric content of the food. We hypothesize that aroma exposure during eating affects bite size for the following two reasons. Firstly, perceived flavor intensity may be regulated adaptively by bite size to maintain moderate intensities; increasing flavor intensity would then elicit smaller bite sizes. Secondly, aromas that signal a creamy, fat-containing dairy product increase the perception of the product's creaminess and thickness on a bite-by-bite basis [8]. By presenting these aromas during a bite, the increase in the perceived creaminess and thickness is also expected to affect bite size. We hypothesized that higher aroma intensities would lead to smaller bite sizes and vice versa.

The present study investigated the dynamics of bite size control by presenting subjects with a series of bites of a semisolid food where the aroma released during oral processing varied from bite to bite. The results indicate whether aroma release can affect the current bite and/or subsequent bites. The formulation of an aroma or taste stimulus will not only affect the evaluation of that stimulus, but also the evaluation of subsequent stimuli [9,10]. If aromas affect bite size, these effects may, therefore, persist over multiple bites. To evaluate the effects of current and previous aroma stimulation

* Correspondence: rene.dewijk@wur.nl
[1]Top Institute Food and Nutrition, Wageningen, The Netherlands
Full list of author information is available at the end of the article

independently, we randomized and balanced subsequent aroma conditions [11,12].

Aroma variations in foods often affect other food properties as well due to physicochemical interactions, such as the food's viscosity [13-15]. Because these effects would hamper the interpretation of the results of this study, an olfactometer was used to present aromas independently of the food. The aroma was presented retronasally so that it follows the same route to the olfactory epithelium as aromas released by the food would follow during normal consumption [8,16].

Results

Bite size averaged 5.95 ml (± 2.1 SD) and did not vary consistently with bite number, indicating that intake-dependent processes such as satiation are not apparent during consumption (see Figure 1, gray bars). However, inspection of individual results showed considerable variation in bite sizes from one bite to the next (see Figure 1, black bars).

The results indicate that bite size decreases with aroma concentration, that is, the high aroma concentration resulted in significantly smaller bite sizes than the stimuli without aroma ($F(1,9) = 5.5$, $P < 0.05$). The low aroma concentration resulted in marginally smaller bite sizes than the stimuli without additional aroma ($F(1,9) = 3.0$, $P = 0.11$) (see Figure 2, aroma condition of current bite N-1 ≤ N ≤ 30, where N is the current bite and N-1 is the last bite). A similar effect was observed for the aroma condition of the previous bite (see Figure 2, aroma condition of bite N-1), however this effect was too small to be significant. The size of a bite is also affected by the aroma condition of bite N-2 (second to

last bite); if bite number N-2 has no aroma, the bite size of bite number N is smaller than when bite number N-2 has a weak aroma ($F(1,9) = 27.1$, $P = 0.01$) or a strong aroma ($F(1,9) = 6.1$, $P = 0.04$) (see Figure 3, aroma condition of bite N-2).

Discussion

Previous findings have demonstrated that bite sizes vary with how familiar the food is, and with its hedonic and textural properties. In addition the findings showed that smaller bite sizes are more satiating and that bite sizes become smaller as the consumer becomes more satiated. These studies indicate that bite size control is sensitive to general food properties as well as to the internal state of the consumer.

The results of the present study demonstrate that bite size control is sensitive to food sensations (aroma intensity) that vary from bite to bite, even at aroma concentrations below or near the perception threshold. In line with our hypothesis, the bite size was smallest for the highest aroma intensity. This result suggests a rapid feedback mechanism in which the aroma is perceived during the filling of the mouth, and where the outcome of this evaluation is used to terminate the bite. This feedback loop takes no more than a few seconds.

A likely reason for reducing the bite size when sensations become more intense is that consumers self-regulate their sensations via bite size, whereby weak sensations are intensified via larger bite sizes and stronger sensations are weakened via smaller bite sizes. This hypothesis is supported by other recent results from our laboratory that demonstrated that higher salt intensities were associated with smaller bite sizes in soups that were designed to be

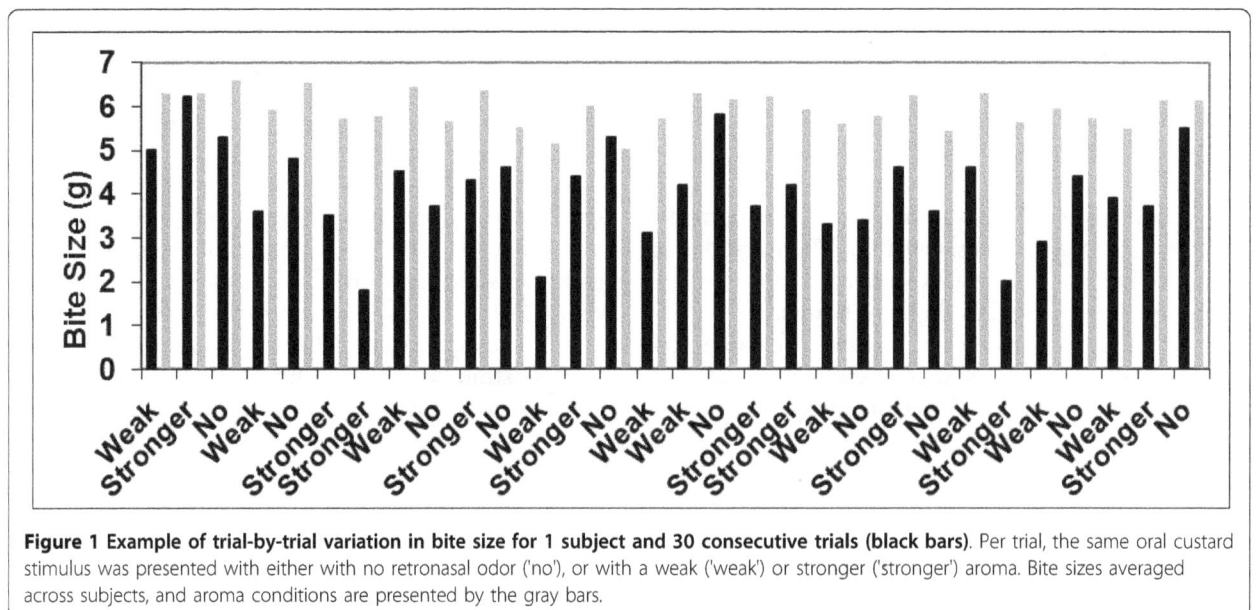

Figure 1 Example of trial-by-trial variation in bite size for 1 subject and 30 consecutive trials (black bars). Per trial, the same oral custard stimulus was presented with either with no retronasal odor ('no'), or with a weak ('weak') or stronger ('stronger') aroma. Bite sizes averaged across subjects, and aroma conditions are presented by the gray bars.

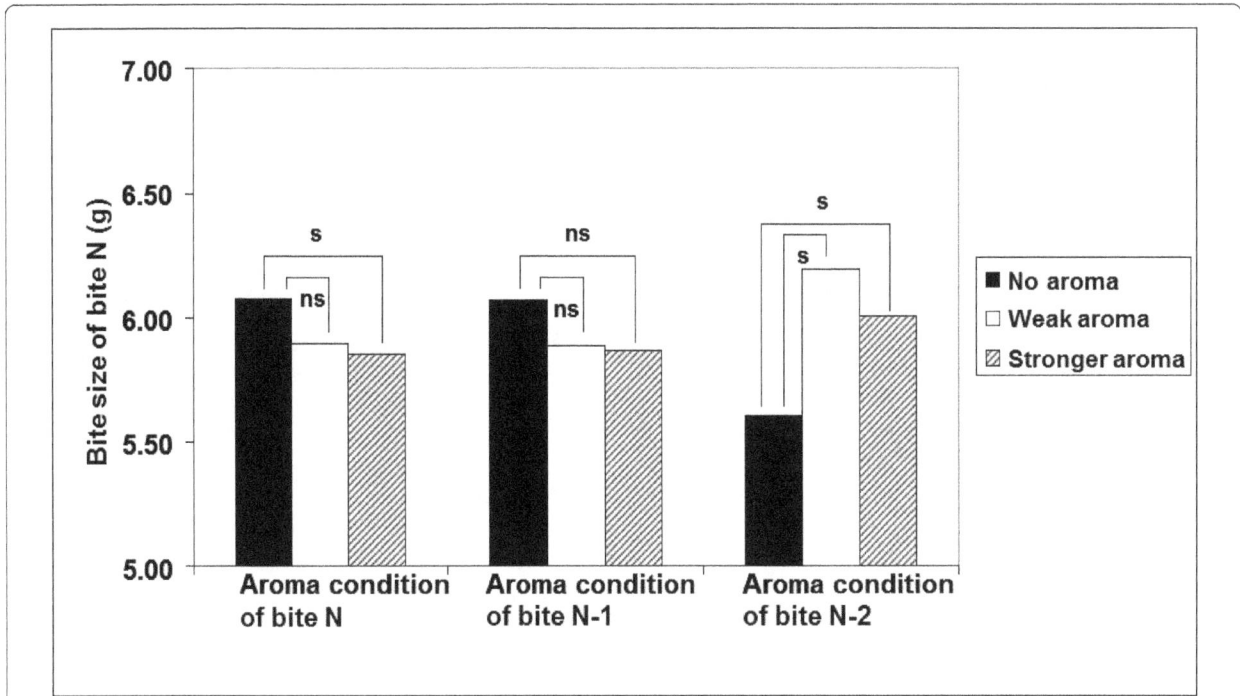

Figure 2 Bite size as a function of aroma condition of current bite (left), last bite (middle) and second to last bite (right). Significances of the differences are indicated in the figure (S = significant, P < 0.05; NS = non-significant).

equally pleasant to eat [17], The bite size reduction caused by taste was approximately 5%, that is, it was similar to the reduction found in the present study for smell. The intensity effect on bite size may not be limited to smell and taste but may also extend to texture sensations, as demonstrated by the fact that thicker textures resulted in smaller bite sizes than thinner ones [2].

Alternatively, stronger cream aromas make the custard seem thicker and more creamy (see [8]), and therefore possibly higher in calories.

The effect of aroma on bite size is not limited to the corresponding bite but also extends to subsequent bites. In fact, the bite size was most affected by the aroma conditions of the second-to-last bite. What is even more

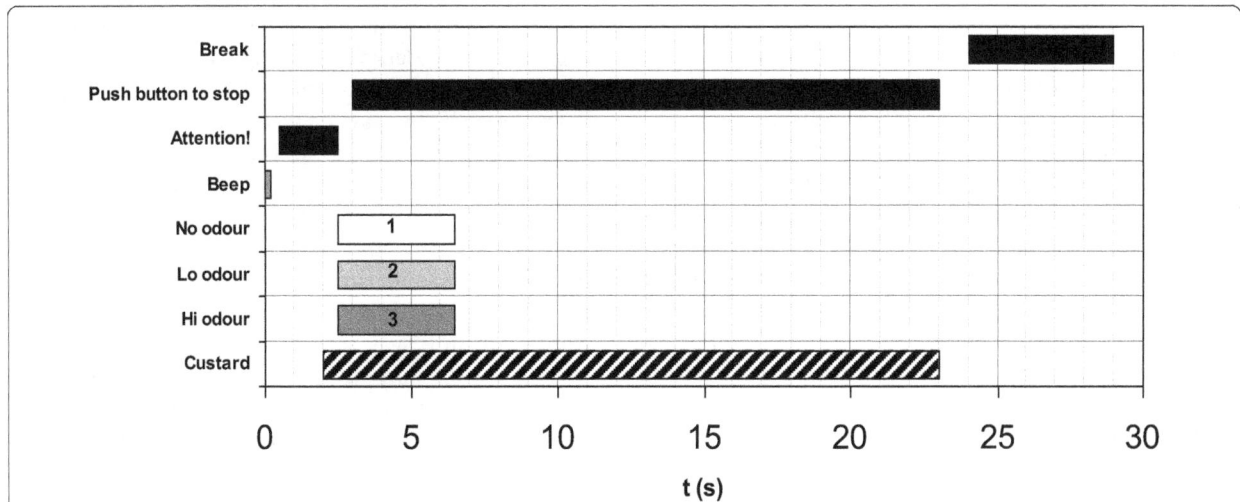

Figure 3 Sequence of events in each 30 s trial as function of time. Custard presentation at 60 g/min (striped bar, maximum 21 s, subject-terminated), screen instructions (black bars), a 400 Hz beep (gray bar) and one of three aroma conditions (no aroma (1), white bar; 10% aroma concentration (weak) (2); light gray bar; 100% aroma concentration (stronger) (3), dark gray bar).

interesting is that the effect of aroma on bite size from this bite is the opposite of the effect observed for the corresponding bite. This study design is not able to provide explanations for this unexpected reversal, but it demonstrates that bite size control is a complex process. One possible explanation, which would require further investigation, is that food intake may be approximately stable over multiple bites, whereby deviations on individual bites are compensated by subsequent bites.

The reduction in bite size as a result of aroma is relatively small, but may be relevant in normal consumption if it is not compensated by larger numbers of bites, that is, when the meal size is also decreased. In that case a reduction of approximately 5% to 10% in intake, as found in the present study based on the single-bite results, is already considerable. An additional study with normal eating is needed to verify this hypothesis.

The set up used in this study, that is, the combination of olfactometer and pump that was stopped by the subject, was developed specifically to investigate the effects of aroma on bite size. In normal eating, bite size is typically more or less determined by the utensil used. A spoon is used to eat a liquid or semisolid food, and a fork is used to eat a solid. During one bite, there is virtually no opportunity to adjust the bite size based on food sensations. Instead, adjustment of intake may take place during subsequent bites, or in the total number of bites, as suggested by a recent study in our laboratory that demonstrated that soups with a high salt content were consumed in fewer bites than low salt soups [18].

Bite size control may not be the only mechanism for self-regulation of food intake. Other possible mechanisms include the way foods are processed in the mouth. Previous findings have demonstrated that the intensity of food sensations varied directly with the way foods are processed. For example, creaminess becomes more intense when the food is not only compressed between tongue and palate but is also sheared via lateral movements of the tongue along the palate [19]. Weaker sensations may elicit more elaborative oral movements than stronger ones. Similarly, foods eliciting weaker sensations may be processed in the mouth for a longer period of time than foods eliciting stronger sensations. To the best of our knowledge, such mechanisms have not yet been studied.

Conclusions

In summary, increasing the aroma intensity reduces the bite size. This result fits into a growing body of literature that suggests that bite size control plays an important role in the self-regulation of food sensations.

Methods

Subjects

Ten subjects (six male, four female), aged between 26 and 50 years and with a normal sense of smell, participated in the experiment. Their average age was 36 years old (SD = 9), and their body mass index (BMI) ranged from 22 to 30.0 kg/m^2. The subjects did not have an aversion to vanilla custard dessert. All subjects had participated in earlier sensory experiments but were unaware of the aim of the present study. All subjects were paid for participation and gave informed consent. The research was approved by the Medical-Ethical Review Committee in Wageningen (File no. NL16918.081.07).

Food and food presentation

A commercially available fresh vanilla custard dessert (Friesche Vlag Halfvolle Vanille vla, 1.5% w/w fat) was used as the test food. A peristaltic pump (HR flow inducer, MHRE 200; Watson-Marlow, Falmouth, UK) with a silicon tube (length 1.5 m, inner diameter 10 mm) was used to deliver the product into the subject's mouth. The pump flow rate was fixed to control the eating rate (60 g/min). Each delivery of food was stopped by the subject via a push button, enabling the subject to control the size of the bite. Bite size was calculated by measuring the difference in weight of the food before and after each bite using a computerized weighing scale. Subjects were not aware of the weighing and received no visual cues as to the amount of food ingested.

Aroma and aroma presentation

At the beginning of each day of experimentation, 5 g of commercially available cream aroma (Butter Buds Asia; Butter Buds, Racine, WI, USA) was dissolved in 1000 g of demineralized water and shaken until fully dissolved. Aromas were presented via an apparatus (Olfactometer OM2s; Burghart Instruments, Wedel, Germany) that allowed application of chemical stimuli without causing concomitant mechanical or thermal sensations [20]. This was achieved by embedding a natural cream aroma in a constantly flowing air stream (8 l/min). The air saturation chamber of the olfactometer was filled with 10 ml of fresh aroma solution before each experimental session. The 'air'/'saturated odorized air' dilution of odorous stimuli was set to 16:1 for the weaker and stronger concentration. The weaker concentration had 10% of the strength of the stronger concentration, and was produced by presenting the stronger concentration for a 100 ms burst once every second. The temperature and humidity of the air stream were kept constant (36.5°C, > 80% relative humidity). The rise time of the odorant concentration was less than 20 ms.

The aromas were released into the epipharynx through a tube positioned under endoscopic control by a licensed doctor [21]. A tube of approximately 23 cm was cut from a sterile suction catheter made from soft polyvinyl chloride. This tube was placed inside the nose under endoscopic control so that the opening of the tube was at the level of the soft palate (approximately 8 cm from the naris). The other end of the tube was connected to the outlet of the olfactometer.

Food presentation, aroma presentation and screen instructions were controlled by a single computing unit running dedicated software written in Delphi [22]. The order of aroma/food combinations was randomized per subject, while keeping combinations of consecutive aroma conditions balanced.

Procedure

Subjects were seated in an upright position in a dentist's chair. The outlets of the olfactometer and pump were positioned in front of the mouth of the subject and mounted on a stable support. This allowed for a comfortable application of the mouthpeices and nosepieces. Food/aroma presentations were repeated every 30 s. All trials consisted of a 400 Hz auditory warning signal, screen instructions, ad lib custard presentation and concurrent retronasal aroma presentations at 0%, 10% or 100% of the concentrations prepared by the olfactometer. A delay of 500 ms between custard and aroma presentations was chosen to mimic the normal eating situation, where there is also a brief delay between the moment food enters the mouth and the start of aroma release (see Figure 3). The aroma concentrations were selected so that the strongest concentration was detected by approximately 50% of the subjects (that is, a perithreshold concentration), whereas the weakest concentration was virtually never detected (that is, a subthreshold concentration). The choice of these concentrations was based on a previous study that demonstrated that sensory integration of food and aromas takes place for relatively weak aromas that do not stand out relative to the flavor of the product [8].

Each of the three aroma concentrations (no aroma/low/high) was presented ten times to each subject in combination with the same oral stimulus. Hence, each subject was presented with 30 test stimuli, preceded by 3 practice stimuli that were not included in the analysis. Each session lasted 16.5 min. To balance consecutive aroma conditions, the 30 stimuli were presented in 10 blocks of 3 stimuli with different aroma concentrations in randomized order within each block. This design also minimizes the influence of day-to-day fluctuations in the subject's nutritional baseline, and of distractions or ambient factors that may interfere with the effect being measured, namely that of aroma on bite size. The total custard consumption did not exceed that of an average dessert. Analysis of how the bite size varies with the number of bites taken indicates whether processes related to bite number, such as satiety, take place.

Statistical analysis

The effects of aroma concentrations on bite size were analyzed using repeated measures analysis of variance (ANOVA; SPSS, SPSS Inc., Chicago, Il, USA) and intra-subject *post hoc* contrast tests to verify the significant effects of the aroma on the bite size of the current bite, and of the two subsequent bites. Data are presented as mean averages.

Acknowledgements

We thank Dr Mari Wigham for her help with editing the manuscript.

Author details

[1]Top Institute Food and Nutrition, Wageningen, The Netherlands. [2]Food & Biobased Research, Consumer Science & Intelligent Systems, Wageningen, The Netherlands. [3]Hospital 'De Gelderse Vallei', Ede, The Netherlands. [4]Nizo Food Research, Ede, The Netherlands.

Authors' contributions

RAdW conceived the idea for the study and drafted the manuscript. JFHB developed the necessary hardware and software for the study IAP participated in the design of the study and carried out the statistical analysis. WB and SC set up the study protocol, conducted the study, and prepared the data for further analyses. All authors were involved in developing the first draft of the manuscript into the final version suitable for publication.

Competing interests

The authors declare that they have no competing interests.

References

1. Prinz JF, de Wijk RA: **Perceptual effects of ingested volume in semi-solid foods.** *J Sens Studies* 2007, **22**:273-280.
2. de Wijk RA, Zijlstra N, Mars M, de Graaf C, Prinz JF: **The effects of food viscosity on bite size, bite effort and food intake.** *Physiol Behav* 2008, **95**:527-532.
3. de Wijk RA, Engelen L, Prinz JF, Weenen H: **The influence of bite size and multiple bites on oral texture sensations.** *J Sens Studies* 2003, **18**:423-435.
4. Linforth RST, Blissett A, Taylor AJ: **Differences in the effect of bolus weight on flavor release into the breath between low-fat and high-fat products.** *J Agr Food Chem* 2005, **53**:7217-7221.
5. Weijzen PLG, Liem GD, Zandstra LH, de Graaf C: **Sensory specific satiety and intake, The difference between nibble- and bar-size snacks.** *Appetite* 2008, **50**:435-442.
6. Zijlstra N, de Wijk RA, Mars M, MS , Stafleu A, de Graaf C: **Effect of bite size and oral processing time of a semi-solid food on satiation".** *The American Journal of Clinical Nutrition, Am J Clinical Nutrition* 2009, **90**:269-275.
7. De Wijk RA, Prinz JF, Polet IA, van Doorn RM: **Amount of ingested custard as affected by its color, smell, and texture.** *Physiol Behav* 2004, **82**:397-403.
8. Bult JHF, de Wijk RA, Hummel T: **Investigations on multimodal sensory integration, texture, taste, and ortho- and retro nasal olfactory stimuli in concert.** *Neurosci Lett* 2007, **411**:6-10.
9. Lawless HT: **A sequential contrast effect in odor perception.** *B Psychonomic Soc* 1991, **29**:317-319.
10. Schifferstein HNJ, Frijters JER: **Contextual and sequential effects on judgements of sweetness intensity.** *Percept Psychophys* 1992, **52**:243-255.
11. Jesteadt W, Luce RD, Green DM: **Sequential effects in judgments of loudness.** *J Exp Psychol Human* 1977, **3**:92-104.

12. DeCarlo LT, Cross DV: Sequential effects in magnitude scaling: models and theory. *J Exp Psychol Gen* 1990, **119**:375-396.

13. Nueslli J, Conde-Petit B, Trommsdorff UR, Escher F: **Influence of starch flavour interactions on rheological properties of low concentration starch systems.** *Carbohyd Polym* 1995, **28**:167-170.

14. Escher F, Nuessli J, Conde-Petit B: **Interactions of flavor compounds with starch in food processing.** In *Flavor Release.* Edited by: Roberts DD, Taylor AJ. Washington DC, USA: American Chemical Society; 2000:230-245.

15. de Wijk RA, Rasing F, Wilkinson CL: **Sensory flavor-texture interactions for custards.** *J Text Studies* 2003, **34**:131-146.

16. Ruijschop RM, Boelrijk AE, de Ru JA, de Graaf C, Westerterp-Plantenga MS: **Effects of retro-nasal aroma release on satiation.** *Br J Nutr* 2008, 99,1140-8.

17. Bolhuis DP, Lakemond CMM, de Wijk RA, Luning PA, de Graaf K: **Both longer oral sensory exposure to and higher intensity of saltiness decrease ad libitum food intake in healthy normal-weight men.** *Accepted for publication in Journal of Nutrition* .

18. Bolhuis DP, Lakemond CMM, de Wijk RA, Luning PA, de Graaf K: **Effect of salt intensity in soup on ad libitum intake and on subsequent food choice.** *Accepted for publication in Appetite* .

19. de Wijk RA, Prinz JF, Engelen L: **The role of intra-oral manipulation in the perception of sensory attributes.** *Appetite* 2003, **40**:1-7.

20. Kobal G, Hummel T: **Cerebral chemosensory evoked potentials elicited by chemical stimulation of the human olfactory and respiratory nasal mucosa.** *Electroencephal Clin Neurophysiol* 1988, **71**:241-250.

21. Heilmann S, Hummel T: **A new method for comparing orthonasal and retronasal olfaction.** *Behav Neurosci* 2004, **118**:412-419.

22. Borland: *Delphi* Scotts Valley, CA, USA: Borland Software Corporation; 2002.

Place-based taste: geography as a starting point for deliciousness

Joshua Evans[1], Roberto Flore[1], Jonas Astrup Pedersen[1] and Michael Bom Frøst[1,2]*

Abstract

Nordic Food Lab (NFL) is a non-profit, open-source organisation that investigates food diversity and deliciousness. We combine scientific and cultural approaches with culinary techniques from around the world to explore the edible potential of the Nordic region. We are intent on broadening our taste, generating and adapting practical ideas and methods for those who make food and those who enjoy eating. This paper describes some of our methods, using geography as a starting point for the exploration of deliciousness, exemplified in our lunch menu served at the Science of Taste symposium in Copenhagen in August 2014.

Keywords: Deliciousness, Geography, Food diversity, Food systems, Nordic region, Food design, Theoretical framework

Introduction

In November 2004, a symposium for Nordic cuisine was organised in Copenhagen at the then newly opened Nordatlantens Brygge, a cultural house for the North Atlantic parts of the Nordic region. Here a group of chefs and food professionals created a manifesto for a new Nordic cuisine that was signed by chefs from Denmark, Faroe Islands, Finland, Greenland, Norway, Sweden and Åland [1]. The symposium and manifesto crystallised a new Nordic food movement that has since developed the regional cuisines of the Nordic countries and territories beyond what anyone could have imagined.

Nordic Food Lab was founded in 2008 in the same spirit, as a research and development lab with the purpose of exploring food in the Nordic region. Chef René Redzepi and gastronomic entrepreneur Claus Meyer, co-owners of the restaurant Noma in Copenhagen, realised that this investigation could not be undertaken in the restaurant kitchen alone. They saw a need for a space where chefs, scientists, and other researchers could come together to investigate raw materials, traditional processes, and modern techniques more deeply than the pressure of daily service would allow. The outcome of the lab's activities was directed primarily towards the development of

restaurants, but also with the purpose of expanding knowledge in academic and applied contexts.

Since then, Nordic Food Lab (NFL) has helped to bring science and gastronomy closer together in Denmark [2]. Over the years, we have attempted to shift how chefs and scientists work together, from a simple one-way process of chefs asking scientists to help troubleshoot and solve immediate problems in the kitchen, to a more collaborative effort where research questions are developed and investigated together, integrating different methods and types of expertise. One good example is the work by Mouritsen et al. [3], which explored the use of seaweeds in a Nordic culinary context, and demonstrated how the seaweeds sugar kelp and in particular dulse have great potential as ingredients in the new Nordic cuisine to provide flavour and umami. The interests of the chefs and scientists are diverse and none are experts outside their respective fields, so a true collaborative work brings all parties further than any of them would have managed alone.

The experimental methods used at NFL often resemble those of a design studio with iterations of recipes and as thorough an exploration as possible of the sensory space a particular food can occupy [4]. For this reason, we rely on team members who are capable of dismantling the unnecessary division between science and craft, drawing on knowledge from natural sciences, the humanities and the vast world of diverse culinary traditions.

* Correspondence: mbf@nordicfoodlab.org
[1]Nordic Food Lab, Department of Food Science, University of Copenhagen, Rolighedsvej 30, DK-1958 Frederiksberg C, Denmark
[2]Sensory Science Group, Department of Food Science, University of Copenhagen, Rolighedsvej 30, DK-1958 Frederiksberg C, Denmark

Diversity is both our starting point and our goal. It forms a loop of feedback mediated by ecology, necessity, and appetite. There is no single food that can nourish us on its own. The pursuit of good food runs parallel with the pursuit of the biological and cultural diversity upon which truly sustainable food systems rely. Yet infinite choice can be paralysing, and we find creative and investigative freedom in the geographical constraint of our base of our raw materials.

Theoretical framework for deliciousness

In order to create delicious food, it is useful to understand the principles for perception of food and the evaluation of goodness in a food. Creating a new dish or finding a new ingredient to use in our cuisine bears similarities to how we interact with other artefacts of human culture. Looking to theories of human affective response to designed objects or artefacts can thus provide a useful perspective on how similar processes play out in the kitchen and laboratory. Desmet and Hekkert [5] argue that the affective response to a product is a function of three components. First is the immediate perception through our senses, what have previously been termed as the aesthetic experience [6]. Second is the experience of meaning that we ascribe through interpretation and association to assess the personal or symbolic significance of a product experience. The third component in our product experience is the emotional experience that arises from an evaluation of the significance that an experience has for the individual's well-being.

A theory for our interaction with food also needs to take into account the function that food serves for us, the relief of hunger, and the nutritional requirements of our bodies. Norman [7] has formulated that we interact with an object at three distinct levels: First, there is the visceral level, the immediate sensory level. It is how our perception is shaped through the hardwiring of our sensory systems. Second, there is the behavioural level, the function that our interaction with an object serves, such as the needs it satisfies for us. In relation to food, the functional level is the food safety and nutritional aspects, the absence of harmful substances or organisms and the provision of beneficial and necessary nutrients. Third, Norman [7] uses the term 'reflective level' to describe the overall impact a product or object has based on the meaning it gives to us, similar to the meaning level described above by Hekkert and Leder [5]. Figure 1 outlines our interpretation of the three levels of interaction with a food. Here we classify our interaction with food at three overall levels: immediately through our senses, the function the food has, and the reflections we have on the creation of the food.

Perceptual level

We appreciate certain tastes from birth (sweet, fatty, and umami [8]) because they signal the presence of available

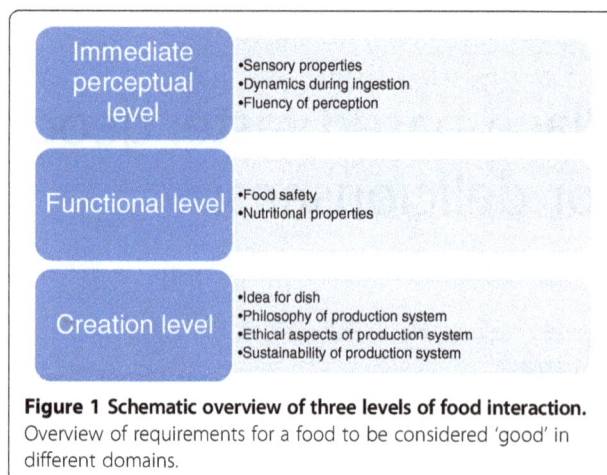

Figure 1 Schematic overview of three levels of food interaction. Overview of requirements for a food to be considered 'good' in different domains.

energy. Appreciation of other sensory properties such as crunchy [9] or creamy [10] is learned from positive consequences through conditioned learning and association [11]. Some sensory properties are more dynamic, and their appreciation is a result of the sensory arc that occurs during ingestion, as we chew and swallow. The main purpose of chewing is to comminute, lubricate and subsequently form food into a bolus that can be swallowed without negative consequences, such as inhalation of small particles into the lower respiratory tract [12]. The success of a food from an oral manipulation point of view depends on the efficiency of comminution, lubrication, and bolus formation. The trajectory of this process has been termed the philosophy of the breakdown path [13].

When we experience foods we implicitly learn some lawful relationship between different sensory properties. For example, we learn to associate the bright orange colour of sea buckthorn with its passion fruit-like aroma and its tangy sourness. After repeated exposures there is fluency in this learnt relationship, which generates intrinsic pleasure as a result of this faster perceptual processing [6]. Gradually, as we become more experienced, our sensory systems can better discern small differences and nuances that in earlier exposures went unnoticed. Gibson [14] suggests that the perceptual development and learning are processes of distinguishing the features of an almost inexhaustibly rich input, hinting at the immense potential to continually develop our senses further. Experienced wine connoisseurs, for example, may be able to distinguish minuscule differences in sensory properties that allow them to correctly identify the vineyard, producer and vintage of a wine and to take great pleasure in analysing and dissecting these sensory inputs of a food.

Functional level

The function of food from a physiological perspective should not be neglected, although it is something that is often taken for granted. Food needs to be safe to eat, i.e.

not cause disease. A good food serves the purpose of providing nourishment, and indeed, the range of intake that provides a person with sufficient macro- and micronutrients is broad. And though nutritional recommendations should be seen as guidelines that can form the basis for nutrition policies, or formulation of diets and foods [15], they are not the be-all and end-all of the complex functionality of food in diets in practice.

Creation level

In relation to food, the parallel to the reflective level or the meaning we ascribe to food is their creation — the production system that brings about the food, or the ideas behind a particular food or dish. A particularly good example of a food that is admired for its idea is Michel Bras' 'chocolat coulant', or chocolate cake with a runny heart that the chef invented in the early 1980s, which for many years has been a signature dish in his restaurant. According to chefs, it is one of the most copied recipes in the world. The ingenuity that was necessary for Michel Bras to develop this particular cake, with a complex preparation that according to legend includes short pieces of a garden hose and freezing the dough before baking, has made it appreciated by his diners for decades, and admired by chefs all over the world. It has helped build Michel Bras' reputation as one of the best chefs in the world (see for instance [16]). Similarly, the artist Olafur Eliasson expresses his admiration for René Redzepi's dish 'Milk skin with Grass', where the grass and the garnishes all originate from the same pasture as the cow that made the milk, and upon which it grazed on, a representation of a particular place at a particular time [17].

A significant part of the appreciation for a food can stem from how it has been created. Several organisations have developed guidelines for goodness in the production system according to their principles. The International Federation of Organic Agriculture Movements (IFOAM) has a set of four principles that form a base for interconnected ethical principles to guide the development of organic agriculture. The four principles are briefly put: health, ecology, fairness and care [18]. The Slow Food movement has a similar succinct statement for their manifesto for good food: good, clean and fair [19,20]. The principles for both these organisations can also be understood in terms of philosophy, ethics and sustainability, as indicated in Figure 1.

These three levels of interaction with a food—perceptual, functional, and creational—help us understand the underlying principles for delicious foods, and can offer explanations for why some foods are indeed delicious.

The menu

Food that excels in the three different domains at the same time is irresistible, as the goodness in the different domains act in synergy with each other. Our pursuit of deliciousness leads us to seek out the delicious potential in as many places and organisms as possible, and often, it is in the neglected, underutilised, forgotten and ignored raw materials that we discover and rediscover unique sources of deliciousness. Similarly, our interest in exploring culinary techniques from both our region and cultures across the world allows us to broaden the culinary potential of these raw materials, by tracing the connections between diverse traditions and translating existing knowledge into our regional context. Combining this biogeographical constraint for raw materials with an openness to all types of knowledge and technique is a starting point for cooking that says something about us and imbues the foods we eat with a connection to this place and this time.

For the Science of Taste symposium, our team developed a menu to both nourish the symposium participants and illustrate how food can be delicious in more than one way. The menu consisted of four dishes served in succession. Figure 2 shows a gallery of images of the different elements of the menu.

Beef heart tartare

We wanted to illustrate the particular qualities of (what are nowadays) underutilised parts of the animal. The heart is a continuously working muscle, which gives it a very different texture than skeletal muscles. Our hearts came from 1-year-old biodynamic calves from Østagergård in Jystrup, Denmark, which we minced while maintaining some structure of the meat. We seasoned the minced heart with black garlic, fresh tarragon, and fig leaf tincture. Black garlic is a product originating in East Asia, and is produced by keeping garlic in a warm environment with little airflow for around 60 days (we seal ours in vacuum bags and keep them at 60°C) [21]. This process denatures the alliinase enzyme responsible for transforming non-volatile alliin into volatile allicin, the pungent sulphurous compound in garlic, especially when its cells are ruptured. Moreover, the low but steady heat creates cascades of low-temperature Maillard reactions, although at a much slower rate than the Maillard reactions commonly experienced in cooking. The finished garlic is characterised by a deep black colour and complex caramelised fragrances.

The tarragon was grown biodynamically at Kiselgården in Ugerløse, Denmark, and provided the freshness to complement the dark richness and acidity of the black garlic.

The Danish island of Bornholm, between Sweden, Germany and Poland at the mouth of the Baltic Sea, has a unique microclimate along its southern coast: soft beaches of fine white sand and an exceptional warmth that lasts later into the fall than is characteristic of the region. This microclimate gives rise to a particular ecology, which includes a robust population of fig trees. In the summer, we made a tincture—a strong infusion of

Figure 2 Gallery of the different elements of the menu. Layout of the tables, crispbread, gin, Peas 'n' Bees, sourdough bread, tongue and koji-chovies, potatoes, cabbage, and koldskål.

high-proof ethanol, which has both gastronomic and medicinal applications—from some of these fig leaves, yielding a concentrated source of their characteristic aroma: part coconut and part coumarin (the sweet-smelling compound in tonka bean, woodruff, and sweet clover, among others). A small amount of tincture provided complex herbal top notes, binding the dish together.

We served the dish with a crispbread laminated with wild mugwort and beach roses, and a chilled shot of fragrant, woodsy gin from the island of Hven in the Øresund.

Peas 'n' Bees

This dish emerged from several sources of inspiration. In June 2014, some of our team visited the island of Livø in the Limfjord in northern Jutland to conduct fieldwork for our insect research. While on the island investigating the European cockchafer, we also obtained some fresh bee larvae from a local beekeeper, along with some very mature lovage stems from her garden. As part of an outdoor experimental cookout we steamed the delicate, fatty larvae inside the lovage stems along with jasmine flowers that at the time were riotously in bloom. The herbal and floral notes of the larvae were enhanced in this rustic and simple preparation, and we wanted to take it further in a more controlled context.

One of us (RF) was reminded of a traditional Italian dish that had a comeback in the 1970s called Risi e Bisi, or risotto with peas. The bee larvae visually reminded RF of the rice. The texture of the dish was enhanced with pearled barley boiled in lovage broth, to create a summery, room-temperature soup of creamed fresh peas and lovage, with some blanched bee larvae, fried bee larvae, fresh lovage, and fermented bee pollen to garnish.

Bee larvae are often a waste product of organic bee-keeping, as the drones are removed periodically through-out the summer months as a strategy to lower the Varroa mite population in the hive [22]. They also happen to be extremely nutritious—around 50% protein and 20% unsaturated fats—and their flavour, like honey, can vary according to the local flora and the time of year. All of this makes them a very exciting product to work with in the kitchen. The bee larvae we used in this dish were obtained from a beekeeper in Værløse, outside of Copenhagen, Denmark.

Along with this course, we served large sourdough loaves made with flour from Øland wheat, an old variety of wheat from the island of Øland in Sweden, and virgin butter—carefully cultured cream churned until just before the butterfat and buttermilk separate, yielding a foamy emulsion with a cloud-like texture and bright acidity.

Tongue and koji-chovies
Here again we wanted to showcase the delicious potential of another less-used cut. We cooked the tongues from the same calves (as used above) whole, sous vide for 4 h at 85°C with lots of aromatics. This was followed by 2 h more at 55°C, with butter added. Then, we sliced them and served them slightly warm with lots of fresh greens and herbs and a bright herb sauce. To go along with the tongue, we boiled some new potatoes and tossed them in an umami-rich sauce of koji-chovies (herring fermented in the style of anchovies [23]) and halved pointy cabbage we had grilled and compressed with shio-koji (a mixture of koji, salt, and water, with powerful enzymatic activity) to break it down and bring out its natural sweetness. Both the koji-chovies and shio-koji are excellent examples of translation of technique from other culinary traditions, taking our love of cured anchovies and applying it to a common small fish of the Nordic region, for example, or using the versatility of koji, grain fermented with the fungus *Aspergillus oryzae*, to enhance our fermentation techniques and other processes [24]. The koji, made mainly on rice in East Asia, produces amylases which saccharify the starches allowing the substrate to be further fermented into alcohol (as is the case with sake, or rice wine), along with proteases and

lipases which can be further used to break down proteins into amino acids and fats into fatty acids. The enzymatic breakdown of proteins is the main mechanism that gives rise to umami taste in many products, such as soy sauce, miso, and their analogues around East and South-east Asia.

With the main course we served a juice made from Danish apples and seasoned lightly with juniper berries.

Koldskål
We finished with our take on a classic Danish summertime dessert—koldskål. It is a buttermilk soup with a base of egg yolk, traditionally aromatised with lemon zest and vanilla, and served with small cookies called 'kammerjunkere' and sometimes with fresh strawberries. In this version we opted for a more herbal profile, infusing the soup with lemon verbena, and serving with a mixture of freeze-dried lingonberries, raspberries and cranberries, and homemade kammerjunkere topped with lemon thyme sugar.

As this dish was served, we sprayed a finely misted tincture of birch buds over each table, a beautifully resinous and enveloping aroma from this underused part of the tree that conjures up forests of this most Nordic of trees.

We offered this variation on a beloved Danish classic to share the delicious Danish summer with our Danish and international guests alike.

Competing interests
The authors declare that they have no competing interests.

Authors' contributions
Josh Evans contributed to the writing and editing of the manuscript, and contributed to the development, creation and execution of the menu. Roberto Flore led the development, creation and execution of the menu, and reviewed the manuscript. Jonas Astrup Pedersen contributed to the development, creation and execution of the menu, and reviewed the manuscript. Michael Bom Frøst contributed to the writing and editing of the manuscript. All authors read and approved the final manuscript.

Authors' information
JE is the lead researcher at Nordic Food Lab, and has a background in the humanities. JE has worked extensively for the last years on food systems and sustainable agriculture. RF is the head chef at Nordic Food Lab. A trained chef, he has focussed in his career on building strong relationships between producers and chefs. JAP is a researcher and product developer at Nordic Food Lab, and has a background in food science, coupled with a longstanding interest in the culinary arts and the restaurant trade. MBF is the director for Nordic Food Lab and associate professor in Sensory Science at University of Copenhagen. He has a background in sensory science, and has worked extensively to connect science and culinary arts to the benefit of both.

Acknowledgements
This work has been funded by the following research and dissemination projects hosted at Nordic Food Lab: Smag for Livet (Taste for Life), a centre without walls that has focus on the taste of foods. (Smag for Livet is funded by the Nordea Foundation), and Discerning taste: creating the gastronomic argument for entomophagy, funded by the Velux Foundation.

References

1. Ny Nordisk Mad: Köksmanifestet. [http://nynordiskmad.org/om-nnm-ii/koeksmanifestet/]
2. Risbo J, Mouritsen OG, Frøst MB, Evans JD, Reade B: **Culinary science in Denmark: molecular gastronomy and beyond.** *J Culin Sci Technol* 2013, **11**:111–130.
3. Mouritsen OG, Williams L, Bjerregaard R, Duelund L: **Seaweeds for umami flavour in the New Nordic Cuisine.** *Flavour* 2012, **1**:4.
4. Evans JD: **Non-Trivial Pursuit - New approaches to Nordic deliciousness.** *Anthropol Food* 2012, **S7**. http://aof.revues.org/7262
5. Desmet P, Hekkert P: **Framework of product experience.** *Int J Des* 2007, **1**:57–66.
6. Hekkert P, Leder H, Aesthetics P: **Product aesthetics.** In *Product Experience.* Edited by Schifferstein HNJ, Hekkert P. Amsterdam: Elsevier; 2008:259–285.
7. Norman DA: *Emotional Design - Why We Love (or Hate) Everyday Things.* New York, NY: Basic Books; 2004:257.
8. Ventura AK, Worobey J: **Early influences on the development of food preferences.** *Curr Biol* 2013, **23**:R401–R408.
9. Szczesniak AS: **Texture is a sensory property.** *Food Qual Prefer* 2002, **13**:215–225.
10. Frøst MB, Janhøj T: **Understanding creaminess.** *Int Dairy J* 2007, **17**:1298–1311.
11. Prescott J: **Chemosensory learning and flavour: perception, preference and intake.** *Physiol Behav* 2012, **107**:553–559.
12. Prinz JF, Lucas PW: **An optimization model for mastication and swallowing in mammals.** *Proc Biol Sci* 1997, **264**:1715–1721.
13. Hutchings JB, Lillford PJ: **The perception of food texture - the philosophy of the breakdown path.** *J Texture Stud* 1988, **19**:103–115.
14. Gibson JJ: *The Senses Considered as Perceptual Systems.* Boston: Houghton Mifflin; 1966.
15. Nordic Council of Ministers (Ed): *Nordic Nutrition Recommendations 2012: Integrating Nutrition and Physical Activity.* 5th edition. Copenhagen: Nordisk Ministerråd; 2012:627.
16. Bras M: **Vivre la cuisine.** In *MAD food Symposium.* 2011 [http://www.madfood.co/michel-bras/]
17. Eliasson O: **Milk skin with grass.** In *Noma - Time and Place in Nordic Cuisine.* Edited by Redzepi R. London: Phaidon; 2010:6–9.
18. **Principles of Organic Agriculture.** [http://www.ifoam.org/en/organic-landmarks/principles-organic-agriculture]
19. **Slowfood - our philosophy.** [http://www.slowfood.com/international/2/our-philosophy]
20. Petrini C: *Slow Food Nation.* New York: Rizzoli Ex Libris; 2007.
21. Evans JD: **Black garlic.** [http://nordicfoodlab.org/blog/2013/2/black-garlic]
22. Evans JD: **The real reason we remove drone brood.** [http://nordicfoodlab.org/blog/2012/12/the-real-reason-we-remove-drone-brood]
23. Reade B: **Koji-chovy.** [http://nordicfoodlab.org/blog/2013/6/koji-chovy]
24. Evans JD: **Koji – history and process.** [http://nordicfoodlab.org/blog/2013/8/koji-history-and-process]

"We only eat what we like" or do we still?

Georges M Halpern

Abstract

We humans only eat what we like, and we died when we could not find or were not given such food. The industry knows that well in affluent societies, and that is why (in part) we do have an epidemic of obesity. Ignoring the basic foundations of physiology (and survival) in the name of "science" perverted into "faith" is the perfect recipe for (criminal) failure! Eating/drinking is one of our basic needs; the others being sex, shelter, family/social support and skills. This did work pretty well in the pre- and early industrial age, but with industrialization of the food supply (agriculture, etc.), based on only limitless profit, we witnessed a tectonic perversion in politics, policies, physiopathology, epigenetics, and ultimately public health. The current quasi-unanimous attitude is to blame the victim (for example, the obese) and/or the messenger (for example, maybe this author).

Keywords: Food, Pleasure, Commensality, Industry, Perversion

When one thinks about food and oral intakes, and intends to share passion sprinkled with knowledge, ego and memories jump to the page. I discovered the pleasure of food when I started missing it during World War II in refugee camps in Switzerland. We were fed the same unappetizing gruel twice daily, and that lasted almost 18 months. Around me, older refugees (>65 years old), formerly rich and hedonistic, discussed recipes from Auguste Escoffier, Ali-Bab, or Curnonsky; these ladies refused to swallow the brew they were served. They said: "We are not pigs. We are civilized, educated human beings with a palate, a taste, a culture". And they died, of self-inflicted starvation. They would only have accepted to eat what they liked. For these months and many, too many years after I thought of food every minute, every second; I dreamt of meals; I woke up craving for food, for fat cheeses, for aromatic sausages, for fresh-out-of-the-oven breads. I was obsessed. It never stopped. We know that we shall always like, love the food we liked, loved before the age of six. The foods, the dishes, the cuisines we pretend to like or even love later in life are very few. They are always judged against the enamored dishes we shared in our first childhood. I did not get these, or they were erased. I discovered the tastes of foods in 1947 in Denmark. Coming from France with its rationed 0%-fat Camemberts, Marshall-plan maize bread I had landed in the Land of Fabulous Foods: smoked eels. Danish blue cheeses, endless charcuteries, real milk, soooo many

breads, the freshest fish and seafood, legs of lamb, and the pastries that were served to the King... YES there was FOOD, endless, diverse, bringing happiness and joy –and health. It started my quest; it never ended; it never will.

What do we like in food? The list is long and open-ended. We obviously follow our senses: sight, smell, taste, texture, diverse sounds. We get messages from our genes, groomed for millennia. We do cherish memories and they rush back fast, very fast. We position food in its current environment, at times pleasantissime, at other times intolerable, but always in context. We eat stuff because we were told, or attracted/lured, brainwashed, coerced, by imitation or begging for inclusion, or even by challenge or curiosity. We also eat because it's time and we are hungry, because we need food if we drink, because of salt, or chili peppers, or gluttony, or..... Sometimes we are satisfied, or full, or happy, or frustrated. There's little rationality, no real rules, no foolproof recipe. It is you, now, then, there, in a given environment; and it will never be the same twice. Food is needed to live and survive. Some of our genes have been honed to perform in a hostile environment, the one that was the lot of the >90% until the twentieth century. Sugar was introduced in the 1500s; fat was synonymous of feast until the 1900s; food was seasonal, and preserving it was difficult, random, expensive. In Warsaw, Poland (where I was born), ice from the Wisła (Vistula) was sliced in March, kept in caves and distributed over the summer and fall until the 1960s. Now, with concentration of production and mass distribution, any failure in the chain

Correspondence: drgeorges@drgeorges.net

The Hong Kong Polytechnic University, Hung Hom, Hong Kong SAR results in decimating food poisoning. The world has changed: Taylorism and industrial production, assisted by expert sociologists, marketers, psychologists, physiologists, nutritionists and other food scientists have managed to globalize food before any other human activity. When we know that the advertizing budget of CocaCola® is bigger than the GNP (gross national product (GNP)) of >100 countries of the United Nations, I feel humbled in composing this Op-Ed. What does the industry sell, why and how? It sells sweet and fat; it sells it at the lowest cost (= mass produced) and maximal profit; it sells it everywhere with minimal differences; it sells it by bribing policy makers, and buying the whole chain of distribution; it sells it with some of the smartest scientists that consult or work there for transient illusion of glory and fortune; it sells it like the tobacco industry sells by targeting children and women [1]. And it works: in 2011, Nestlé was listed No. 1 in the Fortune Global 500 as the world's most profitable corporation; with a market capitalization of $ 200 billion, Nestlé ranked No. 13 in the FT Global 2011 [2]. It has recently added Jenny Craig to its empire, the Jenny Craig that sells meals *plus 2 snacks* including an *Anytime Bar* that packs 110 empty calories in the daily ration of the US customer. The bucket of Kentucky Fried Chicken will deliver ~3,000 kilocalories, mostly fats and carbohydrates, and I know many who eat that by themselves! I wrote that *"the only natural thing in a Diet Coke® is the water"* -if you accept that tap water (for example, Bonaqua®) is strictly natural; the revenue of the Coca Cola Company was US$46.5 billion. Responsiveness to sugars and sweetness has very ancient evolutionary beginnings. Newborn human infants also demonstrate preferences for high sugar concentrations and prefer solutions that are sweeter than lactose, found in breast milk [3]; it also controls pain in preemies and newborns [4,5]. Sweetness appears to have the highest taste recognition threshold, being detectable at ~1 part in 200 of sucrose in solution. Sweetness intensity indicates energy density. The 'sweet tooth' thus has an ancient evolutionary heritage, and while food processing has changed consumption patterns, human physiology remains largely unchanged [3]. Then, recently two discoveries changed our food supply: a by-product of corn/maize, the high-fructose corn syrup (HFCS), and the artificial sweeteners. The HFCS is composed of a mixture of 42 to 55% fructose, 41 to 45% glucose and 0 to 5% glucose polymers depending on the specific blend. In the 1980s it mostly replaced sucrose as the main sweetener of soft drinks. Rates of obesity subsequently rose, paralleling an increase in the consumption of soft drinks in general. In addition, laboratory research suggests a link between consuming large amounts of fructose and various health problems e.g. high blood pressure, elevated blood triglycerides, size and type of low-density lipoproteins, and uric acid levels [6,7]. Worse, HFCS is addictive [8]. The

most elaborate theory of sweetness to date is the *multipoint attachment theory* proposed by Jean-Marie Tinti and Claude Nofre in 1991. This theory involves a total of eight interaction sites between a sweetener and the sweetness receptor, although not all sweeteners interact with all eight sites [9]. This model has successfully helped to develop highly potent sweeteners, including the guanidine family with lugduname, about 225,000 times sweeter than sucrose [3]. But plasma beta-endorphin concentrations were more elevated after an aspartame drink than after the sucrose drink or fasting, and insulin increased after drinking as much with aspartame as with sucrose, meaning that possible addiction and obesity were right there with the sweetener [10]. The problem is that the palatability and enjoyment of foods are often tied to their energy density, and therefore fat content. Energy-dense foods that are rich in fat are more palatable than are many low-energy-density vegetables and fruit. High-fat foods, many containing sugar or salt, have an undeniable sensory appeal and are difficult to resist. There are many explanations for why humans like fat. Several physiological mechanisms have been proposed, many of which are based on the strong links found between fat content, palatability, satiety, and energy density. The orosensory properties of fat or fat "taste" are perceived through specific receptors and a combination of taste, texture, and olfaction. My friend Marian Apfelbaum, MD, a great nutritionist, starts his lectures on fat (and diet) by whispering into the microphone: *"Fat tastes gooood; fat is gooood"*.

Indeed fat tastes and makes food taste very good. Fat is also a concentrated source of energy with rewarding post-ingestive effects. The learning of food preferences may be based on associating sensory attributes with the physiologic consequences of ingestion, such as satiety and well-being [11]. That is why the combination of sweetness and fat in fast or junk food is difficult to resist, and is eventually "as addictive as heroin" [12].

Getting a shot of opioids and dopamine to the brain while on the go, snacking on a dark penis-shaped bar full of industrial fats, and guzzling a supersized HFCS-laden drink is the city dweller's fate in many, and more countries. Then they get glued to the television that brainwashes with seductive ads promoting these products and that lifestyle. These "foods" and beverages are very smartly designed and created to appeal to our *nucleus accumbens*, and hedonic hunger [13].

Meals were and are shared; they are communion, conviviality, commensality (Claude Fischler's great neologism). We eat, not nutrients and calories, but foods, dishes, meals and we do so in specific places, at specific times and with specific people with whom we have interactions and relationships. Eating is not just individual behavior; it also consists of social practices and rituals [14]. In most if not all societies on the planet, eating is done in a social context. Individualization and, as it were, privatization of eating in plethoric societies may carry more liabilities than benefits while there may be long unsuspected benefits associated with the sharing of food in the common meal [14]. Meals are the heart and the hearth of all human groups, from the original *soup*, bread soaked in broth, to the 3-day wedding feasts of Bosnia or India. Eating, sharing together is the quintessential basic human activity –with sex. It has been transmogrified a few decades ago in "feeding" individuals with processed chemicals and flavors, blessed by some nutritionists whose horizon is the lab bench. In none of the 83 countries (and counting) where I have worked is the degradation of the human bonding worse than in the United States; bromides and the omnipresent empty message of "family values" are what politicians and policy makers utter, in a country where one in four children eats alone (watching commercials on junk food); where food stamps do not deliver the healthy foods poor families so badly need, but too often some amongst the worst processed ones; where our daughter Emilie was the only student in her 4th grade class to draw a *real* chicken (beak, feathers and all) while the others sketched a frozen, packaged miserable bird; where most young adults will never know as *fish* anything but Mrs Paul's® fish fillets; where you better not drop a "vine" calibrated tomato on your toe (it will hurt!); where standard sliced sandwich bread has the texture and blandness of a Kleenex® tissue; and where people get lonesome, angry, hopeless, joyless, because they are deprived of sharing the crusty fragrant bread -the meal that we all need. My rant could stop here, but it would barely graze the surface of very complex interactions, many of them have come to light recently, for example the microbiome [15]. I cannot think of any mass-promoted diet that could possibly be beneficial to simply overweight people. These are either chastising –and rapidly abandoned-, or unaffordable, impractical. It is not because these diets are all *bad*; some of them are in fact pretty good, but they address a moment in time, just a given moment in the life of human, social individual, with a complex mixture of pasts, presents and visions of the future. They are not tailored, and revised, adjusted, modified, according to a myriad of interferences. How could they? And they ignore *pleasure*. Pleasure is not an "extra", or bonus bringing a little more soul to certain of our acts; it is a fundamental part of our animal life. It is just as difficult to define as spirit, but nonetheless humans are very conscious of it [16]. Pleasure is a potent drive, inducing forms of behavior adapted to physiological needs, for example temperature regulation and food-andfluid intake; sensory pleasure is an incentive to useful behavior, and maximization of pleasure the answer to physiological conflicts, also known as stress [17]. "The pleasures of the table are for every man, of every land, and no matter of what place in history or society; they can be a part of all other

pleasures and they last the longest, to console us when we have outlived the rest." [18].

Bon Appétit!

Abbreviation

HFCS, High fructose corn syrup.

Competing interests

The author declares that he currently has no competing interests.

Acknowledgements

The author thanks Claude Fischler, PhD for his comments, corrections and suggestions, and the editors of *Flavour* for inviting him to write this Op-Ed; it is obviously subject to changes, inherent to the nature of the subject, and the endless flow of scientific evidence that can only be compared to the tip of an iceberg.

References

1. Dorfman L, Cheyne A, Friedman LC, Wadud A, Gottlieb M: **Soda and tobacco industry corporate social responsibility campaigns: how do they compare?** *PLoS Med* 2012, **9**(6):e1001241. doi:10.1371/journal.pmed.1001241.
2. http://en.wikipedia.org/wiki/Nestl%C3%A9#cite_note-9.
3. *Sweetness* http://en.wikipedia.org/wiki/Sweetness.
4. Mitchell A, Waltman PA: **Oral sucrose and pain relief for preterm infants.** *Pain Manag Nurs* 2003, **4**:62–69.
5. Slater R, Cornelissen L, Fabrizzi L, Patten D, Yoxen J, Worley A, Boyd S, Meek J, Fitzgerald M: **Oral sucrose as an analgesic drug for procedural pain in newborn infants: a randomised controlled trial.** *Lancet* 2010, **376**:1225–1232.
6. http://en.wikipedia.org/wiki/High-fructose_corn_syrup_and_health.
7. Stanhope KL, Havel PJ: **Fructose consumption: considerations for future research on its effects on adipose distribution, lipid metabolism and insulin sensitivity in humans.** *J Nutr* 2009, **139**:1236S–1241S.
8. Lustig RH: **Fructose: metabolic, hedonic, and societal parallels with ethanol.** *J Am Diet Assoc* 2010, **110**:1307–1321.
9. Hayes JE: **Transdisciplinary perspectives on sweetness.** *Chemosensory Percept* 2008, **1**:48–57.
10. Melchior JC, Rigaud D, Colas-Linhart N, Petiet A, Girard A, Apfelbaum M: **Immunoreactive beta-endorphin increases after an aspartame chocolate drink in healthy human subjects.** *Physiol Behav* 1991, **50**:941–944.
11. Drewnowski A, Almiron-Roig E: **Human Perceptions and Preferences for Fat-Rich Foods.** In *Fat Detection: Taste, Texture, and Post Ingestive Effects.* Edited by Montmayeur JP, le Coutre J. Boca Raton FL: CRC Press; 2010. Chapter 11. Frontiers in Neuroscience.
12. Johnson PM, Kenny PJ: **Dopamine D2 receptors in addiction-like reward dysfunction and compulsive eating in obese rats.** *Nat Neurosci* 2010, **13**:635–641.
13. Lowe MR, Butryn ML: **Hedonic hunger: a new dimension of appetite?** *Physiol Behav* 2007, **91**:432–439.
14. Fischler C: **The nutritional cacophony may be detrimental to your health.** *Progr Nutr* 2011, **13**:217–221.
15. Krajmalnik-Brown R, Ilhan ZE, Kang DW, DiBaise JK: **Effects of gut microbes on nutrient absorption and energy regulation.** *Nutr Clin Pract* 2012, **27**:201. doi:10.1177/088453311436116.
16. Vincent JD: **Biology of pleasure.** *Presse Med* 1994, **23**:18711873.
17. Cabanac M: **Preferring the pleasure.** *Am J Clin Nutr* 1985, **42**:1151–1155.
18. Brillat-Savarin J-A: *Physiologie du Goût ou Méditations de Gastronomie Transcendante. Ouvrage théorique, historique et à l'ordre du jour, dédié aux Gastronomes parisiens, par un Professeur, membre de plusieurs sociétés littéraires et savantes.* Paris: Chez A. Sautelet et Cie. Libraires; 1826.

Looking for crossmodal correspondences between classical music and fine wine

Charles Spence[1*], Liana Richards[2], Emma Kjellin[2], Anna-Maria Huhnt[2], Victoria Daskal[3], Alexandra Scheybeler[3], Carlos Velasco[1] and Ophelia Deroy[4]

Abstract

Background: Wine writers sometimes compare wines to pieces of music, a particular musical style or artist, or even to specific musical parameters. To date, though, it is unclear whether such comparisons merely reflect the idiosyncratic matches of the writers concerned or whether instead they reflect more general crossmodal matching tendencies that would also be shared by others (e.g., social drinkers). In our first experiment, we looked for any consensual patterns of crossmodal matching across a group of 24 participants who were presented with four distinctive wines to taste. In our second experiment, three of the wines were presented with and without music and 26 participants were asked to rate the perceived sweetness, acidity, alcohol level, fruitiness, tannin level, and their own enjoyment of the wines.

Results: The results of experiment 1 revealed the existence of a significant agreement amongst the participants in terms of specific classical music - fine wine pairings that appeared to go particularly well (or badly) together. For example, Tchaikovsky's String Quartet No 1 in D major turned out to be a very good match for the Château Margaux 2004 (red wine). Meanwhile, Mozart's Flute Quartet in D major, K285 was found to be a good match for the Pouilly Fumé (white wine). The results of experiment 2 revealed that participants perceived the wine as tasting sweeter and enjoyed the experience more while listening to the matching music than while tasting the wine in silence.

Conclusions: Taken together, the results of the two experiments reported here suggest that people (social drinkers) share a number of crossmodal associations when it comes to pairing wines and music. Furthermore, listening to the appropriate classical music can enhance the overall experience associated with drinking wine. As such, our findings provide prima facie evidence to support the claim that comparing a wine to a particular style of music (as documented in the work of a number of wine writers) might provide the social drinker with useful clues about the sensory properties that they should expect to perceive in a wine should they eventually get to taste it.

Keywords: Crossmodal correspondences, Descriptors, Metaphor, Wine, Classical music

Background

'I have tasted first-attempt Chardonnays that were like Dizzy Gillespie's solos: all over the place. And the color of his trumpet, too. On the other hand a Stony Hill Chardonnay recently had the subtle harmonies and lilting vitality of Bix Beiderbecke. Robert Mondavi's Reserve Cabernets are Duke Ellington numbers: massed talent in full cry. Benny Goodman is a Riesling from Joseph Phelps, Louis Martini's wines have the charm and good manners of Glenn Miller. Joe Heitz, though, is surely Armstrong at the Sunset Café; virtuoso, perverse and glorious.' ([1], p. 253).

This quote is just one amongst many that illustrates the way in which wine writers sometimes resort to describing particular wines in terms of specific musical parameters, pieces of music, or in terms of the musical styles of specific artists. Take, for another example, the following quote:

'Taking a sip of wine, at least a wine worth talking about, is like hearing the sound of a sustained, musical chord.' ([2], p. 27).

* Correspondence: charles.spence@psy.ox.ac.uk
[1]Crossmodal Research Laboratory, Department of Experimental Psychology, University of Oxford, South Parks Road, OX1 3UD Oxford, England
Full list of author information is available at the end of the article

On occasion, comparisons are also made in the opposite direction, with pieces of music being related to wines:

'...it's hard to think of music that is more transparently effervescent than Steve Reich's *Octet and Music for Large Ensemble*. Both have textural aspects strongly reminiscent of Champagne. Bouncing along optimistically, motives advance and recede like the frothy mousse of a freshly poured glass: bubbles forming and popping with little explosive jolts, instantly replaced by others. I've listened to these pieces hundreds of times, and full of perpetual motion, they sparkle along and never sound quite the same.' ([3], pp. 122–123).

Going one step further, some writers have suggested an even closer connection between wine and music[a] [1].

'Red wines need either minor key or they need music that has negative emotion. They don't like happy music...Cabernets like angry music.' ([4], cited in [5]).

The tendency to draw comparisons with experiences from another sensory modality in order to express something that may be difficult to capture in the original sensory domain is a well-attested fact (see [6], for a review). But in the case of wine and music, one can ask whether crossmodal associations reflect anything more than merely the idiosyncratic matches of individual writers? Do they also mean something to the social drinker? If the latter interpretation is correct, the suggestion that emerges is that these metaphors or analogies could perhaps convey some useful information about the sensory qualities of the wines so described.[b] The hope here is that there are certain sensory attributes of a wine that really do share a perceptual affinity with particular musical parameters, and that this might be what the wine writers are picking up on intuitively. However, to date, there has been no empirical test of the claim that such musical associations are shared between individuals within a society. The primary aim of the present study was therefore to determine whether there would be any agreement in a sample of social drinkers concerning which of a pre-selected range of classical music pieces they would select as best matching a selection of four different wines that had been chosen to display a range of distinctive sensory characteristics.[c] [d] In addition, we also investigated whether listening to the appropriate (that is, matching) classical music would affect the experience of drinking wine.

A large part of what people typically describe as the taste or flavor of a wine actually comes from the nose. In fact, some say as much as 80% of flavor is derived from the sense of smell (for example, Dr. Susan Schiffman quoted in the *Chicago Tribune*, 3 May 1990; [7]). Hence, the first thing to check here is whether there is a general crossmodal matching of musical parameters with specific attributes of wines based on their aroma (detected either orthonasally or retronasally; see [8]).[e] In this regard, it is interesting to note that we often describe certain aromas/fragrances in terms of high notes (for example, lychee), while describing others as low notes instead (see [9] for a review). Interestingly, while the use of the same terms - high and low notes - to describe both tones and aromas is widespread and longstanding (compare with [10], for an early example), we still do not have any clear understanding of the underlying reason for this (see [9] on this point).

Crisinel and Spence [11] have conducted a study that was designed to investigate whether participants would reliably match certain of the aromas that are commonly found in wines with particular classes of instrument and/or musical notes (see also [12], for early research in this area). The participants in Crisinel and Spence's study were presented with a selection of aromas from the *Nez du Vin* kit (Brizard & Co, Dorchester, UK); 20 out of the 54 samples from the kit were selected, comprising: almond, apple, apricot, blackberry, caramel, cedar, dark chocolate, cut hay, green pepper, honey, lemon, liquorice, mushroom, musk, pepper, pineapple, raspberry, smoked, vanilla, and violet. The participants had to sniff each aroma orthonasally (from a small bottle) and then match it with one of 13 sustained notes (each note was presented for 1,500 ms), from C2 (64.4 Hz) to C6 (1,046.5 Hz) in intervals of two tones, on a virtual keyboard presented on a computer. By interacting with the digital interface, the participants were able to listen to each note being played by one of four different classes of musical instrument (piano, brass, strings, or wind). Thus, the participants had 52 different sounds (13 notes × 4 instruments) to choose from when selecting a sound to match each of the odors they were presented with.

Crisinel and Spence's [11] results (see Figure 1) revealed that their participants picked notes and instruments in a non-random manner. In particular, the sound of a piano was chosen as a particularly good match for the fruity aromas of apricot, blackberry, and raspberry, as well as for the smell of vanilla (note that these are all in some sense sweet smells; [13]). The sound of a woodwind instrument was chosen as a good match for the smell of apricot, raspberry, and vanilla. By contrast, the rather less pleasant smell of musk was matched with the synthetic sound of a brass instrument instead. In terms of the crossmodal matching of wine aromas to pitch (see Figure 2), it is clear that participants picked much lower pitched sounds as corresponding with the smell of

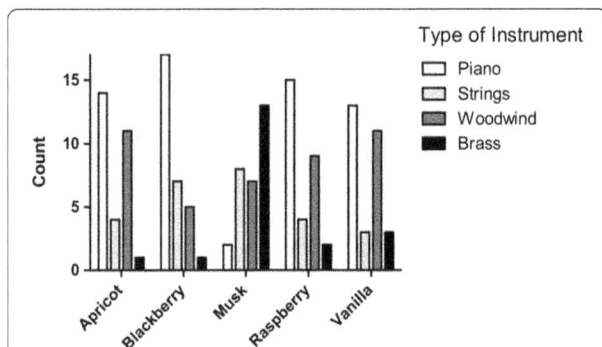

Figure 1 Choice of instrument as a function of the odor presented in Crisinel and Spence's [8] recent study in which the participants matched each of 20 typical wine aromas to a specific class of musical instrument. Only those odors that led to significant preference for a particular class of musical instrument are shown. Note that the total count per category is 30. (Figure reprinted with permission from [8]).

smoke, musk, dark chocolate and cut hay, while generally associating the fruitier aromas (apple, lemon, apricot, raspberry, pineapple, and blackberry) with a higher pitched notes instead.[f]

Over the last few years, a number of other researchers have also investigated the tendency of people to crossmodally match basic tastes (for example, sweet, sour, bitter, and salty) with specific musical parameters (for example, [14-16]; see also [17]). The results of this research have tended to converge on the conclusion that people match sour-tasting foodstuffs with sounds that have a higher pitch while matching bitter-tasting foods and beverages with sounds having a lower pitch (see

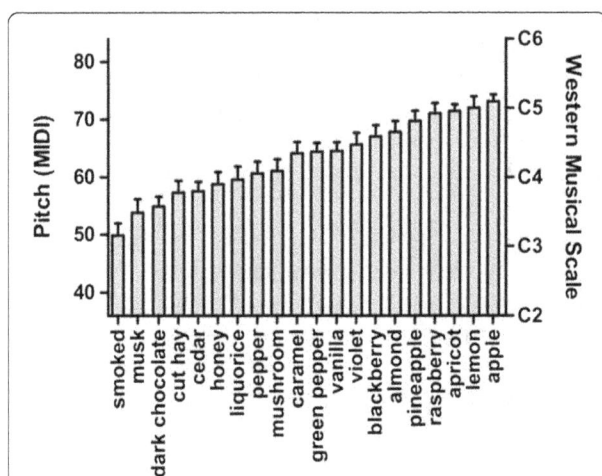

Figure 2 Mean pitch matched to each odor in Crisinel and Spence's [8] study. MIDI (musical instrument digital interface) note numbers were used to code the pitch of the chosen notes. Western musical scale notation is shown on the right-hand y-axis. High-pitched notes were preferred for fruits. (Figure reprinted with permission from [8]).

[18], for a review). The story, as regards sweet tastes, is somewhat more complex, for while people often match them with sounds having a high pitch (for example, [19]) in other studies they have been shown to associate them with lower- pitched sounds (see [17]). In terms of the more complex parameters of music and their association with basic tastes, the available research suggests that 'sour' musical improvisations ought to be high-pitched and dissonant; 'sweet' improvisations are consonant, slow and soft; 'bitter' improvisations are low-pitched and legato; 'salty' improvisations are staccato and dense in wide musical intervals (that is, successive notes of very different pitches). The research that has been published to date clearly demonstrates that certain styles of musical improvisation can reliably connote basic tastes, and possibly also flavors (see [20]).

Here, we present two experiments: In the first, we built on Crisinel and Spence's [11] research documenting the existence of a variety of crossmodal correspondences between specific notes and instrument types and individual aromas, as well as extending the growing body of research showing a crossmodal match between musical parameters and basic tastes and flavors (for example, [18-20]) in order to determine whether any crossmodal correspondences could be documented at the level of more complex pieces of classical music and a selection of quality wines.[g] Should such crossmodal correspondences be obtained in a group of social drinkers then these results would at the very least provide some *prima facie* evidence that when wine writers associate particular wines and specific pieces, or styles, of music then they might be picking up on the crossmodal correspondences that are common to us all ([21]). Note that when assessing the responses of participants in this study, there are no objectively correct answers in terms of which piece of music matches with a specific wine. Instead, what we are looking for is some 'consensuality' or uniformity in terms of the choices that people make. Any such consensuality would be expected to show up in terms of patterns of matching that deviate from what one might expect by chance. In our second experiment, we move on to assess any influence that the musical pieces may have in the perception of a variety of wine attributes; namely sweetness, acidity, alcohol level, fruitiness, tannin level, and the taster's enjoyment of the experience when compared to the ratings of the same wine made in silence.

Experiment 1
Methods

Ethics approval: the experiment was reviewed and approved by the Central University Research Ethics Committee of the University of Oxford, and complied with the Helsinki Declaration.

Participants: 24 participants (15 females, 8 males, and 1 who failed to specify their gender); age (mean ± SD) 34.7 ± 14.5 years, range 24 to 42 years) verbally agreed to participate in the study after the experimental procedure had been explained.

Venue: the study was conducted at the premises of The Antique Wine Company (AWC) Wine Academy in central London. All of the participants were seated in a professional wine tasting room. Each participant was seated at a table and had four pre-poured numbered glasses of wine laid out in front of them. Around 2 cm of each wine was poured into standard tasting glasses. No refills were provided and hence the participants were repeatedly told to conserve their samples for the duration of the experimental test (which lasted for about 45 minutes, and which informal observation revealed that participants were able to do). When the participants arrived at the AWC premises they were held in the reception until everyone had arrived. They were then led into the tasting room as a group.

Wine selection: four wines were chosen for crossmodal matching in the present study. They were selected to present a distinctive array of sensory characteristics (including acidity, fruit, tannins, and sweetness) that previous research suggested could perhaps be matched to distinctive musical attributes. More specifically, the wines (together with the relevant tasting notes provided by the AWC were:

1) Demonstrating balanced acidity: wine: Domaine Didier Dagueneau, Pouilly Fumé Silex 2010. Grape: Sauvignon Blanc. Tasting notes: very fine, pure and sophisticated. Clean with razor sharp acidity which is balanced by fresh blackcurrant leaf flavors.

2) Demonstrating purity of fruit: wine: Domaine Ponsot, Clos de la Roche 2009. Grape: Pinot Noir. Tasting notes: luscious and fruit packed with heaps of fresh cherry, spice, earth and game on the nose. Well-structured with layers of rich, naturally sweet red fruits and mouth coating flavors that explode on the formidably long finish. This is a classy wine with absolutely superb complexity, impeccable balance and almost uncanny presence, all delivered with grace and power.

3) Demonstrating supple tannins: wine: Château Margaux 2004. Grape: Cabernet Sauvignon. Tasting notes: a classic Bordeaux vintage. There is great finesse and above all purity on the nose; in this subtle combination of floral, fruit and spice aromas, all are clearly present but no single aroma dominates. On the palate, the tannic structure is tight-knit, fine, and tender. The general impression is of balance, precision, freshness, and grace.

4) Demonstrating pleasurable sweetness: wine: Château Climens Sauternes 2001. Grape: Semillon. Tasting

notes: attractive nose with hazelnuts, vanilla and apricot. Lots of dried orange peel and honey in this excellent sweet wine. Dense and very sweet on the palate with bright acidity and wonderful purity of fruit. Soft, well-balanced and very elegant long finish.

Musical selections

The London Symphony Orchestra (LSO) selected the following range of pieces of classical music - all involving some combination of wind and string instruments: this constraint in the musical selections was dictated by the fact that this research was conducted as a precursor to a much larger private event in which the plan was that a quartet of wind and string musicians from the LSO would play a selection of musical pieces to accompany the selection of wines tasted in the present study. The final event did indeed take place in LSO St. Luke's, London on 24th October, 2013 in front of an audience of more than 100 people, and with a quartet from the LSO playing the musical selections that had been identified as best-matching by the present research.

1) Mozart, Flute Quartet in D major, K285 - Movement 1, Allegro. This piece of music is melodic, lively, consonant, several themes, a sonata form movement. The flute has concerto-like prominence.

2) Tchaikovsky, String Quartet No 1 in D major - Movement 2, Andante cantabile. This piece of music is poignant and has a melancholic melody based on a folk tune (it moved Tolstoy to tears, apparently).

3) Ravel, String Quartet in F major - Movement 1, Allegro moderato, très doux. This piece has a sonorous, calm first subject followed by a haunting second theme.

4) Debussy, *Syrinx*. Exotic and meandering. Major and minor intervals create a minor dissonant feel which then moves into a major sound and changes between the two. This piece is for solo flute and woodwind.

5) Ravel, String Quartet in F major - Movement 2, Assez vif. This piece of music is playful, dramatic, and has pizzicato features.

6) Mozart, Flute Quartet in D major, K285 - Movement 2, Adagio. This piece of music is slow, melancholic, with pizzicato strings.

7) Tchaikovsky, String Quartet No 1 in D major - Movement 3, Scherzo, allegro non tanto e non fuoco. This piece of music is bold and lively and in a minor key, but skips to a lively dance-like rhythm.

8) CPE Bach, Solo Sonata in A minor - Movement 2, Allegro. Fast, upbeat, and virtuosic. This is a piece for solo flute, woodwind.

Procedure: the participants were presented with the four glasses of wine in a purpose built tasting room.

They were first given a five to ten minute introduction to wine tasting, and the various attributes that they should look out for in a wine. The participants were given a small sample of rosé to practice tasting with. The participants then spent ten minutes filling in a one page questionnaire concerning each wine (see Figure 3; these results were not analyzed). This questionnaire was designed to encourage the participants to think about some of the pertinent sensory qualities of each of the wines. The eight pieces of music were then played, from a portable hi-fi from loudspeakers at a comfortable listening level, in the order listed above. Each piece was presented for approximately three minutes. During that time, the participants were instructed to taste each of the wines and assign it a score from 0 to 10 depending on how well they thought the wine matched the music (with 0 = not at all, and 10 = a perfect match) using the form shown in Figure 4. Thus, the maximum score that any combination of wine and music could achieve was 240 (that is, 24 participants all giving the pairing a maximum rating of 10).

Wine No __:

What pitch would you associate with this wine?

Very low → Very high

Would you associate a slow or fast piece of music with this wine?

Very low → Very high

What musical instrument would you associate to this wine?

Circle (one of): Flute, Violin, Cello, Viola

Rate the following attributes of the wine (put a mark anywhere on the scale)

Ripe red fruits:

Not at all → Very

Vanilla:

Not at all → Very

Apricot:

Not at all → Very

Tannins:

Not at all → → Very

Figure 3 Questionnaire that participants completed (one sheet for each wine) at the start of experiment 1. Note that a C2 (64.4 Hz) note was played in order to anchor the participants' judgment of a low note and a C6 (1,046.5 Hz) note was played in order to anchor their judgment of what was meant by a high note.

The participants were were discouraged from comparing notes. They were instructed to remain silent, and were instructed not to go back and change their answers after going to the next played music. After each wine, the participants had a glass of water to cleanse their palate should they so desire. After having rated each wine for its match with each of the eight pieces of music, the experimenter debriefed the participants concerning the purposes of the study, and summarized the results of previous research on the pairing of music and wine (see [8], for a review). The participants were also encouraged to ask any questions that they might have. The whole session lasted for about 90 minutes.

Results

The numerical values given by the 24 participants were analyzed using a repeated measures analysis of variance (ANOVA) with the factors of wine (four levels) and music (eight levels). This analysis (using the Greenhouse Geisser correction) revealed a significant interaction (F (8.401, 193.212) = 11.429, $P < .001$) between the factors showing that participants' ratings of the match between the music and wines were non-random (see Figure 5).

A closer analysis of the data revealed that participants judged certain of the wines to be a particularly good (or bad) match for specific pieces of the classical music that had been chosen for use in the present study. Separate repeated measures ANOVAs were then performed in order to assess any differences in the matching ratings between musical pieces for each wine and the results of these analyses are presented below.

Domaine Didier Dagueneau, Pouilly Fumé Silex 2010

A significant difference between the ratings for the different musical pieces was observed ($F(7, 161) = 12.116$, $P < .001$). In particular, pairwise comparisons (with the Bonferroni correction) revealed that participants rated Mozart's Flute Quartet in D major, K285 - Movement 1, to be a much better match for this white wine, than Tchaikovsky's String Quartet No 1 in D major - Movement 2 ($P < .001$), Ravel's String Quartet in F major - Movement 1 ($P = 018$), and Tchaikovsky's String Quartet No 1 in D major - Movement 3 ($P = .017$). The participants also rated Debussy's *Syrinx* ($P < .001$), Ravel's String Quartet in F major - Movement 2 ($P < .001$), Mozart's Flute Quartet in D major, K285 - Movement 2 ($P < .001$), and CPE Bach's Solo Sonata in A minor ($P < .001$), as offering a better match with this wine than Tchaikovsky's String Quartet No 1 in D major - Movement 2. CPE Bach's Solo Sonata in A minor was also rated as more congruent with the Pouilly Fumé than Tchaikovsky's String Quartet No 1 in D major- Movement 3 ($P = .018$).

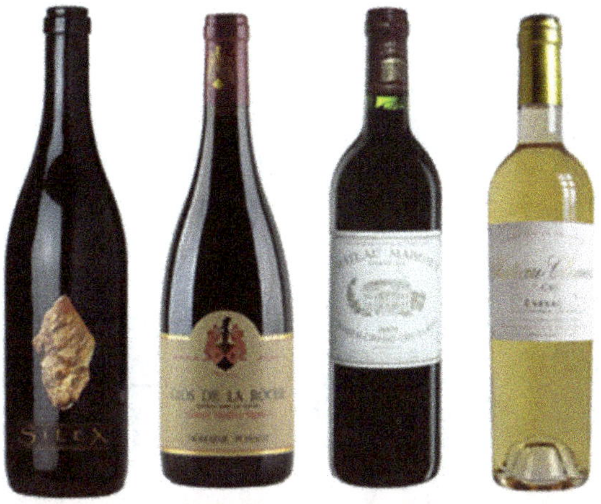

	Wine1	Wine2	Wine3	Wine4
Piece 1				
Piece 2				
Piece 3				
Piece 4				
Piece 5				
Piece 6				
Piece 7				
Piece 8				

Please give each piece a score from 1-10 for how well it suits each wine, with 10 being best suited, and 1 being least well suited. Please add a score to each box

Figure 4 The questionnaire used to assess the degree of match between the four wines and each of the eight pieces of classical music tested in experiment 1. The participants were instructed to give a score between 0 (no match) and 10 (a perfect match) in each cell to indicate how well they felt that the music matched the wine.

Domaine Ponsot, Clos de la Roche 2009

Once again, a significant difference between the ratings for the various musical pieces was observed ($F(7, 161) = 6.691$, $P < .001$). Pairwise comparisons revealed that the participants rated Ravel's String Quartet in F major as offering a better musical match for this wine than Mozart's Flute Quartet in D major, K285 - Movement 1 ($P < .001$), Debussy's *Syrinx* ($P < .001$), and Mozart's Flute Quartet in D major, K285 - Movement 2 ($P = .002$). Additionally, Tchaikovsky's String Quartet No 1 in D major - Movement 2 was rated as a better match for the Clos de la Roche than was Mozart's Flute Quartet in D major, K825 - Movement 1 ($P = .043$). Finally, Tchaikovsky's String Quartet No 1 in D - Movement 3 was rated as more congruent with this red than was Debussy's *Syrinx* ($P = .022$).

Château Margaux 2004

A significant difference between the ratings for the various musical pieces was obtained ($F(7, 161) = 19.710$, $P < .001$). Pairwise comparisons revealed that participants rated Tchaikovsky's String Quartet No 1 in D major - Movement 2 as more congruent with red wine, than any of the other musical pieces ($P < .05$ for all comparisons) except Tchaikovsky's String Quartet No 1 in D major - Movement 3, which was rated as more congruent with the Château Margaux than any of the remaining musical pieces ($P < .05$), with the exception of Tchaikovsky's String Quartet No 1 in D major - Movement 2, and Ravel's String Quartet in F major. Additionally, Ravel's String Quartet in F major was rated as more congruent with the Château Margaux, than either Mozart's Flute Quartet in D major, K285 - Movement 1 ($P = .004$), or CPE Bach's Solo Sonata in A minor ($P = .044$).

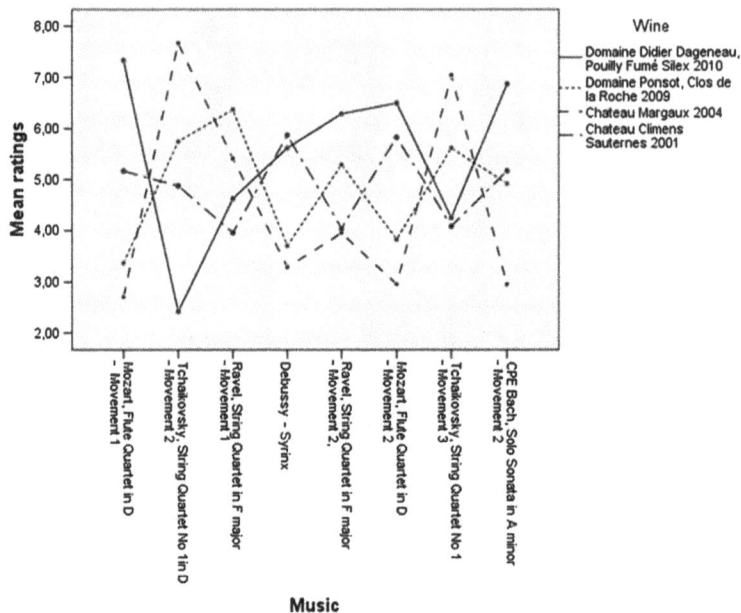

Figure 5 Summary of results from the classical music wine-tasting matching study (experiment 1). The ratings were obtained by asking the participants to rate from 0 to 10 how well they thought the wine matched the music (with 0 = not at all, and 10 = a perfect match).

Château Climens Sauternes 2001

No significant difference between the ratings on each musical piece was found (Greenhouse Geisser corrected, $F(4.127, 94.932) = 12.116$, $P = .064$). One possibility that is worth considering here is that populations normally segment into sweet-likers and sweet-dislikers (see [22]), and hence there may not have been a uniform response to this wine amongst the population tested (unfortunately, however, we did not collect any hedonic ratings for the wines).

Based on the findings of our first experiment, in which evidence was obtained to support the idea that people match certain wines with some pieces of classical music more than with others, we moved on, in experiment 2, to determine whether playing the matching music would actually influence a taster's/drinker's rating of the sensory-discriminative and/or hedonic properties of the wine as compared to the ratings obtained when tasting in silence.[h] Suggestive evidence that such a pattern of results might, in fact, be obtained comes from a number of recently-published studies (for example, [23,24]; though see also [25]). So, for example, Prof. Adrian North [23] has demonstrated that students rate wines (one red, a Chilean Cabernet Sauvignon, and the other a white, Chardonnay) as more powerful and heavy when music that had been rated as 'powerful and heavy' (for example, *Carmina Burana* by Orff) is played in the background; while playing music that has been categorized as 'zingy and refreshing' (*'Just Can't Get Enough'* by Nouvelle Vague) brings out the same qualities in a white wine. Interestingly, however, playing one style of music versus another

did not influence the students' rating of how much they liked the wine in North's study (see [25], on this point).

In addition, several other studies have now demonstrated that changing the music that happens to be playing in the background affects people's rating of everything from whisky ([26]; see also [27]) through to a bittersweet chocolate dessert [28], and from gelato [29] through to functional and dietetic foods [30], not to mention beer [31-33]. What we hear, then, can exert a profound influence on what we taste - and no less importantly on what we think we taste (see [34], for a review).[i]

In experiment 2, we build on the results of experiment 1 (and previous research suggesting that music can influence food and drink perception), in order to assess whether playing the best-matching music (as rated by the participants in our first experiment) would have any effect on people's perception of sweetness, acidity, alcohol level, fruitiness, tannin, and their enjoyment of the wines relative to when the wines were tasted without music. These attributes were chosen, in part, because four of them were the attributes that the wines had been selected to emphasize.

Experiment 2
Methods

Twenty six participants (aged 21 to 60 years) verbally agreed to take part in the study after the experimental procedure had been explained to them. The experiment had been approved by the Central University Research Ethics Committee of the University of Oxford. The tasting event lasted for approximately 90 minutes.

The wines were the same as those used in experiment 1, while the musical selection included four of the pieces used in experiment 1: Mozart's Flute Quartet No 1 in D major, K 285 - Movement 1, Allegro; Ravel's String Quartet in F major - Movement 1, Allegro moderato - très doux; Tchaikovsky's String Quartet No 1 in D major - Movement 2, Andante cantabile; and Debussy's *Syrinx*.

Procedure

Participants were invited to take part in a tutored wine tasting with musical accompaniment. On arrival, the participants were given a glass of Champagne. When they had all arrived, they were escorted as a group into the purpose-built tasting room. The procedure was just as for experiment 1 with the following exceptions. The participants tasted each of the four wines twice. For the first three wines, the second tasting of the wine (which took place about five minutes after the first) was accompanied by the playing of a pre-recorded piece of music at a comfortable listening level. By contrast, the two ratings of the fourth wine were made without any musical accompaniment. The music that, according to the results of experiment 1, best matched the wine was played over the hi-fi via loudspeakers after the participants had made their second rating. The distinctive features of the wines were described to the participants as the experiment proceeded.

Results

In order to determine whether the participants in the present study rated any of the attributes of the wines differently when listening to the music as compared to when tasting in silence, independent paired-samples *t*-tests were performed on the ratings made by participants with and without music (that is, for wines 1 to 3). In addition, repeated measures ANOVAs were performed in order to assess any overall effect of the presence of music, the wine, or the interaction between the two, on the ratings for each attribute, and for each wine.

The results of experiment 2 are shown in Table 1. They highlight a trend toward the participants rating the Domaine Didier Dagueneau as less acidic in the presence of music and as having higher alcohol content. Crucially, this wine was rated as being significantly more enjoyable when tasted together with the pre-selected matching music than when tasted in silence.

By contrast, there were no significant differences between the ratings with and without music, when it came to the Domaine Ponsot, Clos de la Roche 2009. Interestingly, the participants also rated the Château Margaux 2004 as significantly more enjoyable when tasted while listening to the matching music. Figure 6 present the difference between the ratings with and without music.

Based on the results shown in Table 1, further analyses of the data were conducted. In particular, a two-way

Table 1 Summary of paired-samples *t*-tests performed on the ratings for each attribute without and then with music in experiment 2

Wine	Attribute	*t*	df	Sig. (two-tailed)
Domain Didier Dagueneau, Pouilly Fumé Silex 2010	Sweetness	−1.10	25	0.28
	Acidity	1.90	25	**0.07**
	Alcohol	1.91	25	**0.07**
	Fruit	−0.21	24	0.83
	Tannin	−0.44	25	0.66
	Enjoy	−2.82	25	**0.01**
Domain Ponsot, Clos de la Roche 2009	Sweetness	−1.15	24	0.26
	Acidity	0.89	25	0.38
	Alcohol	−0.82	25	0.42
	Fruit	1.25	25	0.22
	Tannin	0.00	25	1.00
	Enjoy	−1.63	25	0.12
Chateau Margaux 2004	Sweetness	−1.41	24	0.17
	Acidity	0.21	24	0.83
	Alcohol	1.59	24	0.13
	Fruit	−0.59	23	0.56
	Tannin	0.75	24	0.46
	Enjoy	−2.68	24	**0.01**

The bolded *P* values either reached significance (*P* < .05), or else were near significant, suggesting the existence of a trend (see acidity in Domaine Didier Dagueneau - two-tailed- and enjoyment, sweetness, and alcohol in Domaine Ponsot, Clos de la Roche 2009 - one-tailed). Here, t stands for the t value, df for degrees of freedom, and sig. for significance.

repeated measures ANOVA was performed with the factors of wine (three levels) and music (absent versus present) for each of the attributes that were rated by participants for the first three wines (that is, those that were rated the second time around while listening to music).

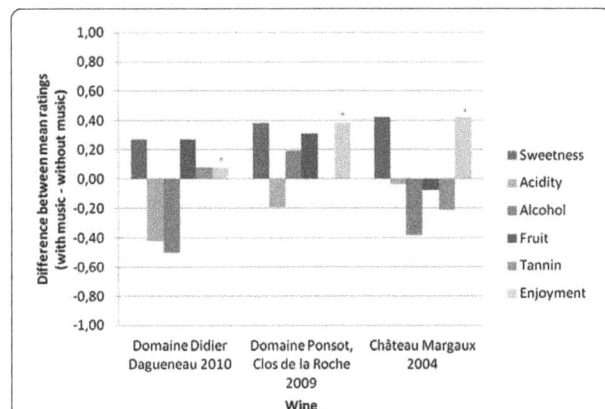

Figure 6 Mean difference between the first and second ratings of the three wines that were evaluated first without music, followed by an evaluation in the presence of the putatively matching music in experiment 2. The significant results (*P* < .05) obtained through the paired-samples *t*-tests are indicated with a *.

Sweetness

There was a significant main effect of music ($F(1, 23) = 4.466$, $P = .046$). Pairwise comparisons (using the Bonferroni correction) revealed that sweetness ratings were higher overall when tasting while the participants listened to the music ($P = .046$, without music M = 2.64, with music M = 2.93).

Acidity

There was a significant main effect of the wine ($F(2, 48) = 6.869$, $P = .002$), and a trend toward a main effect of music was also documented ($F(1, 24) = 3.417$, $P = .077$). Pairwise comparisons revealed that the Domaine Didier Dagueneau 2010 received higher sourness ratings as compared to the Château Margaux 2004 ($P = .002$). The participants also rated the wine as tasting somewhat more acidic while listening to the music that had been selected as matching in the preceding experiment (though this difference just failed to reach significance; p = .077).

Alcohol

There was a significant main effect of the wine ($F(2, 48) = 3.478$, $P = .039$). Pairwise comparisons revealed that Château Margaux 2004 was rated as higher in alcohol than the Domaine Didier Dagueneau ($P = .039$).

Fruitiness

No significant terms were found for the analysis of this attribute.

Tannin

There was a significant main effect of the wine ($F(2, 48) = 81.219$, $P < .001$). Perhaps unsurprisingly, the Château Margaux 2004 was rated as more tannic than either of the other two wines, followed by Domaine Ponsot, Clos de la Roche 2009, and finally the Domaine Didier Dagueneau ($P < .001$ for all comparisons).

Enjoyable

Crucially, the analysis of the data from experiment 2 revealed a highly-significant main effect of music ($F(1, 24) = 8.167$, $P < .001$), with the participants rating the wines as being more enjoyable while listening to the music (M = 5.65) than without (M = 5.19) ($P < .001$). See Figure 2 for the mean ratings when tasting in the absence of music and while listening to the music.

The key result for present purposes is the significant main effect of matching music on our participants' overall enjoyment of the wine ($P < .001$, see Figure 7). Note that one limitation with the present study was that the music was pre-recorded and played over a fairly low quality hi-fi system. One might reasonably anticipate therefore that the beneficial effects of playing the matching music on participants' enjoyment of wine would be even more pronounced were the matching music be played live, as anticipated for the third and final tasting.

Discussion

The results of the two experiments reported in the present study demonstrate that people do indeed reliably associate certain kinds of classical music with particular wines. These results therefore add new evidence concerning how music can influence participant's perception of wine. For example, Tchaikovsky's String Quartet No 1 in D major turned out to be a particularly good match for the Château Margaux 2004 (red wine), while presenting a particularly bad match for the Domaine Didier Dagueneau, Pouilly Fumé Silex 2010 (white wine). Meanwhile, Mozart's Flute Quartet in D major, K285 was found to be a particularly good match for the Pouilly Fumé but a bad match for the Château Margaux. On the basis of these results, it appears that wines displaying prominent acidity offer a good match for musical pieces that incorporate the flute, while red wines appear to offer a crossmodal match for string quartets. That we obtained any significant results in experiment 1 is all the more impressive once it is realized that we selected a much narrower range of wines and musical styles/excerpts than have contributed to previous research in this area. Note here that all four wines were from the same region (France), and that the music selections only involved wind and string instruments (that is, brass and piano were absent from all of the musical selections, despite their being often chosen as the instruments that best match particular taste/flavor attributes; [11]).

The results of experiment 2 demonstrated the significant effect that listening to music that has been pre-selected (in experiment 1) to match the wine can have on how enjoyable people find it to taste wine. Participants rated each of four wines twice: for the first three wines, the second rating took place while the participants listened to the most appropriate musical selection, based on the results of our first experiment. We specifically investigated whether presenting the appropriate music would actually enhance participants' ratings (that is, enjoyment) of the wines when the two were experienced together. The results provided clear evidence in support of such a claim. Mean enjoyment ratings were significantly higher (approximately 4.6%) when listening to the matching music (M = 5.65) than when rating the wines in silence (M = 5.19).

Having demonstrated in experiment 1 that social drinkers experience particular pieces of classical music as corresponding with specific fine French wines, one question for future research will be to determine whether sensory characteristics or semantic or emotional associations govern this matching. Our results provide some modest support for Gray's [4] suggestion of an emotional mediation with which we started this piece:

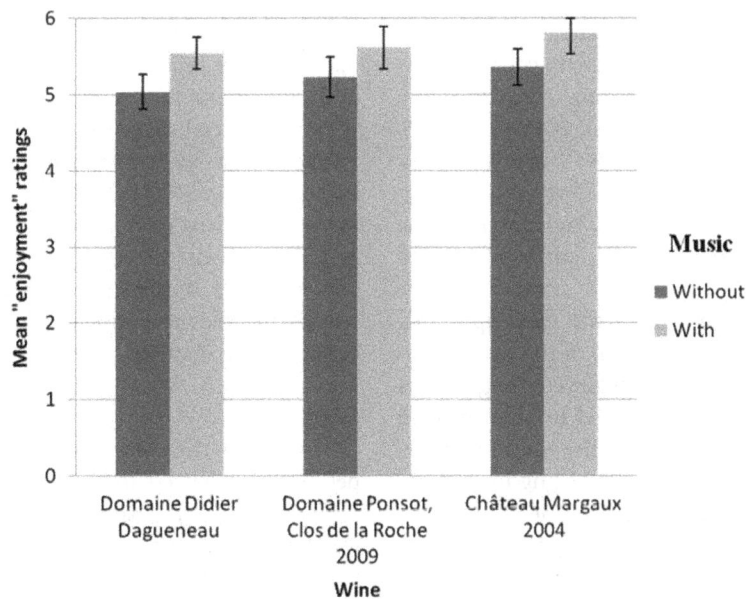

Figure 7 Mean ratings of how enjoyable the wines were when rated without music versus when rated a second time while listening to the pre-recorded musical selections in experiment 2.

'Red wines need either minor key or they need music that has negative emotion. They don't like happy music...Cabernets like angry music'. In particular, the music with the highest scores for the Cabernet Sauvignon (Chateaux Margaux 2004) were the Andante cantabile (Movement 2) and Scherzo (Movement 3) of Tchaikovsky's String Quartet No 1 in D major (musical Selections 2 and 7, respectively). While the Andante is slow in tempo with a melancholic melody, the Scherzo is more vigorous and, at times, has a discordant sound to it which could be described as heated and possibly even angry with its energetic bow strokes.

The hypothesis of an emotional mediation account is different from the explanation that one might simply match the stimuli presented in the different modalities based on how much one likes them individually (see [19]). Such an account has also been tested in the domain of color and music matching (see [35,36]). Here, though, it would seem that the complexity of the matched object might involve other matching principles, noticeably of the kind of sensory correspondences between flavor/aromas and sounds evidenced elsewhere (for example, [37]; see [9] for a review). It would seem possible that people might match stimuli crossmodally in terms of their intensity, or brightness, both putatively amodal stimulus properties ([38]; though see also [39]). However, it is important to note that these are but a few of the various accounts for crossmodal matching (or correspondences) that have been put forward over the years (see [9,38,40,41], for others). Further research is clearly needed in order to determine the most appropriate

explanation for these findings. Relevant to answering this particular question, researchers will need to investigate whether the consensual crossmodal matches documented here in Western participants would also be deemed appropriate by those who come from a very different cultural background, and who are thus perhaps more familiar with a very different musical repertoire [21-48].

One potential concern regarding the results of experiment 2 is that our participants might simply have given a higher numerical rating the second time that they evaluated the wines, regardless of whether or not any music happened to have been playing in the background. Now, while this certainly does not seem to have been the case for the sensory-discriminative attributes of the wine (see Table 1), it seemed possible that hedonic ratings (for example, of enjoyment) might be different (see also [26]) - especially given evidence concerning the mere exposure effect (for example, [49,50]). However, the results of the comparison between the first and second ratings of the fourth wine (the dessert wine), did not show any such significant effect (M = 5.48 on first tasting versus 5.60 on second tasting - thus only a very slight trend was observed in that direction).[j] Hence, it can be concluded that it was the presence of the music that made a difference to the ratings obtained during the second tasting of the first three wines.

One other possibility that we cannot evaluate with the present dataset concerns how much of the beneficial effect of the music depends on its crossmodally corresponding to the wine. Thus, we know from previous research that simply listening to music or noise can impact on taste sensitivity and the higher level semantic judgments that people make about a wine's qualities (see [23-34,36-41,43-51]). It would therefore be interesting in future research, to compare people's rating of a wine when the putatively matching music is played versus when mismatching music is played instead. So, for example, the results of experiment 1 clearly demonstrate that people thought that for Tchaikovsky's String Quartet No 1 in D major, offered a very bad match for the Domaine Didier Dagueneau, Pouilly Fumé Silex 2010 (white wine).

One more abstract concern about the most appropriate interpretation, or rather extension, of the present results, for example, to more commercial contexts relates to the question of whether people need to be primed to think about the relationship between the music they are listening to and the wine that they are tasting in order to make the connection between one and the other. Certainly, the results of North's [23] study suggest that this need not be the case. A second concern, more generally, is about the role of any experimenter expectancy effects in this kind of experimental design (involving, as it does, a public tasting event). Here, one might wonder what role the experimenter's hopes, or for that matter the participants' own opinions regarding what is being tested and how they might be expected to respond could influence the pattern of results obtained (see for example, [52-57]). While such concerns are by no means specific to the crossmodal matching or/influence of music on wine (see, for example, [58,59]), the public and interactive environments in which such data are collected presumably leave themselves open to such indirect forms of influence. That said, a number of the participants from experiment 1 came away from the event convinced of the veracity of the phenomenon (suggesting what we were picking up on was more than merely experimenter expectancy effects).

Having demonstrated that people enjoy tasting quality wines more while listening to matching music, one further question to be addressed by future research will be to determine whether the crossmodal influences documented here also operate in the opposite direction - that is, can people's appreciation of classical music also be enhanced by drinking wine? It may perhaps be that the influence is bidirectional, so it would be interesting to see whether the wines may also enhance the appreciation of these pieces of music [25]. In addition, further research is needed in order to clarify the potential mechanisms that explain these matchings.

Conclusions

The results presented here provide some of the first evidence for the idea that people follow similar crossmodal correspondences when pairing wines and classical music. In addition, the results reveal that classical music can enhance the overall experience of drinking wine. All together, these findings suggest that music-wine pairings may not only reflect wine writers' idiosyncratic matches [1-3], but more general crossmodal associations that are shared by social drinkers, which also provide the social drinker with useful information about the expected sensory properties of a wine. These results may be explained in terms of the potential crossmodal correspondences that exist between the various features of music and wines [8-38,42,43,60]; nonetheless, research is still needed in order to clarify the potential mechanisms that explain these matches.

Endnotes

[a]Note also that wine is not the only foodstuff that people have started to match with music: a number of chefs have recently started to suggest musical pairings for their recipes too (for example, see [61,62]). Or take the following from A A Gill ([63], p. 174): 'If heaven is eating *foie gras* to the sound of trumpets, then purgatory is a hamburger consumed to the sound of the same shrill, flat note being blown on a harmonica over and over' (see also [64,65]).

[b]Paul White [3], p. 122 captures the concern here when he says that: '...I've rarely resorted to describing wines through musical terminology (staccato, crescendo, rubato, riff, and so on) or made direct associations between tunes and individual wines: 'This Riesling is so middle-period Nirvana...' That's not to say those aren't valid expressions; it's just not the way I've sensed wine and tried to lay it out in words. To be frank, I've always feared how easily that sort of discussion can end up sounding trite or pretentious or simply slink off into esoteric nonsense'.

[c]Finally, beyond the occasional use of classical music metaphors when describing fine wines, these two classes of stimuli have also been shown to have something of an affinity for one another in previous marketing research. So, for example, Areni and Kim [66] have demonstrated that customers in a North American wine cellar spend significantly more on a bottle of wine classical music happens to be playing in store than when 'Top-40' pop music was played instead. Such a pattern of results have been taken to suggest that both expensive wine and classical music may share a semantic association with the notion of high quality or classiness.

[d]Charters and Pettigrew [67], p. 126 also note that: '[...] informants considered that the consumption of wine shows some similarities to the appreciation of 'pure' art forms - especially music [...]'.

[e]The orthonasal olfactory system is associated with the inhalation of external odors as when we sniff a wine in the glass, while the retronasal system (involving the posterior nares), is associated with the detection of the

olfactory stimuli that emanate when odors are periodically forced out of the nasal cavity when, for example, we swallow a mouthful of wine.

[f]While some writers have been tempted to explain such surprising crossmodal matches in terms of synesthesia (for example, [14,60]), this is not an explanation that we find particularly useful (see [9]). That said, we are rather partial to the name 'oenesthesia', coined by Jo Burzynska (MW), and used to describe the widespread tendency for people to match sounds, instruments, and music to the tastes, aromas, and flavors that are present in wine.

[g]It should be noted here there is a very recent precedent for studying crossmodal correspondences involving pieces of classical music in the work of Palmer, Schloss, Xu, and Prado-León [35]. These researchers documented some surprisingly robust crossmodal correspondences between pieces of classical music and color patches in both Californian and Mexican participants.

[h]What we are looking for here is captured by the following from James John, Director of the Bath Wine School when he says of the combination of Mozart's *Laudate dominum*, and Chardonnay: '[...] Just as the sonant complexity is doubled, the gustatory effects of ripe fruit on toasted vanilla explode on the palate and the appreciation of both is taken to an entirely new level' (quote from [42]). Of course, there are alternative views of the interaction of wine and music as captured by the following from Doug Frost (MW): 'So I don't want music and wine to match up; I want them to talk to each other. They may agree; they may argue. Sometimes they don't speak at all; they just yell past each other. That's cool too' (quote also taken from [42]).

[i]Emile Peynaud ([43], p. 104) was perhaps prescient in this regard then when he advised the professional wine taster some years back that: 'The sense of hearing can interfere with the other senses during tasting and quiet has always been considered necessary for a taster's concentration. Without insisting on absolute silence, difficult to obtain within a group in any case, one should avoid too high a level of background noise as well as occasional noises which can divert the taster's attention'.

[j]Furthermore, any slight trend toward the fourth wine being rated more highly the second time round might in part be explained by a few of the participants subsequently admitting that they had not rated this wine, when instructed to do so in silence, but had instead waited until the putatively matching music had been played.

Abbreviations
AWC: The Antique Wine Company; LSO: The London Symphony Orchestra.

Competing interests
The authors declare that they have no competing interests.

Authors' contributions
CS, LR, EK, AMH, VD, and AS developed the idea of the research project and coordinated the logistics, together with The Antique Wine Company (who selected and provided the wine), and The London Symphony Orchestra (who chose the musical selections). CS, LR, EK, AMH, VD, and AS collected the data. CS, CV, LR EK and AMH conducted the analysis of the data. CS, OD, and CV interpreted the data, and drafted the manuscript. All authors read and approved the final manuscript.

Acknowledgments
We would like to thank The Antique Wine Company for providing the testing facilities and wine used in the present study. We would also like to thank The London Symphony Orchestra for providing the musical excerpts utilized in the present study. This project was not funded by a funding agency.

Author details
[1]Crossmodal Research Laboratory, Department of Experimental Psychology, University of Oxford, South Parks Road, OX1 3UD Oxford, England. [2]London Symphony Orchestra, Barbican Center, Silk Street, EC2Y 8DS London, England. [3]The Antique Wine Company (AWC), London, England. [4]Center for the Study of the Senses, School of Advanced Study, University of London, London, UK.

References
1. Johnson H: *Wine: A Life Uncorked*. London: Weidenfeld & Nicolson; 2005.
2. Bach K: **Knowledge, wine, and taste: what good is knowledge (in enjoying wine)?** In *Questions of Taste: The Philosophy of Wine*. Edited by Smith BC. Oxford: Oxford University Press; 2007:21–40.
3. White P: **Food of love: wine and music.** *The World of Fine Wine* 2008, 21:120–123.
4. Gray WB: **Music to drink wine by: vintner insists music can change wine's flavors.** *San Francisco Chronicle* 2007, 11:2.
5. Lehrer A: *Wine & Conversation*. 2nd edition. Oxford: Oxford University Press; 2009.
6. Cacciari C: **Crossing the senses in metaphoric language.** In *Oxford Handbook of Synesthesia*. Edited by Simmer J, Hubbard ED. Cambridge UK: Cambridge University Press; 2008:425–443.
7. Martin GN: **A neuroanatomy of flavor.** *Petits Propos Culinaires* 2004, 76:58–82.
8. Spence C: **Wine and music.** *The World of Fine Wine* 2011, 31:96–104.
9. Deroy O, Crisinel A-S, Spence C: **Crossmodal correspondences between odors and contingent features: odors, musical notes, and geometrical shapes.** *Psychonom Bull Rev* 2013, 20:878–896.
10. Piesse CH: *Piesse's Art of Perfumery*. 5th edition. [http://www.gutenberg.org/files/16378/16378-h/16378-h.htm]
11. Crisinel AS, Spence C: **A fruity note: crossmodal associations between odors and musical notes.** *Chem Senses* 2012, 37:151–158.
12. Belkin K, Martin R, Kemp SE, Gilbert AN: **Auditory pitch as a perceptual analogue to odor quality.** *Psychol Sci* 1997, 8:340–342.
13. Stevenson RJ, Boakes RA: **Sweet and sour smells: learned synesthesia between the senses of taste and smell.** In *The Handbook of Multisensory Processing*. Edited by Calvert GA, Spence C, Stein BE. Cambridge: MA: MIT Press; 2004:69–83.
14. *Synesthesia: Smells like Beethoven*. [http://www.economist.com/node/21545975]
15. Crisinel AS, Spence C: **Implicit association between basic tastes and pitch.** *Neurosci Lett* 2009, 464:39–42.
16. Mesz B, Sigman M, Trevisan MA: **A composition algorithm based on crossmodal taste-music correspondences.** *Front Hum Neurosci* 2012, 6:1–6.
17. Simner J, Cuskley C, Kirby S: **What sound does that taste? Cross-modal mapping across gustation and audition.** *Perception* 2010, 39:553–569.
18. Knöferle KM, Spence C: **Crossmodal correspondences between sounds and tastes.** *Psychonom Bull Rev* 2012, 19:992–1006.
19. Crisinel AS, Spence C: **The impact of pleasantness ratings on crossmodal associations between food samples and musical notes.** *Food Qual Prefer* 2012, 24:136–140.
20. Bronner K, Bruhn H, Hirt R, Piper D: **What is the sound of citrus? Research on the correspondences between the perception of sound and flavor.** In *Proceedings of the 12th International Conference of Music Perception and*

Cognition and the 8th Triennial Conference of the European Society for the Cognitive Sciences of Music: 23 to 28 July 2012; Thessaloniki, Greece. Edited by Cambouropoulos E, Tsougras C, Mavromantis P, Paastiadis K. 2012:142–148.

21. Rader CM, Tellegen A: An investigation of synesthesia. J Pers Soc Psychol 1987, 52:981–987.

22. Prescott J: Taste Matters: Why we Like the Foods we do. London: Reaktion Books; 2012.

23. North AC: The effect of background music on the taste of wine. Brit J Psychol 2012, 103:293–301.

24. Reid M: How Guns & Roses can change your tune on wine. TimesOnline 2008:.

25. Spence C, Deroy O: On why music changes what (we think) we taste. i-Perception 2013, 4:137–140.

26. Velasco C, Jones R, King S, Spence C: Assessing the influence of the multisensory environment on the whisky drinking experience. Flavour 2013, 2:23.

27. Stafford LD, Fernandes M, Agobiani E: Effects of noise and distraction on alcohol perception. Food Qual Prefer 2012, 24:218–224.

28. Crisinel A-S, Cosser S, King S, Jones R, Petrie J, Spence C: A bittersweet symphony: systematically modulating the taste of food by changing the sonic properties of the soundtrack playing in the background. Food Qual Prefer 2012, 24:201–204.

29. Kantono K, Hamid N, Sheperd D, Yoo MJY, Grazioli G: Effect of music genre on pleasantness of three types of chocolate gelati. Food Qual Prefer 2013. Submitted.

30. Silva DW, Bolini HMA: Influence of music genres on consumer's perception of a functional and dietetic food, Poster presented at the 10th Pangborn Sensory Science Symposium. Brazil: Rio de Janeiro; 2013.

31. Brown P: Ale, ale, rock and roll! Word Magazine 2012, 28:28–29.

32. Holt-Hansen K: Taste and pitch. Percept Mot Skills 1968, 27:59–68.

33. Holt-Hansen K: Extraordinary experiences during cross-modal perception. Percept Mot Skills 1976, 43:1023–1027.

34. Spence C: Auditory contributions to flavor perception and feeding behavior. Physiol Behav 2012, 107:505–515.

35. Palmer SE, Schloss KB, Xu Z, Prado-León LR: Music-color associations are mediated by emotion. Proc Natl Acad Sci U S A 2013, 110:8836–8841.

36. Schifferstein HNJ, Tanudjaja I: Visualizing fragrances through colors: the mediating role of emotions. Perception 2004, 33:1249–1266.

37. Seo H-S, Hummel T: Auditory-olfactory integration: congruent or pleasant sounds amplify odor pleasantness. Chem Senses 2010, 36:301–309.

38. Spence C: Crossmodal correspondences: a tutorial review. Atten Percept Psychophys 2011, 73:971–995.

39. Spence C, Deroy O, Bremner A: Questioning the utility of the concept of amodality: towards a revised framework for understanding crossmodal relations. Multisens Res 2013, 26:57.

40. Stevenson RJ, Rich A, Russell A: The nature and origin of cross-modal associations to odors. Perception 2012, 41:606–619.

41. Walker P, Walker L: Size-brightness correspondence: crosstalk and congruity among dimensions of connotative meaning. Atten Percept Psychophys 2012, 74:1226–1240.

42. Sachse-Weinert M: Wine & Musik: 2 + 2 = 5. Vortrag im Rahmen der Ringvorlesung 'Weinwissenschaft' an der Johannes Gutenberg-Universität Mainz im Sommersemester. 2012. Presentation given on 4 July.

43. Peynaud E: The Taste of Wine: The Art and Science of Wine Appreciation (Translator M Schuster). London: Macdonald & Co; 1987.

44. Bremner A, Caparos S, Davidoff J, de Fockert J, Linnell K, Spence C: Bouba and Kiki in Namibia? Western shape-symbolism does not extend to taste in a remote population. Cognition 2013, 126:165–172.

45. Henrich J, Heine SJ, Norenzayan A: The weirdest people in the world? Behav Brain Sci 2010, 33:61–135.

46. Walker R: The effects of culture, environment, age, and musical training on choices of visual metaphors for sound. Percept Psychophys 1987, 42:491–502.

47. Wang QJ: Music, Mind, and Mouth: Exploring the interaction between music and flavor perception. Cambridge, MA: Manuscript submitted in partial fulfillment of the requirements for the Degree of Master of Science in Media Arts and Sciences at The Massachusetts Institute of Technology; 2013.

48. Woods AT, Spence C, Butcher N, Deroy O: Fast lemons and sour boulders: testing the semantic hypothesis of crossmodal correspondences using an internet-based testing methodology. i-Perception 2013, 4:365–369.

49. Gallace A, Spence C: Touch with the Future: The Sense of Touch from Cognitive Neuroscience to Virtual Reality. Oxford: Oxford University Press. 2014.

50. Monahan JL, Murphy ST, Zajonc RB: Subliminal mere exposure: specific, general, and diffuse effects. Psych Sci 2000, 11:462–466.

51. Deroy O, Spence C: Why we are not all synesthetes (not even weakly so). Psychon Bull Rev 2013, 20:1–22.

52. Rosenthal R: Experimenter outcome-orientation and the results of the psychological experiment. Psych Bull 1964, 61:405–412.

53. Rosenthal R: Experimenter Effects in Behavioral Research. New York: Appleton-Century-Crofts; 1966.

54. Rosenthal R: Covert communication in the psychological experiment. Psych Bull 1967, 67:356–367.

55. Levitt SD, List JA: Was there really a Hawthorne effect at the Hawthorne plant? An analysis of the original illumination experiments. Am Econ J Appl Econ 2011, 3:224–238.

56. Mayo E: Hawthorne and the Western Electric Company. The Social Problems of an Industrial Civilization. London: Routledge; 1949.

57. Zdep SM, Irvine SH: A reverse Hawthorne effect in educational evaluation. J Educ Psychol 1970, 8:89–95.

58. Doyen S, Klein O, Pichon C, Cleeremans A: Behavioral priming: it's all in the mind, but whose mind? PLoS One 2012, 7(1):e29081.

59. Farah MJ: Is visual imagery really visual? Overlooked evidence from neuropsychology. Psychol Rev 1988, 95:307–317.

60. Rudmin F, Cappelli M: Tone-taste synesthesia: a replication. Percept Mot Skills 1983, 56:118.

61. Gagnaire P, Gonzales C: Bande Originale. 175 Recettes & Une Heure de Musique (Bande Originale. 175 Recipes and an Hour of Music). Paris: Flammarion; 2010.

62. Pelaccio Z: Eat with your Hands. New York: Ecco; 2012.

63. Gill AA: Table Talk: Sweet and Sour, Salt and Bitter. London: Weidenfeld & Nicolson; 2007.

64. Peralta E: The Sounds of Asparagus, as Explored Through Opera. [http://www.npr.org/blogs/thesalt/2012/05/29/153950254/the-sounds-of-asparagus-as-explored-through-opera]

65. Steinberger M: Au Revoir to all That: The Rise and Fall of French Cuisine. London: Bloomsbury; 2010.

66. Areni CS, Kim D: The influence of background music on shopping behavior: classical versus top-forty music in a wine store. Adv Consum Res 1993, 20:336–340.

67. Charters S, Pettigrew S: Is wine consumption an aesthetic experience? J Wine Res 2005, 16:121–136.

5

Can you find the golden ratio in your plate?

Ophelia Deroy[1*] and Charles Spence[2]

Abstract

A scientific approach to plating needs to be based on perceivers' responses and anticipate possible cultural and individual differences. It cannot just follow common sense principles, whose validity remain untested and only attract journalists' attention, like the claim that people will prefer food composition based on the golden ratio.

Keywords: Plating, Aesthetics, Colour, Composition, Psychology, Visual perception

Introduction

According to the latest scientific research, the visual appearance of a dish can affect how much diners like it, and even how they rate its overall flavor [1]. Whereas the majority of the research that has been published to date has focused on how specific properties, such as the colour of individual ingredients, can influence people's evaluation, the understanding of the visual appeal of a dish [2], such as a curry, with sauce, vegetables, and rice, undoubtedly needs to take into account the composition of the various elements on the plate [3-6]. According to Dr Hadley, a physicist at Warwick University's Department of Physics (Coventry, UK), applying a mathematical formula can tell you exactly how much rice and curry will look appealing to all consumers. A plate size of 27 cm, with a 23 cm diameter bed of rice precisely 5 mm thick, and supporting a low dome of curry with a diameter of 14 cm and a maximum height of 2.4 cm, represents, or at least so Dr Hadley would like to have us believe, the perfect presentation for a plate of curry, and should be liked by everyone. Other variants of the perfect curry can be envisaged, with equal precision, for a hungry student or an aficionado of nouvelle cuisine [7]. The press have certainly jumped enthusiastically onto this story [8,9].

The calculations applied here are supposed to satisfy what is known as the golden ratio, the ratio whereby the relation of the greater part to the sum of the two parts equals that of the two parts. To have the most aesthetically pleasing curry, then, the supposition is that the rice must be approximately 1.61 times wider than the circle of curry that is laid on top ($(\sqrt{5} + 1)/2$, to be exact). Of course, here one needs to grant Dr Hadley the right to extend the concept of the golden ratio, usually meant to apply to rectangles and ellipses, to the relation between the radiuses of two circles. It is also not meant to explain the aesthetics of three dimensional objects, and apply to the height of objects, like here.

Is a claim that plates should obey the golden ratio science or enigmatic calculation? Are we not sprinkling some mysticism back onto the plate? In the past, a number of authors have supposed that the golden ratio dictates a viewer's preference for certain architectural achievements or visual displays (not to mention to be a proof of God's mathematical skills [10-12]). However, careful scientific research has demonstrated that this magical number actually explains little of the sense of balance and harmony that people typically attribute to shapes or composition. Despite some optimistic early results with rectangles ([13-16], but see [17]), Fechner, one of the fathers of modern psychology, failed to find any evidence that people actually preferred ellipses built on the golden ratio rather than others [18]. Furthermore, specialists in the field of experimental aesthetics have since demonstrated that many other factors bear on what people think has a balanced composition or harmonious figure [19,20].

What such research has demonstrated is that what counts as a balanced composition depends to a large degree on what is represented or presented, and preferences will differ for different shapes, colours, and objects [21,22]. In this sense, preferences for curry and rice might not be the same depending on whether it is a green or a red curry. Aesthetic preferences, if one chooses to extend them to the plating of food, will also vary with context [23] and present individual differences.

* Correspondence: ophelia.deroy@sas.ac.uk
[1]Centre for the Study of the Senses, University of London, London, UK
Full list of author information is available at the end of the article

If this is true for simple geometrical shapes, with different groups of individuals expressing markedly different preferences when it comes to judging the most beautiful of rectangles [20,24], it is also likely to be true when it comes to a plate of curry. Instead of the myth of the golden ratio, the reality will likely depend on the diner's cultural background, whether they happen to be hungry or not, the types of plating arrangements that they have been exposed to previously, and perhaps even their personality (see McManus *et al.* for a recent investigation of simple compositions [25]). The first thing to do, of course, would be to measure these preferences directly, by asking people to select the most pleasing or appetizing plate of curry (online testing might be useful here). This can be done at minimal cost, by any reasonable marketing department. If science needs to be involved, it will be to make sure that the presented pictures systematically vary along key dimensions, and that important interactions between, for example, size, colour, orientation, and so on, are not ignored.

Ignoring which science is relevant to the choice of the perfect plating, or the elaboration of the perfect meal, is perhaps the biggest problem at the present time. A plate needs to be resistant, smooth, and perhaps shiny, and physics and design may certainly be relevant. Students of physics might also be able to describe how to provide an equal distribution of weight in the plate if it is to be filled with a certain quantity of low-density rice and high-density curry. But when it comes to the preferences of diners, equations and simple premises are just not that relevant. It is rather the discovery of the fundamental premises themselves which is at the core of the work in this area, or at least it should be. As mentioned, many studies in experimental psychology were needed to show that the golden ratio does not necessarily represent a useful guide to people's visual preferences. Many more experiments are being performed, even today, in order to try and gain a better understanding of what governs the sense of balance or harmony in visual composition. It will take many more psychologists and careful testing to demonstrate what drives the preferences of consumers once the visual composition is also supposed to be eaten, as it is the case for a combination of curry and rice on a plate.

Take, for example, Hadley's claim that diners want a clear rim of at least 2 cm around the food on the plate. This is certainly not the case in the most admired Michelin-starred plating styles, where beautifully arranged sauces and spices cover all the plate's surface; and why would it be 2 cm rather than 3 cm or 5 cm? Do preferences depend on the color, the size, or even the shape of the plate, and the type of food that it contains? Is it the same for desserts, coming at the end of the meal, and starters? Dr Harvey's research [26] certainly rests on intuitive aesthetic principles, which explain why the final result (the 'perfect curry') will indeed look appealing enough to a large body of individuals.

Given the role played by intuition in this research, one would rather trust the chef's intuitive sense of presentation. Claims that chefs and cooks 'have been getting curry all wrong,' as reported in *The Times*, is not just provocative, but totally misplaced [8]. The development of a scientific approach to the presentation of food, in all its cultural, aesthetic, and individual complexity [3], is a noble prospect. For this very reason, we need more than merely the feeding of intuitions into complex equations. We certainly need more experimental rigor and to put the diners at the centre of our scientific investigation of the aesthetics of plating [27]. Not magic numbers.

Competing interests
The authors declare that they have no competing interests.

Authors' contribution
OD and CS contributed equally to this manuscript. Both authors read and approved the final version of the manuscript.

Acknowledgements
OD and CS are supported by an Arts and Humanities Research Council (AHRC) grant in the 'Science in Culture' theme (re-thinking the senses).

Author details
[1]Centre for the Study of the Senses, University of London, London, UK.
[2]Crossmodal Research Laboratory, Department of Experimental Psychology, Oxford University, Oxford, UK.

References
1. Spence C, Piqueras-Fiszman B, Michel C, Deroy O: **Plating manifesto (II): the art and science of plating.** *Flavour.* in press.
2. Lyman B: *A Psychology of Food, More Than a Matter of Taste.* New York, NY: Avi, Van Nostrand Reinhold; 1989.
3. Zampollo F, Wansink B, Kniffin KM, Shimuzu M, Omori A: **Looks good enough to eat: how food plating preferences differ across cultures and continents.** *Cross Cult Res* 2012, **46**:31–49.
4. Zellner DA, Lankford M, Ambrose L, Locher P: **Art on the plate: effect of balance and color on attractiveness of, willingness to try and liking for food.** *Food Qual Prefer* 2010, **21**:575–578.
5. Zellner DA, Siemers E, Teran V, Conroy R, Lankford M, Agrafiotis A, Ambrose L, Locher P: **Neatness counts. How plating affects liking for the taste of food.** *Appetite* 2011, **57**:642–648.
6. Reisfelt HH, Gabrielsen G, Aaslyng MD, Bjerre MS, Møller P: **Consumer preferences for visually presented meals.** *J Sens Stud* 2009, **24**:182–203.
7. Hadley M: **The perfect curry.** In *Unpublished manuscript commissioned by Tilda Rice.* Rainham: Tilda; 2013.
8. de Bruxelles S: **Scientists find formula for the perfect curry.** *The Times,* 1 November:45.
9. Griffiths S: **Calculate your perfect CURRY: physicist creates formula to ensure the perfect ratio of rice to sauce when tucking into a korma.** *Mail Online* 2013. [http://www.dailymail.co.uk/sciencetech/article-2481071/Physicist-creates-formula-perfect-ratio-rice-sauce-tucking-curry.html]
10. Alberti LB: *On the Art of Building.* Cambridge, MA: MIT Press; 1989.
11. Ghyka M: *The Geometry of Art and Life.* New York, NY: Dover Publications; 1977.
12. Green CD: **All that glitters: a review of psychological research on the aesthetics of the golden section.** *Perception* 1995, **24**:937–968.
13. Fechner GT: *Zur Experimentalen Aesthetik [On Experimental aesthetics].* Leipzig: Hirzel; 1871.
14. Fechner GT: *Vorschule der Aesthetik. [Introduction to Aesthetics].* Leipzig: Breitkopf & Haertel; 1876.

15. Fechner GT: **Various attempts to establish a basic form of beauty: experimental, aesthetics, golden section, and square.** *Empir Stud Arts* 1876, **15:**115–130. Translation of chapter XIV.
16. Lalo C: *L'esthetique Experimentale Contemporaine. [Contemporary Experimental Aesthetics].* Paris: Alcan; 1908.
17. Godkewitsch M: **The 'golden section': an artifact of stimulus range and measure of preference.** *Am J Psychol* 1974, **87:**269–277.
18. Boselie F: **The golden section has no special aesthetic attractivity!** *Empir Stud Arts* 1992, **10:**1–18.
19. McManus IC: **The aesthetics of simple figures.** *Br J Psychol* 1980, **71:**505–524.
20. McManus IC, Weatherby P: **The golden section and the aesthetics of form and composition: a cognitive model.** *Empir Stud Arts* 1997, **15:**209–232.
21. Boselie F: **The golden section and the shape of objects.** *Empir Stud Arts* 1997, **15:**131–141.
22. Koneni VJ: **The vase on the mantelpiece: the golden section in context.** *Empir Stud Arts* 1997, **15:**177–207.
23. Benjafield J, McFarlane K: **Preference for proportions as a function of context.** *Empir Stud Arts* 1997, **15:**143–151.
24. Macrosson WDK, Strachan GC: **The preference amongst product designers for the golden section in line partitioning.** *Empir Stud Arts* 1997, **15:**153–163.
25. McManus IC, Cook R, Hunt A: **Beyond the golden section and normative aesthetics: why do individuals differ so much in their aesthetic preferences for rectangles?** *Psychol Aesthet Creativity Arts* 2010, **4:**113.
26. Anon: **Academic reveals the secret to the perfect curry.** *Press release* 2013.
27. Deroy O, Michel C, Piqueras-Fiszman B, Spence C: **The plating manifesto (I): from decoration to creation.** *Flavour.* in press.

6

The emerging science of gastrophysics and its application to the algal cuisine

Ole G Mouritsen[1,2]

Abstract

This paper points to gastrophysics as an emerging scientific discipline that will employ a wide range of the most powerful theoretical, simulational and experimental biophysical techniques to study the empirical world of cooking and gastronomy. Gastrophysics aims to exploit recent advances in the physical sciences to forward the scientific study of food, its raw materials, the effects of processing food and quantitative aspects of the physical basis for food quality, flavour and absorption into the human body. It suggests the use in cooking of a class of raw materials little used in the Western world, the marine macroalgae or seaweeds, as a laboratory for defining, characterizing and shaping the emerging scientific discipline of gastrophysics. In relation to gastronomy, seaweed materials are virtually unexplored scientifically by physical experimentation and theory. The sea is one of the last resorts for humankind to exploit the ability to obtain more food to feed a hungry world, because world fisheries can no longer meet the need for healthy seafood. Hence, seaweeds offer a rich and virtually unexploited source of primary marine foodstuff in the Western world. To explore the full gastronomical potential of this resource, there is a need for fundamental research into the gastrophysics of seaweeds.

Keywords: gastrophysics, molecular biophysics, algae, seaweeds, science of cooking

Gastronomy and gastrophysics

The father of gastronomy, Jean Anthelme Brillat-Savarin, defined 'gastronomy' in his 1825 masterpiece, *Physiologie du goût* (*The Physiology of Taste*) [1], as the knowledge and understanding of everything that relates to man as he eats. Its purpose is to ensure the conservation of men, using the best food possible. Nowadays, 'gastronomy' is used as a broad term that covers the art and science of good cooking, including aesthetics, the qualities of raw materials, food preparation and cooking techniques, flavour and the cultural history of cooking. A wide range of established scientific disciplines relate to gastronomy, in particular food chemistry and technology, sensory sciences and human nutrition. A recent trend, pioneered by the British-Hungarian physicist Nicolas Kurti [2] and the French chemist Hervé This [3,4] and fuelled by a close collaboration between scientists and chefs, has led to the new term 'molecular gastronomy'. A closer inspection of what molecular gastronomy is claimed to be shows that in most cases it tends to involve few if any quantitative molecular considerations. Despite the somewhat misleading terminology, or maybe exactly because of the use of this terminology, molecular gastronomy has received a lot of publicity among the general public and celebrated chefs. Some prominent chefs, including Heston Blumenthal, from restaurant *The Fat Duck*, and Ferran Adrià, from restaurant *El Bulli*, have issued disclaimers stating that their art and style of cooking cannot be described by the concept of molecular gastronomy [5].

In a recent review of molecular gastronomy, the British physicist Peter Barham and his colleagues [6] made a serious attempt to define what molecular gastronomy is and how it differs from gastronomy. These authors advocate the very pragmatic viewpoint that molecular gastronomy "should be considered as the scientific study of why some food tastes terrible, some is mediocre, some good, and occasionally some absolutely delicious" (p. 2315). In this way, they strike a delicate balance between, on the one hand, considering food simply as materials with certain properties described by food chemistry, and, on the other hand, taking into account the empirical fact that there are very passionate feelings associated with food and

Correspondence: ogm@memphys.sdu.dk
[1]MEMPHYS, Center for Biomembrane Physics, Department of Physics, Chemistry and Pharmacy, University of Southern Denmark, Campusvej 55, DK-5230 Odense M, Denmark
Full list of author information is available at the end of the article

eating. Although according to this definition there seems to be no specific reference to why the term 'molecular' is invoked, it is presupposed that the field is based on the utilisation of molecularly based sciences such as chemistry for the scientific and systematic study of all aspects of food, including raw materials, cooking procedures, flavour, acceptance of food and systematisation of recipes and cooking procedures. In contrast, aspects of human nutrition and health are less emphasised. The Barham *et al.* review ultimately concludes that "perhaps the most important objective of [molecular gastronomy] should be to delineate the essential principles that underpin our individual enjoyment of food" [6] (p. 2361).

Molecular gastronomy heavily relies on well-established sciences such as food chemistry, general food science and food-processing technology, which are themselves, of course, of seminal importance for the food industry [7-9], although it has been claimed that there is very little science in the food industry [10]. A related focus on the chemistry of food is internalized in a less well-defined discipline called 'culinary chemistry', a term directly inspired by the title of an 1821 book by the German chemist Friedrich Accum, who wrote about the practical aspects of applying scientific principles, in particular chemistry, to cooking [11].

The application of physical principles to study foods from a materials science perspective is well-established in research on food physics and food biophysics [12], with a focus on physical and physicochemical properties such as texture, foam stability, emulsification properties, phase transformations, the physical principles underlying cooking processes and so on. These approaches are traditionally separated from sensory sciences, and they are often less concerned with gastronomical considerations. Several popular science books are embodiments of this category [13-18]. The most celebrated monograph on the science of cooking is Harold McGee's by now classic and encyclopaedic monograph, *On Food and Cooking: The Science and Lore of the Kitchen* [19].

In recent years, the collaboration between famous chefs and scientists has materialized in cookbooks that contain a significant element of scientific explanations, such as Heston Blumenthal's *The Fat Duck Cookbook* [20], and books that focus on the science of particular types of food, such as the science of ice cream [21], chocolate [22,23], pizza [24] and sushi [25]. The most recent development in writing on the science of cooking is embodied in the monumental five-volume tome by Nathan Myhrvold and colleagues on what is called 'Modernist Cuisine' [26]. To date, Myhrvold's books provide the most comprehensive accounts of the physical aspects of food and cooking and could well qualify as a gastrophysical body of work, although he and his colleagues do not use the term 'gastrophysics' in their five-volume series. In fact,

very little, if anything, that can be called 'scientific' is written about gastrophysics [27,28], and there appear to be no published scientific papers about gastrophysics.

It is the contention of the present paper that gastrophysics has as its goal the demonstration that fundamental principles of physics, in particular soft matter physics, biophysical chemistry and molecular biophysics, can and should be brought together in scientific work dealing with food and should focus on molecular aspects, as well as scientific mechanisms and explanations, and their relation to gastronomy to a much higher degree than traditional food chemistry and food physics. The perspective on the plethora of possible research activities within gastrophysics should be gastronomy itself. Gastrophysics takes gastronomy as its departure point of inspiration and uses gastronomical problems as a driving force. The intrinsic dynamics of this way of doing science is very similar to what has been witnessed in recent decades in the case of another emerging field of science, biological physics [29], which deals with biologically inspired physics problems. Biological physics not only solves problems in biology but also contributes to physics in general by revitalizing it and opening up a new empirical world for physicists. In a similar vein, gastrophysics uses gastronomical problems as a motor. Gastrophysics hence places itself as an emerging new scientific and molecularly based subdiscipline at the borders between soft matter physics and chemistry, culinary sciences, food chemistry and molecular gastronomy.

It is to be expected that a sufficiently original, ambitious and courageous research endeavour should be able to make significant advances in defining the emerging and little developed field of gastrophysics and demonstrate where it could lead. This is not to say that the various experimental, computational and theoretical methodologies, typical of the three pillars of modern natural science, which should be applied to gastrophysics, are in a similarly less advanced and less well-defined stage. On the contrary, the idea is to use, so to speak, the 'sharpest knives in the kitchen' to carve out the future of the field. These 'knives' include molecular simulation, single-molecule studies of taste, high-resolution micro- and nanoscale visualisation of macromolecular aggregates of food molecules and particles, measurements of physical forces controlling the stability of food formulations, molecular-structural characterization of raw food materials and their transformation during preparation, cooking, ageing, drying, conservation and fermentation, as well as the design of physical model systems tailored to mimic, for example, the absorption of food molecules and food particles in the intestines. All of these approaches are highly sophisticated and advanced and are likely to make a strong impact on the emerging field of gastrophysics.

The present paper suggests that a unique and clear-cut opportunity exists to carve out the future for gastrophysical research by focusing attention on a specific class of materials, the marine macroalgae or seaweeds, which are virtually unexplored in a quantitative gastronomical setting. The bidirectional value of the approach, as implied by the analogy described above regarding biological physics, can then be assessed by investigating (1) how gastrophysics contributes to developing the algal cuisine and (2) how a focus on algae helps to frame gastrophysics as an emerging science.

Use of gastrophysics to explore and gastronomically chart a new territory: the kingdom of algae

The food we eat is derived from the mighty biological kingdoms of animals, plants and the fungi, as illustrated in Figure 1. Each kingdom contributes to our food and food cultures, with each their varieties in species, nutritional value and taste. Yet another big kingdom is out there which we hardly use for food: the algae. In terms of organic production on Earth, the algae are by far the most important, and because algae also perform photosynthesis, they are the main suppliers of atmospheric oxygen [30]. Algae are not plants, although the green algae are believed to be the ancestors of the higher (flowering) plants. Algae

are at the bottom of the food chain, hence they are the primary source of many of the valued nutrients, as well as flavour compounds, that end up in, for example, marine animals such as fish and shellfish.

The algal kingdom is a very diverse group of organisms, many of which are only very distantly related [30,31]. The unicellular algae constitute the largest group. Some of them are related to plants, whereas others are bacteria, such as the green-blue algae, which are cyanobacteria. The multicellular algae that live in the sea are called 'marine macroalgae' or 'seaweeds'. Seaweeds are traditionally classified into three categories: brown, red and green seaweeds. The classification scheme is the subject of much dispute, and it is noteworthy that the brown and green seaweeds are, loosely speaking, genetically as distinct as plants and animals [32] (Figure 1). There are about 10,000 different species of seaweeds, and seaweeds are found in all climatic regions of the globe. It is interesting to note that whereas many plants and fungi are poisonous and unsuitable for human consumption, all seaweeds harvested in clean water are not poisonous, except for a few species in the tropics. Viewed in this light, seaweeds comprise a rich new territory for gastronomical exploitation and innovation which is almost

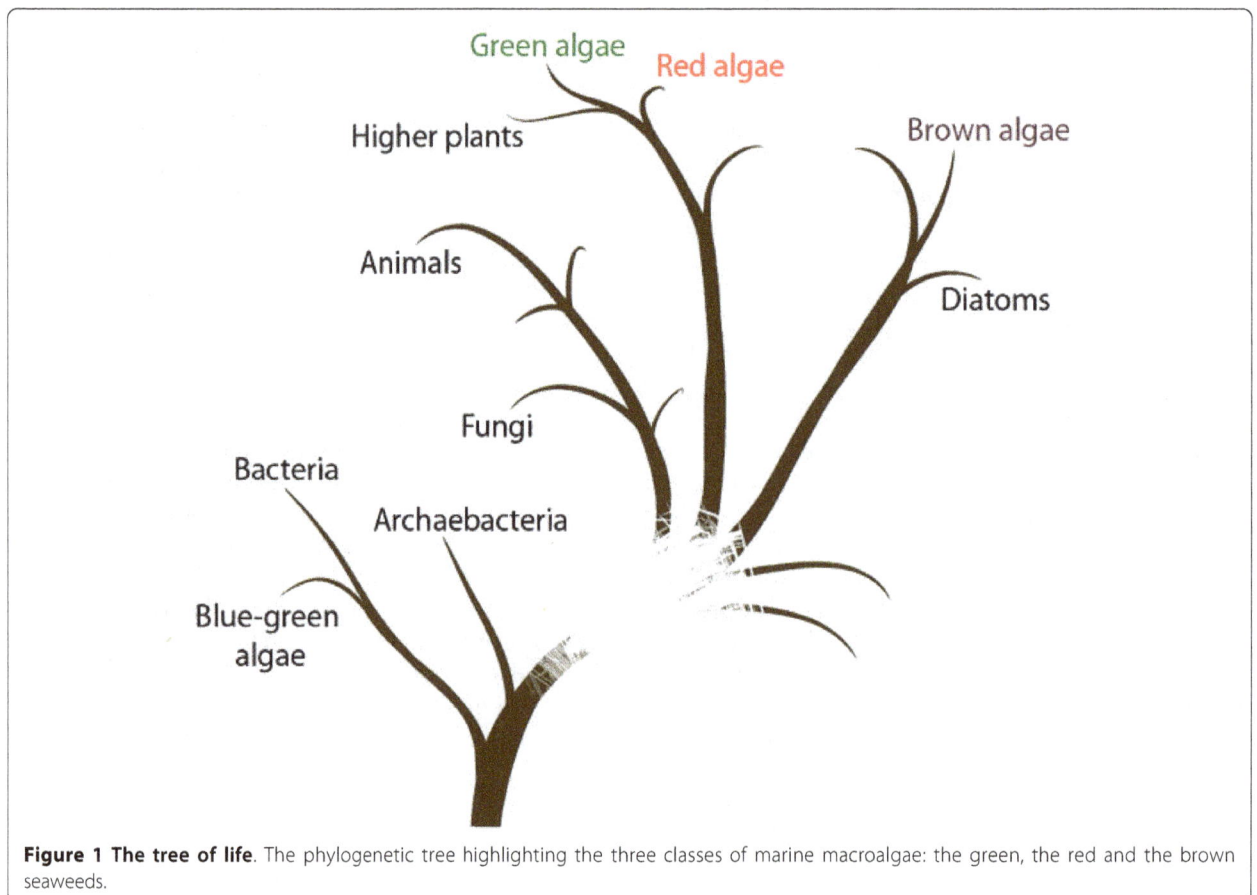

Figure 1 The tree of life. The phylogenetic tree highlighting the three classes of marine macroalgae: the green, the red and the brown seaweeds.

comparable in richness to the addition of the plant kingdom to a hypothetical cuisine based only on animal products.

The marine macroalgae are an almost unexploited resource for primary foodstuff in the Western world, whereas it is an essential part of the daily diet in the East [33]. In the West, however, and in Europe in particular, seaweeds are exploited mostly as extracts for additives and stabilizers in processed food, specifically alginate, carrageenan and agar (E400-E407). There is, however, a growing interest in North America and some parts of Europe in using raw and processed forms of seaweeds as foods and food supplements, and one also notices that an increasing interest in healthy Asian food stimulates a local demand for seaweeds such as nori, wakame and konbu. In Europe and North America, small companies in Iceland, Brittany, Wales, Ireland, Maine, British Columbia and California produce and market various forms of dried and salted seaweed products for human consumption.

To put the potential of an algal cuisine into the context of sustainability, it should be remarked that the sea is one of the last resorts for humankind to exploit for obtaining more food to feed a hungry world. The domestication of species in the ocean is now increasing at a much more rapid speed than terrestrial species [34], and the ocean holds the greatest potential for obtaining more food. Furthermore, food from the ocean, that is fish, shellfish and marine algae, is known to be beneficial for human nutrition as well as for general health and well-being [35]. However, fisheries can no longer meet the needs of the world population [36].

Seaweeds are in many ways optimal for human nutrition because they contain a bounty of important minerals, trace elements, vitamins, proteins, iodine and polyunsaturated fatty acids, which are listed in Table 1 [35,37-46]. Moreover, seaweeds have plenty of soluble and insoluble dietary fibres (45% to 75%) and hence contain few calories. Their actual composition depends very much on the seaweed species, age, location of growth, time of harvesting and storage conditions. The protein content varies generally from 7% to 35%, and it involves most of the essential amino acids. Nori (*Porphyra* spp.) is one of the seaweeds with the highest amount of protein. The mineral content of seaweeds is generally ten times higher than that of terrestrial plants. It is noteworthy that the salty taste of many types of seaweed is due to a large extent to potassium salts, which generally outbalance the sodium salts. This makes seaweeds a good salt substitute in food and may be a way of combating the increasing global problem of hypertension.

Although the fat content of seaweeds is generally low (about 2% to 3% on a dry weight basis), it is noteworthy

Table 1 Important nutritional elements in seaweeds[a]

Nutrients in seaweeds	Composition
Proteins, essential amino acids	7% to 35%
Dietary fibres (soluble, insoluble)	45% to 75%
Vitamins	A, B (B_1, B_2, B_3, B_6 and B_{12} of folate), C, E
Iodine	Variable large amounts in some brown species
More K^+ salts than Na^+ salts	
Minerals	Fe, Ca, P, Mg, Cl
Trace compounds	Zn, Cu, Mn, Se, Mb, Cr
Essential fatty acids	(2% to 5%), ω-3 (EPA, no DHA), ω-6 (AA)

[a]AA, arachidonic acid; DHA, docosahexaenoic acid; EPA: eicosapentaenoic acid. The actual composition depends very much on the species, its age, the location of growth and the time of harvesting [35,37-46]. Percentages refer to dry weight.

that generally more than one-half of the fat is made up of unsaturated fatty acids, the majority of which are essential polyunsaturated ω-6 and ω-3 fatty acids [35]. An important group of ω-6 and ω-3 fatty acids are the very long and superunsaturated fatty acids, specifically arachidonic acid (AA) in the ω-6 family and eicosapentaenoic acid (EPA) and docosahexaenoic acid (DHA) in the ω-3 family. Whereas microalgae contain all three of these fatty acids, it is noteworthy that seaweeds contain only AA and EPA and very little, if any, DHA. An interesting observation is that the ratio of the amount of ω-3 to ω-6 fatty acids in seaweeds is in the range of unity or even greater in certain cases [35,41]. This should be contrasted to a typical European or North American diet, which has ratios of about 0.1 and 0.05, respectively [47]. Nutritionists generally recommend that the diet be equally balanced with regard to ω-3 and ω-6 fatty acids [47]. The severe imbalance toward too much ω-6 in the Western diet is believed to be one of the main causes of the burden of ill health, in particular in relation to heart and coronary diseases, obesity, diabetes and certain mental diseases [48-50]. Because seaweeds contain large amounts of natural antioxidants such as polyphenols [37], one can speculate that these antioxidants are instrumental in protecting the polyunsaturated fatty acids in the seaweeds when they are used as whole foods. Little is known to this effect, and a study of these important fatty acids, their amounts, their flavours and their bioavailability in foods could well turn out to be a rewarding gastrophysical project.

Seafood, including seaweeds, was probably the most important component of our ancestors' diet and part of the basis for the evolution of our brain and neural system, which are characterized by a large content of the superunsaturated ω-3 and ω-6 fatty acids at a proportion close to 1:1 [50]. Seaweeds may therefore be considered to be brain food, and it is their polyunsaturated fatty acid content and their perfect balance between ω-3 and ω-6 fatty acids that are particularly beneficial for our health.

Seaweeds can be harvested in the wild or grown in the sea in large amounts in a sustainable fashion. Undoubtedly, seaweeds will become a larger part of our daily diet in the future. Importantly, these 'vegetables of the sea' will help us to renew and balance our diet and may contribute to combating the increased incidence of Western lifestyle-related diseases, in particular cardiovascular and heart diseases, such as hypertension, cancer and obesity, as well as mental disorders [47,49].

A 2010 international seaweed symposium [51] rallied key actors in the science and use of seaweeds for foodstuff and also attracted practitioners in the fields of aquaculture, the food industry, gastronomy, nutrition and the health sciences. It became clear at the conference that there is a very strong and growing interest in seaweeds as food, and it was concluded that there is an urgent need for more research and the establishment of scientifically based knowledge about the uses of seaweeds for human consumption in people's ordinary diets, in futuristic functional foods and in gastronomy at all levels.

Therefore, to introduce the algal cuisine in the Western world, there is a need for a gastronomically inspired approach to using algae in cuisine. It is the contention of the present paper that the emerging science of gastrophysics can be defined and can show its potential by being applied to algal cuisine.

Traditional uses of algae in cooking

Whereas many microalgae, in particular spirulina [52] and *Chlorella* [53], are widely used for human consumption and food supplementation, in particular because of their high content of proteins, essential fatty acids and various minerals, these algae are not particularly interesting for gastronomical uses, except possibly for their unique ability to impart a green hue to dishes and drinks such as 'smoothies'. Microalgae generally have no interesting taste and texture. In contrast, macroalgae are very interesting, in particular because of their flavours and textures.

Although brown, red and green seaweed species have been eaten by all coastal peoples since prehistoric times [54], the regular consumption of seaweeds has predominantly survived to the present primarily in the contemporary cuisines of Asian countries such as Japan, Korea, China and the Philippines. In Japan especially, seaweeds are both fully integrated into the daily diet and used to create gastronomic specialties. Elsewhere seaweeds are now overwhelmingly linked to the preparation of Asian-style meals. Nevertheless, the globalization of foods such as sushi has raised their profile in North America and, increasingly, in Europe.

The traditional use of seaweeds as part of the normal diet has almost died out in the Western world, but it can still be encountered here and there. The pattern of consumption is on an upward trend, however, no doubt driven by an awareness of its possible beneficial effects on overall health. In Europe, and particularly in Ireland [55], there are signs of a revival of interest in cooking with seaweeds, although this is still considered quirky and exotic. Many of today's leading chefs, however, have become fascinated with marine algae and are experimenting with their tastes, textures and colours in familiar dishes and in exciting, new gastronomic creations.

In some of the coastal areas of North America and northern Europe, the traditional consumption of seaweeds has been preserved and has become more prominent in the past few decades. This is especially the case in California and Maine in the United States, in British Columbia and Nova Scotia in Canada, and in Iceland, where dried seaweeds are now available in many supermarkets. In Europe, seaweeds still have a place in the popular cuisines of Brittany, Wales and Ireland. In Iceland, dulse is eaten in dried form as a snack and is mixed into salads, bread dough or curds, just as it was in the time of the sagas. Chefs in Scandinavia are actively finding creative ways to use seaweeds in fusion dishes as part of a revitalization of the Nordic food cultures of former times and the rise of the New Nordic Cuisine [56].

Seaweeds in gastronomy

Whereas there is a large body of scientific literature about the nutritional composition of various seaweeds, in particular with respect to protein, fibres, carbohydrates, minerals and fatty acids (see [35,37-46]), virtually no scientific studies of the basic physical and physicochemical properties of seaweeds in the context of gastronomy and gastronomical innovation have been undertaken. This is surprising, considering that the world market for seaweeds for human consumption amounts to almost $10 billion annually. Over the past few years, the interest among the public and chefs up to the Michelin star level has been on a steep increase. At the same time, some European aquaculturists have expressed an interest in farming seaweeds for human consumption.

It is the seaweeds' content of high amounts of minerals, proteins and essential amino acids, trace elements, vitamins, iodine, low-calorie dietary fibres and polyunsaturated fatty acids which underlie their many nutritional qualities as human foodstuff. This is well-recognized in the East, such as in China, Korea and Polynesia, and particularly in Japan, where seaweeds are an essential part of the classic cuisine and highly regarded for their nutritional value, flavour and texture [33].

In the Western world, the interest in using seaweeds for cooking was stimulated by the macrobiotic diet movement in the 1960s and 1970s. Judith Cooper Madlener's classic seaweed cookbook was published during this period [57]. Later the globalization of Asian food, in particular sushi from Japan and its revival in Californian cuisine, introduced seaweeds such as konbu, wakame and particularly nori to the Western palate [25]. A number of cookbooks focused on seaweeds were subsequently published, many of them with reference to the macrobiotic and health food movements [58-66], others with a firm basis in the experience of individual seaweed harvesters [67-70] and a few books penned by scientists [35,55,71].

Seaweed as a novel raw material for foods has a unique potential for gastronomical and sensory innovation. Although it has been used for centuries in other parts of the world, a serious and dedicated research program has to be carried out to develop and exploit this potential in a Western context and for the Western palate. It is noteworthy that whereas a very substantial body of scientific literature exists on seaweed farming, processing and nutrition, there is hardly any science-based sensory and gastronomical knowledge about seaweeds available on the international scene. Prominent exceptions are studies of the large brown seaweed konbu as a source of *umami* related to its contents of monosodium glutamate [56,72-75]. Together with 5'-ribonucleotides such as inosinate and guanylate, glutamate promotes synergistic flavour enhancement effects (*umami*) [76] and may have applications in salt and sugar reduction in a variety of foods, including meats [75,77,78].

Probably no other ingredients used for cooking are as versatile as seaweeds. They can be eaten raw, cooked, baked, toasted, puréed, dried, granulated or deep-fried. They can be eaten on their own or combined in countless ways with other cold or hot ingredients. In most instances, the seaweeds are believed to retain the greater part of their nutritional value in unchanged form, although much research along these lines is needed.

One can turn for inspiration to the greater attention that has been paid to marine algae during the past few years by many serious practitioners of 'haute cuisine'. This has been driven to a certain extent by a felicitous combination of factors: the locavore movement, a much greater consciousness of promoting sustainable food sources and the desirability of eating healthy, pesticide-free foods.

For the professional chef, the challenge of how to use raw materials is on a completely different plane from the one faced in the home kitchen. Like art and the search for new knowledge, gastronomy is driven by a combination of craft, desire for renewal, self-criticism, vision and delight in playing with new ingredients and ideas. For most chefs in Europe and North America,

seaweeds are a relative novelty, which inspires them to rework traditional recipes in daring ways and to devise pioneering gastronomic experiences.

In inventive gastronomy, the goal is to invent fabulous new dishes in which seaweeds have been incorporated as essential, and often surprising, elements, which elevate the resulting creation to something unique and possibly sublime, something that calls out for both sensory and aesthetic appreciation. Well-known chefs have already embraced the use of seaweeds in their kitchens and restaurants to capitalise on the profusion of tastes, textures, shapes and colours that are characteristic of marine algae. Often this involves making excursions to harvest fresh seaweeds from their local surroundings, a natural extension of the quest for wild, edible plants and herbs in forests and fields. Others have drawn inspiration from classic Asian cuisine, where seaweeds have retained a prominent role.

One might wonder how the patrons of high-end expensive restaurants react to finding seaweeds on the menu. It seems that part of the appeal lies in the ingredients themselves. The diners appear to be intrigued by the idea of eating seaweeds and take an interest in learning more about them and their places of origin. In this way, tasty and flavourful marine algae may win a permanent place on the menu and, in the future, be featured more prominently, which would represent a vindication for the food of our ancestors.

Research questions for gastrophysics applied to the algal cuisine

The present paper has advocated the viewpoint that there is a need to apply concepts, methodologies and state-of-the-art methods from physics and biophysical chemistry to solve gastronomically inspired problems. This approach which we call 'gastrophysics' employs a wide range of the most powerful theoretical, simulational and experimental biophysical techniques to study the structure, dynamics and functional properties of raw and processed foodstuff with a focus on small-scale, macromolecular physical properties and how they relate to flavour and texture.

Among many others, the following could be some of the specific focal areas of the gastrophysical study of seaweeds: (1) visualization and characterization of structure on the small scale of raw and processed seaweeds; (2) determination of the stability and micromechanical properties of foodstuff derived from seaweeds and optimisation of these properties with respect to enhanced taste, mouthfeel and flavour; (3) measurement of the release of flavour compounds as well as the absorption and transport of algal food particles and nutrients via physical models of the intestinal barrier; (4) assessment of the biophysical properties of the precious ω-3 fatty acids and sterols synthesised by algae and the effect of seaweed bioactive compounds on cell membranes; (5) characterization of the physics of the taste sensory system stimulated by glutamate from seaweeds using single-molecule receptor taste compound binding assays; and (6) investigations of synergies in taste by computer simulation of the *umami* receptor.

The plethora of physical studies should be fully integrated to define and characterise the emerging science of gastrophysics with seaweeds as the test case. *En route*, collaboration with food chemists, phycologists, sensory scientists, science communicators, high-level chefs and gastronomical entrepreneurs will help to shape the research directions and provide the means for implementation of scientific achievements in tomorrow's cuisine.

Conclusion and future outlook

The empirical world of cooking and gastronomy is a promising new territory for application of state-of-the-art methods used in physics and biophysical chemistry that may not only transform our study of this empirical world into a new science, 'gastrophysics', but also revitalize the physics of the kind of soft matter of which foodstuffs are made, potentially leading to new fundamental insights that can be translated into a more scientifically inspired approach to gastronomy. The field of gastrophysics will employ a wide range of the most powerful theoretical, simulational and experimental biophysical techniques to study the structure, dynamics and functional properties of raw and processed foodstuff materials with a focus on small-scale and macromolecular physical properties and how they relate to flavour and texture.

This paper presents a vision of a gastrophysical journey into an overlooked and neglected territory of foodstuff, the seaweeds, well-known in the Asian cuisine but only beginning to enter Western cuisine. Seaweeds are basically unexplored scientifically in relation to gastronomy. For reasons related to the world's future supplies of food, a low-carbon imprint of sustainable food production, human health and nutrition, and gastronomical innovation, seaweeds appear to be an ideal subject of study for the emerging science of gastrophysics.

The future use of seaweeds and seaweed products for human consumption will be a highly innovative enterprise because the market is basically virgin in Europe and North America. There is currently a large import of seaweeds from Asia, and individuals make purchases via the internet, but the potential for seaweed use is increasing steeply because of consumers' growing interest in healthy foods that are produced in an environmentally sustainable way. Many leading chefs and gastronomical entrepreneurs are spearheading this trend, and food writers, nutritionists and food scientists are fuelling the interest with novel ideas and factual information.

Consumers have started asking for products available in restaurants and stores. This underscores an urgent need for scientifically based information on which consumers, producers, and companies can base their purchases and strategies for gastronomical development, respectively.

Hence the time is ripe for gastrophysical studies of seaweeds, and it can be expected that the results of such studies may have an impact on decision-making among producers, manufacturers and gastronomical entrepreneurs. Last but not least, knowledge-based information communicated to the public will provide the best support for consumers' ability to navigate tomorrow's food market, which invariably will involve more seaweed products than we have ever seen before. It is worth pointing out that a recently launched plan in the Nordic countries for a special *terroir*-based 'New Nordic Diet' does indeed now recommend regular consumption of seaweed products [79].

Application of gastrophysical studies of seaweeds are hence timely and should serve as a nucleus for the growth of a sector comprising the production and processing of seaweeds for human consumption, with a clear aim of creating gastronomical value and, not least, flavour. The anticipated results of gastrophysical investigations of seaweeds will provide a valuable source of knowledge for (1) future recommendations regarding the development and production of select, high-quality seaweeds for human consumption (fresh, dried and processed); (2) best practice designs of drying and storage procedures; (3) selection of algal species with maximal health benefits and desirable sensory perceptions; (4) inspiration of gastronomical entrepreneurs and producers and distributors of seaweed products; and (5) information for the general public about the various benefits of a diet with a certain content of seaweed products.

Wider informed use of seaweeds and seaweed products in the diet is likely to lead to a number of general health benefits and may aid in combating diet-related lifestyle-related diseases. Gastronomical innovation based on seaweeds is expected to create a demand for seaweeds as foodstuff or food ingredients and hence to fuel the creation of new industries and small businesses, as well as to rejuvenate the commercial potential for existing producers of daily foodstuff. The development of farmed seaweed production, possibly with organic certification, will provide a new and sustainable use of the resources of local waters.

Abbreviations
AA: arachidonic acid; DHA: docosahexaenoic acid; EPA: eicosapentaenoic acid.

Acknowledgements

The author gratefully acknowledges many discussions with Per Lyngs Hansen and Amy Rowat about the science and practicalities of cooking and for coining and clarifying the term 'gastrophysics'. Felix Göni is thanked for references to papers on the science of cooking in the Spanish magazine *SEBBM*. Mariela Johansen is gratefully acknowledged for careful translation, from Danish into English, of some of the author's writing on seaweeds. MEMPHYS was supported as a centre of excellence by the Danish National Research Foundation for the period 2001 to 2011. The Nordic Food Lab is an independent institution supported by Danish government sources, external funds and private companies, as well as by foundations such as NordeaFonden. This work was supported by a grant (J.nr. 3414-09-02518) from the Danish Food Industry Agency.

Author details

[1]MEMPHYS, Center for Biomembrane Physics, Department of Physics, Chemistry and Pharmacy, University of Southern Denmark, Campusvej 55, DK-5230 Odense M, Denmark. [2]Nordic Food Lab, 93 Strandgade, DK-1401 Copenhagen K, Denmark.

Competing interests

The author declares that he has no competing interests.

References

1. Brillat-Savarin JA: In *Physiologie du goût [The Physiology of Taste: Or, Meditations on Transcendental Gastronomy]*. Edited by: Drayton A, transl. London: Penguin Books; 1994:.
2. Kurti N: **The physicist in the kitchen.** *Proc R Inst G B* 1969, **42**:451-467.
3. This H: In *Traité élémentaire de cuisine [Molecular Gastronomy: Exploring the Science of Flavor]*. Edited by: DeBevoise MB, transl. New York: Columbia University Press; 2006:.
4. This H: In *Les secrets de la casserole [Kitchen Mysteries: Revealing the Science of Cooking]*. Edited by: Gladding J, transl. New York: Columbia University Press; 2007:.
5. Adrià F, Blumenthal H, Keller T, McGee M: **Statement on the 'new cookery'.** *Observer (Lond)* 2006, [online article] http://www.guardian.co.uk/uk/2006/dec/10/foodanddrink.obsfoodmonthly.
6. Barham P, Skibsted LH, Bredie WLP, Bom Frøst M, Møller P, Risbo J, Snitkjær P, Mørch Mortensen LM: **Molecular gastronomy: a new emerging scientific discipline.** *Chem Rev* 2010, **110**:2313-2365.
7. Belitz HD, Grosch W, Schieberle P: In *Food Chemistry*. 3 edition. Edited by: Burghagen MM, transl. Heidelberg: Springer; 2004:.
8. **Fennema's Food Chemistry (Food Science and Technology).** In *Boca Raton, FL: CRC Press* Edited by: Damodaran S, Parkin KL, Fennema OR , 4 2007.
9. Coultate TP: *Food: The Chemistry of Its Components*. 4 edition. Cambridge, UK: Royal Society of Chemistry; 2002.
10. This H: **Hay muy poca ciencia en la industria alimentaria [in Spanish].** *SEBBM* 2010, **166**:25-29.
11. Accum FCA: *System of Theoretical and Practical Chemistry* Philadelphia: Kimber and Conrad;, 1808.
12. Barham P: *The Science of Cooking* Heidelberg: Springer; 2001.
13. Wolke RL: *What Einstein Told His Cook: Kitchen Science Explained* New York: W.W. Norton & Co; 2002.
14. Wolke RL: *What Einstein Told His Cook 2: Further Adventures in Kitchen Science-The Sequel* New York: W.W. Norton & Co; 2005.
15. Hillman H: *The New Kitchen Science: A Guide to Knowing the Hows and Whys of Fun and Success in the Kitchen* New York: Houghton Mifflin; 2003.
16. Joachim D, Schloss A: *The Science of Good Food: The Ultimate Reference on How Cooking Works* Toronto: Robert Rose; 2008.
17. Potter J: *Cooking for Geeks. Real Science, Great Hacks, and Good Food* Sebastopol, CA: O'Reilly; 2010.
18. Vega C, Ubbibk J, van der Linden E: *The Kitchen as Laboratory. Reflections on the Science of Food and Cooking* York: Columbia University Press; 2012, .
19. McGee H: *On Food and Cooking: The Science and Lore of the Kitchen* New York: Scribner; 2004, Revised edition.
20. Blumenthal H: *The Fat Duck Cookbook* London: Bloomsbury; 2008.
21. Clarke C: *The Science of Ice Cream* Cambridge: Royal Society of Chemistry; 2004.
22. Beckett ST: *The Science of Chocolate* Cambridge: Royal Society of Chemistry; 2000.
23. Rowat AC, Hollar K, Rosenberg D, Stone HA: **The science of chocolate: phase transitions, emulsification, and nucleation.** *J Chem Educ* 2011, **88**:29-33.
24. Rowat AC, Hollar K, Rosenberg D, Stone HA: **The science of pizza: the molecular origins of cheese, bread, and digestion using interactive activities.** *J Food Sci Educ* 2010, **9**:106-112.
25. Mouritsen OG: *Sushi: Food for the Eye, the Body, & the Soul* New York: Springer; 2009.
26. Myhrvold N, Young C, Bilet M: *Modernist Cuisine: The Art and Science of Cooking* Bellevue, WA: The Cooking Lab; 2011, 5 vols.
27. Parkers K: **Recipe for success: teachers get inspiration from 'gastrophysics'.** *Phys Educ* 2004, **39**:19.
28. Ogborn J: **Soft matter: food for thought.** *Phys Educ* 2004, **39**:45-51.
29. *Physics of Biological Systems*. Heidelberg: Springer Edited by: Flyvbjerg H, Hertz J, Jensen MH, Mouritsen OG, Sneppen K 1997.
30. Barsanti L, Gualtieri P: *Algae: Anatomy, Biochemistry, and Biotechnology* Boca Raton, FL: CRC Press; 2006.
31. Andersen RA: **Diversity of eukaryotic algae.** *Biodivers Conserv* 1992, **1**:267-292.
32. Braune W, Guiry MD: *Seaweeds: A Colour Identification Guide to the Common Benthic Green, Brown and Red Algae of the World's Oceans* Königstein, Germany: Koeltz Scientific Books; 2011, Revised and translated by Guiry MD.
33. Arasaki S, Arasaki T: *Low Calorie, High Nutrition Vegetables from the Sea to Help You Look and Feel Better* Tokyo: Japan Publications Inc; 1983.
34. Duarte CM, Marbá N, Holmer M: **Rapid domestication of marine species.** *Science* 2001, **316**:382-383.
35. Mouritsen OG: *Seaweeds. Edible, Available & Sustainable [in Danish]* Copenhagen: Nyt Nordisk Forlag Arnold Busck; 2009.
36. Food and Agriculture Organization of the United Nations: **The State of World Fisheries and Aquaculture 2010.** [http://www.fao.org/docrep/013/i1820e/i1820e00.htm].
37. Løvstad Holdt S, Kraan S: **Bioactive compounds in seaweed; functional food applications and legislation.** *J Appl Phycol* 2010, **23**:543-597.
38. Pereira L: **A review of the nutrient composition of selected edible seaweeds.** In *Seaweed: Ecology, Nutrient Composition and Medicinal Uses.* Edited by: Pomin VH. Happauge, NY: Nova Science Publishers; 2012:.
39. Colombo ML, Risè P, Giavarini F, De Angelis L, Galli C, Bolis CL: **Marine macroalgae as sources of polyunsaturated fatty acids.** *Plant Foods Hum Nutr* 2006, **61**:67-72.
40. Khotimchenko SV, Vaskovsky VE, Titlyanova TV: **Fatty acids of marine algae from the Pacific Coast of North California.** *Botanica Marina* 2002, **45**:17-22.
41. Dawczynski C, Schubert R, Jahreis G: **Amino acids, fatty acids, and dietary fibre in edible seaweed products.** *Food Chem* 2006, **103**:891-899.
42. Rupérez P: **Mineral content of edible marine seaweeds.** *Food Chem* 2002, **79**:23-26.
43. Teas J, Pino S, Critchley A, Braverman LE: **Variability of iodine content in common commercially available edible seaweeds.** *Thyroid* 2004, **14**:836-841.
44. van Netten C, Hoption Cann SA, Morley DR, van Netten JP: **Elemental and radioactive analysis of commercially available seaweed.** *Sci Total Environ* 2000, **255**:169-175.
45. Sánchez-Machado DI, Lopez-Cervantes J, López-Hernández J, Paseiro-Losada P: **Fatty acids, total lipid, protein and ash contents of processed edible seaweeds.** *Food Chem* 2004, **85**:439-444.
46. MacArtain P, Gill CIR, Brooks M, Campbell R, Rowland IR: **Nutritional value of edible seaweeds.** *Nutr Rev* 2007, **65**:535-543.
47. Simopoulos AP: **The importance of the ratio of ω-6/ω-3 essential fatty acids.** *Biomed Pharmacother* 2002, **56**:365-379.
48. Valentine RC, Valentine DL: *Ω-3 Fatty Acids and the DHA Principle*. Boca Raton, FL: CRC Press 2010.
49. **Poly-unsaturated fatty acids, neural function and mental health.** In *Biol Skr Dan Vid Selsk* Edited by: Mouritsen OG, Crawford MA 2007, **56**:1-87.
50. Cunnane SC, Stewart KM: *Human Brain Evolution: The Influence of Freshwater and Marine Food Resources* Hoboken, NJ: Wiley-Blackwell; 2010.
51. Seaweed Symposium: **Seaweeds for Human Consumption, Bioactive Compounds, and Combating of Diseases.** An international interdisciplinary

symposium, Carlsberg Academy, Copenhagen, Denmark; 2010 [http://www.tangbog.dk/?page_id=597].

52. *Spirulina in Human Nutrition and Health.* Boca Raton, FL: CRC Press Edited by: Gershwin ME, Belay A 2008.

53. Ley BM: *Chlorella: The Ultimate Green Food* Golden Valley, MN: BL Publications/NHL Ministries; 2003.

54. Dillehay T, Ramírez DC, Pino M, Collins MB, Rossen J, Pino-Navarro JD: **Monte Verde: Seaweed, food, medicine, and the peopling of South America.** *Science* 2008, **320**:784-786.

55. Rhatigan P: *Irish Seaweed Kitchen: The Comprehensive Guide to Healthy Everyday Cooking with Seaweeds* Co Down, Ireland: Booklink; 2010.

56. Mouritsen OG, Williams L, Bjerregaard R, Duelund L: **Seaweeds for *umami* flavour in the New Nordic Cuisine.** *Flavour* 2012, **1**(1):4.

57. Cooper Madlener J: *The Seavegetable Book: Foraging and Cooking Seaweed* New York: Clarkson N Potter; 1977.

58. Babel K: *Seafood Sense: The Truth About Seafood Nutrition & Safety* Laguna Beach, CA: Basic Health Publications; 2005.

59. Bradford P, Bradford M: *Cooking with Sea Vegetables: A Collection of Naturally Delicious Dishes Using to the Full the Bountiful Harvest of the Oceans* New York: Thorson Publishers; 1985.

60. Chavannes CD: *Algues: Légumes de la mer* Sète, France: Editions La Plage; 2002.

61. Cooksley VG: *Seaweed: Nature's Secret Balancing Your Metabolism, Fighting Disease, and Revitalizing Body & Soul* New York: Stewart, Tabori and Chang; 2007.

62. Ellis L: *Seaweed, A Cook's Guide: Tempting Recipes for Seaweed and Sea Vegetables* San Francisco: Fisher Books; 1998.

63. Fryer L, Simmons D: *Food Power from the Sea: The Seaweed Story* Austin, TX: Acres U.S.A; 1977.

64. Gusman J: *Vegetables from the Sea. Everyday Cooking with Sea Greens* New York: William Morrow Cookbooks; 2003.

65. Huston F, Milne X: *Seaweed and Eat It: A Family Foraging and Cooking Adventure* London: Virgin Books; 2008.

66. Maderia CJ: *The New Seaweed Cookbook: A Complete Guide to Discovering the Deep Flavors of the Sea* Berkeley, CA: North Atlantic Books; 2007.

67. Lewallen E, Lewallen J: *Sea Vegetable Gourmet Cookbook and Wildcrafter's Guide* Mendocino, CA: Mendocino Sea Vegetable Co; 1996.

68. Erhart S, Cerier L: *Sea Vegetable Celebration* Summertown, TN: Tennessee Book Publishing Co; 2001.

69. McConnaughey E: *Sea Vegetables: Harvesting Guide & Cookbook* Happy Camp, CA: Naturegraph Publishers; 2002.

70. Harbo RM: *The Edible Seashore: Pacific Shores Cookbook & Guide* Surrey, BC: Hancock House; 1988.

71. Druehl L: *Pacific Seaweeds: A Guide to Common Seaweeds of the West Coast* Madeira Park, BC: Harbour Publishing; 2000.

72. Ninomiya K: ***Umami*: a universal taste.** *Food Rev Int* 2002, **18**:23-38.

73. Kasabian A, Kasabian D: *The Fifth Taste: Cooking with Umami* New York: Universe Publishing; 2005.

74. Blumenthal H, Barbot P, Matsushisa N, Mikuni K: *Dashi and Umami: The Heart of Japanese Cuisine* London: Eat-Japan, Cross Media; 2009.

75. Mouritsen OG, Styrbæk K: *Umami: The Gourmet Ape & the Fifth Taste [in Danish]* Copenhagen: Nyt Nordisk Forlag Arnold Busck; 2011.

76. Zhang F, Klebansky B, Fine RM, Xu H, Pronin A, Liu C, Tachdjian C, Li X: **Molecular mechanism for the *umami* taste synergism.** *Proc Natl Acad Sci USA* 2008, **105**:20930-20934.

77. Strauss S: **Parse the salt, please.** *Nat Med* 2010, **16**:841-843.

78. López-López I, Cofrades S, Jiménez-Colmenero F: **Low-fat frankfurters enriched with *n*-3 PUFA and edible seaweed: effects of olive oil and chilled storage on physicochemical, sensory and microbial characteristics.** *Meat Sci* 2009, **83**:148-154.

79. Mithril C, Dragsted LO, Meyer C, Blauert E, Holt MK, Astrup A: **Guidelines for the New Nordic Diet.** *Pub Health Nutr* 2012, **17**:1-7.

7

Comparative biology of taste: Insights into mechanism and function

Gary K Beauchamp[*] and Peihua Jiang

Abstract

Each animal lives in its own sensory world that is coordinated with its diet. In this brief review, we describe several examples of this coordination from studies of the sense of taste, particularly from species of the order Carnivora. This order includes species that are obligate carnivores (e.g., *Felis* species), omnivores, and strict plant eaters. Many of the obligate carnivores have lost function for sweet taste, presumably through relaxation of selection for eating sugars from plants. In contrast, the giant panda, which feeds almost exclusively on bamboo, retains sweet taste function but may have lost amino acid (umami) taste perception. Finally, mammals that have "returned" to the sea, such as sea lions, have experienced even more extensive taste loss, presumably as a consequence of adaptations to a diet of fish and other sea creatures swallowed whole. Future comparative studies will surely reveal important relationships between diet and molecular, cellular, and behavioral taste adaptations that will shed light on how evolution moulds sensory structure and function.

Keywords: Taste, Taste receptors, Comparative studies, Carnivora, Cats, Giant panda, Sea lion, Evolution

Each animal species lives in a separate sensory world that is coordinated with its behavioral ecology. A dramatic example of this occurs for the sense of taste [1] where sensory perception and diet choice are intimately intertwined.

The evolutionary basis for the existence of a small number of primary taste qualities (sweet, bitter, sour, salty, umami, and perhaps a few others) is that these qualities evolved to detect and motivate consumption of critical nutrients and detect and avoid potential poisons. It is widely believed that sweet taste evolved in animals that eat plants to detect energy-rich simple sugars such as glucose, fructose, and sucrose. In contrast, bitter taste presumably functions to insure that an animal avoids poisons; most poisons are bitter and most bitter substances are harmful although this relationship is not perfect. Salty taste is thought to enable detection of sodium, an absolutely essential mineral. When some species of animals become deficient in sodium—usually this occurs in herbivorous animals—a powerful appetite for salty taste is aroused. And for many species, salt is consumed even when there is no apparent need. For sour taste,

many have suggested that it is involved in the detection of the ripeness of fruits. Finally, the fifth basic taste, umami or savory, probably serves to signal amino acids and protein. This however remains speculative. Other classes of compounds may also interact with the taste system (e.g., fatty acids, calcium, starch), but they do not give rise to the (to humans) strong qualitative percept that the other five do.

To obtain a clearer understanding of the functional significance for these basic taste qualities, we have studied the order Carnivora. Our goal is to understand how taste receptors and taste perception in different species are related to different feeding ecologies with a particular focus on sweet compounds. For example, some Carnivora species are obligate carnivores (e.g., cats), whereas others are almost completely herbivorous, sometimes feeding on virtually a single plant (e.g., giant panda). If the function of sweet taste is to detect simple sugars in plants, we predict that animals that do not consume plants would not need/have sweet taste perception. By examining sweet taste perception across a number of species in this order, we can put this prediction to the test.

Many years ago, we [2] demonstrated that domestic and wild cats (*Felis* and *Panthera* species) are indifferent to all sweeteners tested but are highly responsive to

* Correspondence: beauchamp@monell.org
Monell Chemical Senses Center, 3500 Market Street, Philadelphia, PA 19104, USA

certain amino acids and fats. We speculated that these species may not have the ability to perceive sweet (to humans) sugars. Following the discovery of the major sweet taste receptor, the T1R2 + T1R3 heterodimer (review: [3]), we demonstrated that the cat's indifference to sweeteners can be explained by the pseudogenization of the *Tas1r2* gene which encodes the T1R2 receptor. That is, the sweet taste receptor of the domestic cat as well as closely related wild cats such as lions and tigers has accumulated numerous germ-line mutations of the *Tas1r2* gene, thereby rendering the sweet receptor non-functional [4].

We next reasoned that other exclusively meat-eating species might also have an inactive form of this gene. Sequencing of the entire coding region of the *Tas1r2* gene from 12 Carnivora species revealed that seven of these species, all exclusive meat eaters, had independently fixed a defective *Tas1r2* allele [5]. Since these disabling mutations occurred at different places within the *Tas1r2* gene in each species, this loss of sweet taste function in multiple species in the Carnivora has occurred independently and thus repeatedly during their evolution. Behavioral tests of two of the genotyped species, the Asian otter (defective *Tas1r2*) and the spectacled bear (intact *Tas1r2*), were consistent with the genetic findings: The former showed no preference for sweet-tasting compounds, while the latter preferred sugars and some non-caloric sweeteners. These results indicate that the independent loss of a functional *Tas1r2* is widespread among obligate carnivores. We suggest that this loss is a consequence of the relaxation of selective pressures maintaining receptor integrity.

A striking study with birds provides additional support for the hypothesis that sweet taste exists to detect simple sugars. All birds apparently lack a homolog for the *Tas1r2* gene; this loss likely occurred as the non-avian reptile and bird lines split. Thus, it would seem that birds should not be able to taste sweet sugars. But if this were the case, how can one explain the behavior of avian species that consume sweet sugars such as hummingbirds? Baldwin et al. [6,7] provide one answer: The receptor dimer T1R1 + T1R3, the amino acid or umami receptor in mammals, has been repurposed in these bird species to detect simple sugars thereby opening a novel source of energy not available to many other birds. In sum, these studies provide strong support for the hypothesis that sweet taste perception exists to provide an ability to identify energy-rich sugars.

More recently [8], we conducted behavioral and molecular studies with giant pandas, animals that consume plants, but ones (bamboo) without abundant simple sugars. Would this member of the order Carnivora retain sweet taste perception, or would the absence of a need to find specific plants that taste sweet also result in

relaxed selection for maintenance of receptor function? We found that sweet taste perception is fully functional in giant pandas. Although giant pandas thus retain an avidity for sweet compounds, genetic evidence suggests that this species has lost umami taste perception [9], but as yet we know of no behavioral studies verifying this nor do we understand why this may have occurred and how widespread such loss might be.

Although loss of sweet taste seems common for animals that do not consume plants, are there species that have lost even more of the basic tastes? And if so, how can this be interpreted? Based on genetic studies, we [5] and others [10] have reported that many mammalian species that have returned to the sea (e.g., sea lions, dolphins, whales) may have independently lost function for several, perhaps all, taste quality perception. These genetic studies are consistent with anatomy (many of the species do not have identifiable taste cell structures) and behavior (many eat their food whole, without apparently "tasting" it). The factors responsible for this extensive loss of taste function in marine mammals remain to be determined.

In summary, these data dramatically illustrate how plastic the taste system is and, as illustrated through the sweet taste modality, how it has adapted to changes in diet as species evolved. Similar changes are likely in the other taste qualities. For example, it is likely that species differences in the repertoires of bitter receptors reflect different classes of poisons that these species are likely to confront [11]. Species variation in salt taste perception is also likely to be coordinated with diet. For example, it is possible that strict carnivores may not perceive NaCl in the same way as do herbivorous mammals since carnivores' all-meat diet likely provides sufficient Na$^+$. Finally, as a third example, the human umami or amino acid receptor responds to only a few compounds (glutamate and a few others). However, this receptor acts as a more general amino acid receptor for rodents and other species. These species differences may also be explained by different feeding ecologies although this remains to be determined. Future comparative research will surely reveal many more interesting and important relationships between taste function, food choice, and diet.

Competing interests
The authors declare that they have no competing interests.

Authors' contributions
This review was written by both authors. Both authors read and approved the final manuscript.

References
1. Kare MR, Beauchamp GK, Marsh RR: **Special senses II: taste, smell and hearing.** In *Duke's Physiology of Domestic Animals, Eleventh Edition.* Edited by

Swenson MJ, Reece WO. Ithaca and London: Comstock Publishing Associates; 1993:816–835.

2. Beauchamp GK, Maller O, Rogers JG Jr: **Flavor preferences in cats (*Felis catus* and *Panthera* sp.).** *J Comp Physiol Psychol* 1977, **97**(5):1118–1127.

3. Bachmanov AA, Beauchamp GK: **Taste receptor genes.** *Annu Rev Nutr* 2007, **27**:389–414.

4. Li X, Li W, Wang H, Cao J, Maehashi K, Huang L, Bachmanov AA, Reed DR, Legrand-Defretin V, Beauchamp GK, Brand JG: **Pseudogenization of a sweet-receptor gene accounts for cats' indifference toward sugar.** *PLoS Genet* 2005, **1**:27–35.

5. Jiang P, Josue J, Li X, Glaser D, Li W, Brand JG, Margolskee RF, Reed DR, Beauchamp GK: **Major taste loss in carnivorous mammals.** *Proc Natl Acad Sci U S A* 2012, **109**:4956–496.

6. Baldwin ME, Toda Y, Nakagita T, O'Connell MJ, Klasing KC, Misaka T, Edwards SV, Liberles SD: **Evolution of sweet taste perception in hummingbirds by transformation of the ancestral umami receptor.** *Science* 2014, **345**:929–933.

7. Jiang P, Beauchamp GK: **Sensing nectar's sweetness.** *Science* 2014, **345**:878–879.

8. Jiang P, Josue-Almqvist J, Jin X, Li X, Brand JG, Margolskee RF, Reed DR, Beauchamp GK: **The bamboo-eating giant panda (*Ailuropoda melanoleucap*) has a sweet tooth: behavioral and molecular responses to compounds that taste sweet to humans.** *PLoS One* 2014, **9**:e93043.

9. Li R, Fan W, Tian WG, Zhu H, He L, Cai J, Huang Q, Cai Q, Li B, Bai Y, Zhang Z, Zhang Y, Wang W, Li J, Wei F, Li H, Jian M, Li J, Zhang Z, Nielsen R, Li D, Gu W, Yang Z, Xuan Z, Ryder OA, Chi-Ching Leung F, Zhou Y, Cao J, Sun X, Fu Y, *et al*: **The sequence and de novo assembly of the giant panda genome.** *Nature* 2010, **463**:311–317.

10. Feng P, Zheng J, Rossiter J, Wang D, Zhao H: **Massive losses of taste receptor genes in toothed and baleen whales.** *Genome Biol Evol* 2014, **6**(6):1254–1265.

11. Li D, Zhang J: **Diet shapes the evolution of the vertebrate bitter taste receptor gene repertoire.** *Mol Biol Evol* 2013, **31**(2):303–309.

8

Does the type of receptacle influence the crossmodal association between colour and flavour? A cross-cultural comparison

Xiaoang Wan[1,2]*, Carlos Velasco[2], Charles Michel[2], Bingbing Mu[1], Andy T Woods[3] and Charles Spence[2]

Abstract

Background: We report a cross-cultural study designed to investigate whether the type of receptacle in which a coloured beverage is presented influences the colour-flavour associations that consumers make. Participants from the United States of America (USA) and China were shown photographs of red, green, yellow, blue, orange, brown, and clear liquids in a water glass, a wine glass, a cocktail glass, and a plastic cup.

Results: The two groups of participants exhibited different colour-flavour associations for the green, yellow, orange, and brown drinks when these were presented in the different receptacles, suggesting some interesting interactions between the receptacle, colour, and flavour. Cross-cultural differences were also observed in the colour-flavour associations for red and blue drinks that were independent of the type of container in which the drinks were presented.

Conclusions: These findings highlight the existence of an interaction between contextual factors (the receptacle in which a drink is presented) and the cultural background of our participants (China versus the USA) in terms of colour-flavour associations. Such results raise interesting questions regarding the underlying mechanisms responsible for these effects.

Keywords: Colour-flavour interactions, Tableware, Cross-cultural difference, Contextual factors

Background

Perceiving the flavour of foods and drinks involves a process of multisensory integration [1,2]. The colour of a food or drink item may, for example, provide clues regarding the likely identity and intensity of the flavour [3,4]. Yet, flavour perception is also influenced by cognitive factors such as a person's expectations, which may lead people to generate false expectations and to misidentify flavours and/or odours [5-11].

Colours convey different aesthetic values and meanings in different contexts [12], so it would seem likely that consumers from different cultures might have different expectations when presented with the same coloured beverage. Evidence in support of this suggestion comes from Shankar,

Levitan, and Spence [13]. These researchers conducted a study in which groups of participants from the United Kingdom (UK) and Taiwan viewed coloured and clear drinks and were asked what flavour first came to mind on seeing each drink. The two groups of participants exhibited different colour-flavour associations for the yellow, blue, orange, and brown drinks. For example, the most common flavour associated with the brown drink was cola for the British participants but grape for those from Taiwan. Recently, significant cross-cultural differences have also been reported in the case of colour-odour associations [14].

To date, however, most researchers have tended to argue that colour has a specific meaning regardless of the particular context in which that colour happens to have been presented. However, it would seem more likely that the same colour can have a variety of different meanings depending upon the particular context and country (or culture) in which it happens to have been presented [15,16]. Here, we hypothesized that one such contextual variable that might influence the meaning of

* Correspondence: wanxa@mail.tsinghua.edu.cn
[1]Department of Psychology, School of Social Sciences, Tsinghua University, Qinghua Yuan, Beijing 100084, China
[2]Crossmodal Research Laboratory, Department of Experimental Psychology, University of Oxford, South Parks Road, OX1 3UD, Oxford, UK
Full list of author information is available at the end of the article

the colour of a beverage is the container in which it happens to be presented. A dark purple coloured drink might, for example, make one think of wine if presented in a wine glass, and of grape juice, blackcurrant, or something else entirely if presented in a high ball glass.

A number of studies have already examined how the shape, size, and material of the container can influence the perception and evaluation by consumers of wine, tea, soft drinks, and water (for a recent review, see Spence *et al.* [17]). The shape of a glass can also influence the perceived aroma and odour of wine, at least when people are aware of the dimensions of the glass from which they happen to be tasting [18] (also see [19], for a review). The material from which a container is made has also been shown to affect participants' experience of drinking tea and soft drinks [20]. To date, however, few studies have attempted to assess how contextual variables such as the type of receptacle and culture interact to influence the specific flavours that people may associate with particular colours. The present study was therefore designed to investigate whether the type of receptacle,

specifically different glasses containing a transparent coloured liquid, would influence people's colour-flavour associations. Another question of interest was whether there would be any cross-cultural differences either in terms of the crossmodal colour-flavour associations held by participants from different parts of the world, and/or in terms of the influence of the type of receptacle on the meaning of colour in those different regions.

Given that some cultural differences in colour-flavour associations have been reported previously between those participants from the UK and Taiwan [13], it seemed sensible to extend this study to investigate cross-cultural associations with participants from the United States of America (USA) and from mainland China. In this study, we asked our participants to choose the flavour (from a list) to indicate the flavour that first came to their mind when they saw pictures of drinks in different colours (red, green, yellow, blue, orange, brown, and clear) presented in four different types of receptacles, including the water glass, wine glass, cocktail glass, and plastic cup (see Figure 1 for an illustration). Specifically, we addressed the following

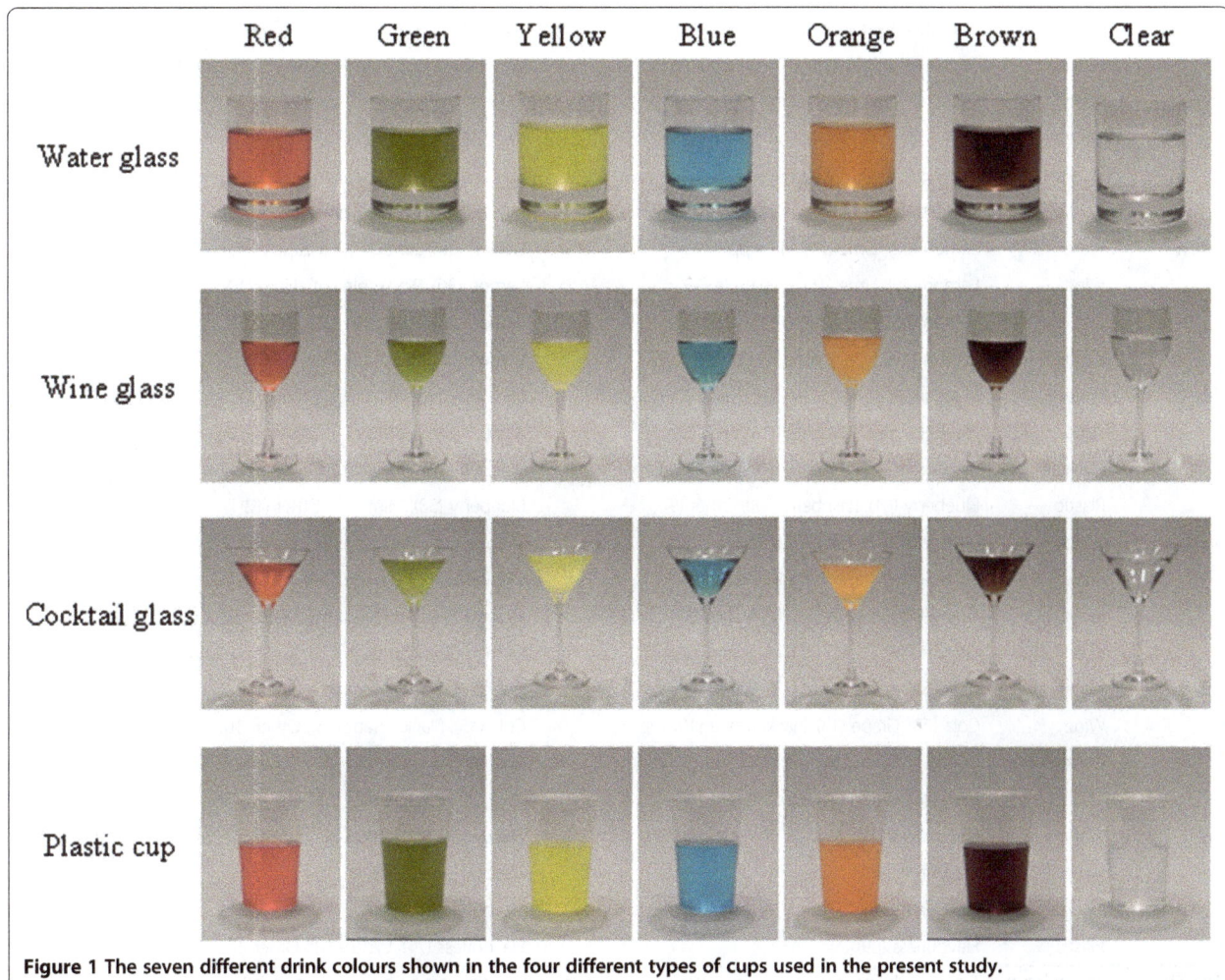

Figure 1 The seven different drink colours shown in the four different types of cups used in the present study.

questions: First, are there any differences in the nature of colour-flavour associations between the participants from these two countries? Second, are such colour-flavour associations affected by the type of receptacle in which the coloured liquid happens to be presented? Third, how, if at all, do these two factors (the type of receptacle and culture) interact? Note also that a substantially larger sample size (N = 200) was tested in the present study compared to that in Shankar *et al.*'s [13] previous study (N = 35).

Results

The colour-flavour responses for each group of participants are highlighted in Table 1. Note that the analysis reported here is based on that used by Shankar *et al.* [13]. One-sample Kolmogorov-Smirnov tests revealed that

the pattern of flavours associated with each coloured drink in each type of receptacle was significantly different from a uniform distribution in both groups of participants, for all $P < .001$, suggesting the existence of robust crossmodal, colour-flavour associations in each group. Fisher's Exact Tests with P values estimated with 100,000 Monte Carlo simulations and Bonferroni's correction for multiple comparisons (see [21] for the rationale for this correction) revealed that the two groups had different colour-flavour associations for (1) red or blue drinks in all type of receptacles; (2) green drinks in the water and wine glasses; and (3) yellow, orange, or brown drinks in the plastic cup, for all $P < .01$. No cross-cultural differences were observed in terms of the flavour associations for (1) the green drink presented in a cocktail glass or plastic cup; (2) the yellow,

Table 1 Top three flavour responses for each coloured drink in each type of cup in the participants from the United States of America (USA) and China (exact count in brackets)

Colour	Cup	Participants from the USA (N = 100)	Participants from China (N = 100)	Difference
Red	Water	Cherry (27), Cranberry (25), Strawberry (19)	Strawberry (27), Watermelon (25), Cherry (12)	**
	Wine	Strawberry (28), Cherry (25), Cranberry (16)	Watermelon (24), Cherry (20), Strawberry (18)	**
	Cocktail	Cherry (31), Strawberry (25), Cranberry (17)	Watermelon (27), Cherry (18), Strawberry (17)	**
	Plastic	Cherry (31), Strawberry (25), Watermelon (14)	Watermelon (35), Strawberry (20), Cherry (19)	**
Green	Water	Kiwi (34), Lime (27), Apple (13)	Kiwi (54), Apple (14), Lime (12)	**
	Wine	Kiwi (31), Lime (30), Melon (12)	Kiwi (40), Apple (20), Lime (16)	**
	Cocktail	Lime (32), Kiwi (27), Apple (19)	Kiwi (33), Lime (23), Apple (17)	
	Plastic	Kiwi (40), Lime (23), Apple (13)	Kiwi (58), Lime (16), Apple (7)	
Yellow	Water	Lemon (32), Lime (24), Pineapple (14)	Lemon (26), Lime (14), Pineapple (13)	
	Wine	Lemon (49), Lime (15), Pineapple (13)	Lemon (36), Pineapple (14), Lime (10)	
	Cocktail	Lemon (45), Pineapple (22), Banana (9)	Lemon (25), Pineapple (19), Pear (10)	
	Plastic	Lemon (49), Pineapple (18), Banana (13)	Lemon (28), Pineapple (26), Pear (11)	**
Blue	Water	Blueberry (63), Raspberry (16), Mint (8), Other (8)	Blueberry (43), Mint (30), Other (12)	**
	Wine	Blueberry (61), Raspberry (14), Mint (9)	Blueberry (35), Mint (31), Other (12)	**
	Cocktail	Blueberry (59), Raspberry (18), Mint (8), Other (8)	Blueberry (32), Mint (30), Other (12)	**
	Plastic	Blueberry (61), Raspberry (18), Other (9)	Blueberry (38), Mint (27), Other (12)	**
Orange	Water	Orange (43), Peach (18), Mandarin (17)	Orange (24), Mandarin (21), Peach (11)	
	Wine	Orange (46), Peach (21), Mandarin (17)	Orange (38), Mandarin (34), Peach (7)	
	Cocktail	Orange (44), Peach (23), Mandarin (23)	Orange (32), Mandarin (31), Melon (10)	
	Plastic	Orange (48), Peach (15), Mandarin (14)	Orange (38), Mandarin (21), Watermelon (13)	**
Brown	Water	Cola (66), Blackcurrant (12), Grape (9)	Cola (46), Blackcurrant (16), Other (14)	
	Wine	Cola (68), Grape (13), Blackcurrant (10)	Cola (43), Blackcurrant (21), Other (10)	
	Cocktail	Cola (63), Blackcurrant (23), Cranberry (4)	Cola (48), Blackcurrant (20), Other (10)	
	Plastic	Cola (71), Grape (14), Blackcurrant (6)	Cola (32), Blackcurrant (27), Other (13)	**
Clear	Water	Flavourless (91)	Flavourless (81), Other (6)	
	Wine	Flavourless (88)	Flavourless (73), Other (10)	
	Cocktail	Flavourless (86)	Flavourless (73), Other (12)	
	Plastic	Flavourless (89)	Flavourless (71), Lychee (8), Other (8)	

Note: Flavour counts of less than 5 are not shown in this table, while responses that tied for third place are both shown (where $**P < .01$).

orange, or brown drinks in the water, wine, or cocktail glass; or (3) the clear liquid in any of the receptacles, for all $P > .01$.

The participants' responses regarding the drink that they associated with each type of receptacle were classified into the following categories: (1) water, (2) alcoholic drink, (3) soda, (4) juice, (5) other, and (6) 'Don't know.' The top three drinks chosen by each group of participants are listed in Table 2. One-sample Kolmogorov-Smirnov tests revealed that the pattern of drinks associated with each type of receptacle was significantly different from a uniform distribution in both groups of participants, for all $P < .001$, thus suggesting that the participants associated each type of receptacle with a certain type of drink. The results of Fisher's Exact Tests with P values estimated with 100,000 Monte Carlo simulations and Bonferroni's correction for multiple comparisons suggested that the participants from the USA and China had associated each type of receptacle with different drinks. As for the water glass, alcohol was the drink that most often came to mind for the participants from the USA, whereas the Chinese participants thought of water instead. As for the wine and cocktail glasses, although alcohol was most often associated for both groups, more of the participants from the USA than from China (90% versus 60%, respectively) provided alcohol as their answer. As for the plastic cup, although water was the most frequent answer for both groups, the second most frequently chosen answer was soda for the participants from the USA, and 'Don't know' for the participants from China.

Discussion
Cross-cultural differences in receptacle-colour-flavour associations

Both the statistics and counts of chosen flavours consistently demonstrated the interaction between the type of

Table 2 Top three drinks associated with each type of cup/glass for the groups of participants from the United States of America (USA) and from China when containing clear water (exact count in brackets)

Cup	Participants from the USA (N = 100)	Participants from China (N = 100)	Difference
Water glass	Alcohol (49), Water (39)	Water (52), Alcohol (16), Juice (10)	**
Wine glass	Alcohol (90), Water (8)	Alcohol (60), Soda (9), Juice (9)	**
Cocktail glass	Alcohol (93)	Alcohol (58), Don't know (17), Water (9), Juice (9)	**
Plastic cup	Water (58), Soda (22)	Water (48), Don't know (18), Soda (12)	**

Note: **$P < .01$.

receptacle and the cultural background of the participants on the flavour associations elicited by the green, yellow, orange, and brown drinks. That is, the cross-cultural difference for yellow, orange, and brown drinks emerged only when the drinks were presented in the plastic cup. This result allowed us to compare these results directly to those reported by Shankar et al. [13]. In the latter study, all of the drinks were presented in plastic cups. Shankar and colleagues also found cross-cultural differences in terms of the particular colour-flavour associations for yellow, orange, and brown drinks. Our results confirmed the cross-cultural difference in the associations between flavour expectations and the yellow, orange, and brown beverage colours. In addition, our results also highlight the possibility that such cross-cultural differences are very likely to occur when a drink is presented in a plastic cup.

As for the green drink presented in a plastic cup, the most common flavour choice was kiwi for all participants. In contrast, Shankar et al. [13] reported that participants from the UK and Taiwan consistently chose the flavour of mint for a green drink when it was presented in a plastic cup, whereas kiwi was mentioned less often. These results might be taken to suggest another cross-cultural difference, but they might also be attributed to the fact that the green drinks presented in these two studies were not exactly the same.

Taken together, the findings of the present study suggest how the type of receptacle, as a contextual factor, can interact with a participant's cultural background in order to influence the flavour expectations they hold for green-, yellow-, orange-, and brown-coloured drinks. According to the associative learning account of crossmodal correspondences, people are sensitive to the repeated exposure to specific combinations of stimuli [22]. What is more, consumers presumably learn crossmodal associations between colours and flavours across the lifespan as a result of their being exposed to specific combinations of stimuli (colours and flavours) that happen to co-occur together. Those combinations may differ across cultures, thus leading different groups of consumers to hold different crossmodal colour-flavour associations. Furthermore, the cross-cultural differences in colour-flavour associations are also presumably based on different expectations generated from the repeated co-occurrence of certain pairings of colour and flavour in everyday life. These cross-cultural differences may also result from a participant's exposure to different commercial products and to differences in agriculture practices in different environments [13]. The influence of the type of receptacle on such cross-cultural differences suggests that participants' expectations might be strengthened, weakened, or otherwise altered by contextual factors, such as the type of container in which a drink is shown (e.g., as in an advertisement).

Cross-cultural differences in colour-flavour associations

The present results revealed a cross-cultural difference in the associations between flavour expectations that were elicited by the sight of a red beverage, which was not influenced by the type of receptacle in which the beverage was presented. For those participants from the USA, cherry flavour was most commonly associated with the red drink in three out of four containers, whereas the most frequent flavour choice for the same drink for the Chinese participants was watermelon. This cross-cultural difference might well be attributed to the fact that the same colour red is repeatedly seen with different flavours in the marketplace of these two countries. Although red is associated with love and happiness in both countries [23], it is reported as being associated with 'good-tasting' in the USA, but not by those from China. It is therefore difficult to determine whether documented cross-cultural differences in the colour-flavour associations arise at a perceptual or at a more semantic level [10,24].

Similarly, the cross-cultural difference in the crossmodal associations between flavour and the blue-coloured beverage was not influenced by the type of receptacle either. In the present study, the cross-cultural difference was shown for the second most common flavour choice. Blueberry was the flavour that was most often associated with the blue-coloured drinks by both groups of participants. The next most common choice was raspberry for the participants from the USA but mint for those from China. This difference might perhaps be attributable to the fact that raspberry is not a particularly common fruit in China. Interestingly, blue was rated as the most liked colour by participants from eight different countries, including the USA and China [15], and they both associate it with high quality [23].

The impact of the container on drink expectations

Interestingly, the results of the present study demonstrate that participants from the USA and China associated the same container with different drinks. Alcohol most often came to the minds of the USA participants' when viewing the water glass, whilst it was water that came to mind for Chinese participants. As for the wine and cocktail glasses, although alcohol was the drink that was most often associated for both groups, more of the participants from the USA than from China suggested alcohol. Therefore, it would seem reasonable to expect that these two groups of consumers might have different expectations regarding the flavour of one and the same coloured drink if shown in an online advertisement, say. For example, when presented in a water glass, an alcoholic drink might be judged as highly 'appropriate' for those from the USA, while not eliciting such a strong feeling in those from China. Given that expectations concerning the flavour of a drink have been shown to

influence people's subjective evaluation of the drink [25], the present findings provide evidence in support of the idea that it is important to choose the type of receptacle in which to present coloured drinks carefully, based on both the colour of the drink and the culture background of the audience.

Limitation and future directions

Further research will be needed in order to overcome the limitations of the present study. First, it should be noted that the two groups of participants tested here differed not only in terms of their culture but also in terms of their age, as those participants who were recruited from the USA were older than those who took part in the study from China. It is possible that their age might also have influenced their familiarity with certain alcoholic drinks or the expectations on the drinks presented to them, but future research is needed to test this possibility. Second, when the participants were asked to indicate the first flavour that came to mind for each drink, they were asked to choose just one flavour from a list. Although this method has commonly been used in the literature [10], it is important to note that it may not accurately represent the colour-flavour associations that a participant might come up with spontaneously (that is, it fails to capture the strength of the association for a given individual). However, that said, it is worth noting that Shankar et al. [13] had their participants freely write down any flavour that they thought of when seeing each coloured drink, and many of the results we obtained with providing a list to choose from were consistent with those freely written down by their participants. Third, the participants from both the U.S.A. and China finished the experiment on their own computers and saw the photos of coloured drinks on their own monitors, so we had no control over the colours that they actually saw. This is, of course, also exactly the same challenge that faces any international marketer wanting to show their latest product offerings online.

Conclusions

In summary, the results of the present study suggest that the type of the container influences the colour-flavour associations for certain beverage colours, and also reveals some cross-cultural differences between participants from the USA and China in terms of their colour-flavour associations. The two groups of participants exhibited different colour-flavour associations for the green, yellow, orange, and brown drinks when presented in the different receptacles, whereas the cross-cultural differences in the colour-flavour associations for red and blue drinks were independent of the type of container used to present the drinks. These findings therefore highlight an interaction between contextual factors and culture in terms of colour-flavour associations. These findings have relevance to the international marketing of drinks. It is worth remembering that many international drinks manufacturers are currently thinking about how best to break into the Chinese market, which undoubtedly represents a huge area of potential growth for many businesses. We predict that online marketing, possibly using images and presentation formats similar to those utilized here (for example, online testing) is going to become increasingly popular and important in the years to come. Critically, previous research has tended to assess the meaning of different colours for flavour perception without necessarily taking into account (or at the very least without giving sufficient consideration to) the receptacle in which the coloured liquid happens to have been presented. The results reported here demonstrate, for what we believe to be the first time, that a given beverage colour can have a very different meaning (or associated flavour) depending on the receptacle in which it is presented - which can be thought of as a kind of context effect [17] - as well as the country in which it is presented. We would argue though that the findings of the present study should be intriguing enough to capture the attention of those marketers who wish to better tailor their products by using appropriate colour-flavour pairs for the consumers from different countries and by using the most appropriate/flattering receptacle.

Methods

Participants

One hundred participants from mainland China (50 women and 50 men, mean age = 27.3 years, SD = 1.8, ranging from 20 to 32 years) were recruited to take part in this study from the subject pool of the Spatial Cognition Lab at the Psychology Department of Tsinghua University; 100 participants from the USA (44 women and 56 men, mean age = 32.9 years, SD = 11.2, ranging from 19 to 61 years) were recruited from Amazon's Mechanical Turk in exchange for a payment of 0.80 US dollars. All of the participants provided informed consent prior to their taking part in the study. The experiment was approved by the Central University Research Ethics Committee of the University of Oxford.

Apparatus and materials

Photos of coloured-water drinks (180 pixels wide × 240 pixels high) were shown to participants. The drinks were made using 400 ml plastic cups (Tesco, Cheshunt, UK) to mix 250 ml of water with the different commercial food colourings (Dr. Oetker, Leeds, UK) to achieve six colours: red (.30 ml red food colouring), green (.45 ml green and .05 yellow food colouring), yellow (.05 ml yellow food colouring), blue (.20 blue food colouring), brown (.15 ml red, .10 ml blue, and .10 ml yellow food

colouring), and orange (.10 ml yellow and .05 ml red food colouring). Clear water was also presented as the seventh condition. As shown in Figure 1, the drinks were presented in four different containers, including a water glass, a wine glass, a cocktail glass, and a plastic cup.

Both groups of participants took part in the study online using their own computer on two websites. The participants from China completed the task in Chinese via Unipark (www.unipark.de), whereas the participants from the USA completed it in English via Xperiment (http://www.xperiment.mobi downloaded on 15 May 13).

Design and procedure

A 2 (Culture: USA versus China) × 4 (Receptacle: water glass, wine glass, cocktail glass, or plastic cup) design was used, with Culture as a between-group factor and Receptacle as a within-group factor. Each participant completed 28 trials (7 colours × 4 types of receptacle) presented in a random order.

Before starting the main experiment, the participants were informed about the general aims of the study, and only those participants who accepted to take part continued. During each trial, a photo of a drink was shown, and the participant had to choose the first flavour or drink that came to mind based on the drink's colour from a list of 24 flavours, including apple, banana, blackcurrant, blueberry, cherry, cola, cranberry, grape, kiwi, lemon, lime, lychee, mandarin, melon, mint, orange, peach, pear, pineapple, raspberry, strawberry, watermelon, flavourless, and other. The list was constructed by combining those flavour options found in the papers of Shankar et al. [13] and Zampini et al. [10], the wine aroma wheel (winearomawheel.com), and several flavours that are commonly found in the Chinese marketplace. These options were presented in a random order in each trial to avoid any possible order or position effects. The participants were instructed to specify which flavour they had in mind if they chose the 'other' option instead[a]. At the end of the experiment, the participants were shown the pictures of four receptacles with clear water in them (in a random order) and asked to specify the drink they associated with the receptacle or answer 'Don't know' if they were unable to provide an answer. The experiment lasted approximately 15 minutes.

Endnotes

[a]After participants were asked to choose the first flavour they thought of when seeing the photo of a coloured drink, and they were also asked to rate the drink for familiarity and how pleasant they expected it to be using a 7-point Likert scale. The data are not reported here because they fall outside the scope of this paper.

Competing interests

The authors declare that they have no competing interests.

Authors' contributions

Each of the listed co-authors made the following contributions to the paper: XW, CS, and CV co-developed the idea for the study and collaboratively designed the study. CM and XW designed the stimuli for the study. BM and AT collected the data and revised the manuscript. XW and CS conducted the data analysis and interpretation of the data, and drafted the manuscript. CV, AT, and BM aided in the initial stages of the manuscript's preparation. All of the authors have read and approved the final version of the manuscript.

Acknowledgements

This research was supported by Tsinghua University Cultural Inheritance and Innovation Program and by the Tsinghua-Santander Young Faculty Program. The authors would like to thank Xi Zhou for her assistance in data collection. Comments concerning this paper should be sent to Dr. Xiaoang Wan at wanxa@mail.tsinghua.edu.cn.

Author details

[1]Department of Psychology, School of Social Sciences, Tsinghua University, Qinghua Yuan, Beijing 100084, China. [2]Crossmodal Research Laboratory, Department of Experimental Psychology, University of Oxford, South Parks Road, OX1 3UD, Oxford, UK. [3]Xperiment, Lausanne, Switzerland.

References

1. Auvray M, Spence C: **The multisensory perception of flavor.** *Conscious Cogn* 2008, **17**:1016–1031.
2. Zellner DA: **Color-odor interactions: a review and model.** *Chem Percept* 2013, **6**:155–169.
3. DuBose CN, Cardello AV, Maller O: **Effects of colorants and flavorants on identification, perceived flavor intensity, and hedonic quality of fruit-flavored beverages and cake.** *J Food Sci* 1980, **45**:1393–1399.
4. Hyman A: **The influence of color on the taste perception of carbonated water preparations.** *B Psychonomic Soc* 1983, **21**:145–148.
5. Morrot G, Brochet F, Dubourdieu D: **The color of odors.** *Brain Lang* 2001, **79**:309–320.
6. Shankar M, Simons C, Shiv B, Levitan C, McClure S, Spence C: **An expectations-based approach to explaining the influence of color on odor identification: the influence of degree of discrepancy.** *Atten Percept Psychophys* 2010, **72**:1981–1993.
7. Stillman JA: **Color influences flavor identification in fruit-flavored beverages.** *J Food Sci* 1993, **58**:810–812.
8. Wood AT, Lloyd DM, Kuenzel J, Poliakoff E, Dijksterhuis GB, Thomas A: **Expected taste intensity affects response to sweet drinks in primary taste cortex.** *Neuroreport* 2011, **22**:365–369.
9. Wood AT, Poliakoff E, Lloyd DM, Dijksterhuis GB, Thomas A: **Flavour expectation: the effect of assuming homogeneity on drinking perception.** *Chem Percept* 2010, **3**:174–181.
10. Zampini M, Sanabria D, Phillips N, Spence C: **The multisensory perception of flavor: assessing the influence of color cues on flavor discrimination responses.** *Food Qual Prefer* 2007, **18**:975–984.
11. Zellner DA, Bartoli AM, Eckard R: **Influence of color on odor identification and liking ratings.** *Am J Psychol* 1991, **104**:547–561.
12. Aslam MM: **Are you selling the right colour? A cross-cultural review of colour as a marketing cue.** *J Marketing Commun* 2006, **12**:15–30.
13. Shankar MU, Levitan C, Spence C: **Grape expectations: the role of cognitive influences in color-flavor interactions.** *Conscious Cogn* 2010, **19**:380–390.
14. Ren J, Woods A, McKenzie K, Ru LX, Levitan CA: **Cross-cultural colour-odour associations.** *Appetite* 2012, **59**:634.
15. Madden TJ, Hewett K, Roth MS: **Managing images in different cultures: a cross-national study of color meanings and preferences.** *J Int Marketing* 2000, **84**:90–107.
16. Piqueras-Fiszman B, Velasco C, Spence C: **Exploring implicit and explicit crossmodal colour-flavour correspondences in product packaging.** *Food Qual Prefer* 2012, **25**:148–155.
17. Spence C, Harrar V, Piqueras-Fiszman B: **Assessing the impact of tableware and other contextual variables on multisensory flavour perception.** *Flavour* 2012, **1**:7.
18. Hummel T, Delwiche JF, Schmidt C, Hüttenbrink KB: **Effects of the form of glass on the perception of wine flavors: a study in untrained subjects.** *Appetite* 2003, **41**:197–202.
19. Spence C: **Crystal clear or gobbletigook?** *World Fine Wine* 2011, **33**:96–101.
20. Schifferstein HNJ: **The drinking experience: cup or content?** *Food Qual Prefer* 2009, **20**:268–276.
21. Maxwell SE, Delaney HD: *Designing experiments and analyzing data: a model comparison perspective.* Mahwah NJ: Erlbaum; 2004.
22. Wilson DA, Stevenson RJ: *Learning To Smell: Olfactory Perception From Neurobiology To Behavior.* Baltimore, MD: The Johns Hopkins University Press; 2006.
23. Jacobs L, Keown C, Worthley R, Ghymn KI: **Cross-cultural colour comparisons: global marketers beware!** *Int Marketing Rev* 1991, **83**:21–31.
24. Spence C: **Crossmodal correspondences: a tutorial review.** *Atten Percept Psychophys* 2011, **73**:971–995.
25. Raudenbush B, Meyer B, Eppich W, Corley N, Petterson S: **Ratings of pleasantness and intensity for beverages served in containers congruent and incongruent with expectancy.** *Percept Motor Skill* 2002, **94**:671–674.

The pleasure of food: underlying brain mechanisms of eating and other pleasures

Morten L Kringelbach[1,2]

Abstract

As all chefs know, great food can have a transformational impact. A great deal of recent research has gone into using the new techniques from molecular gastronomy and gastrophysics to create innovative meals with delicious original textures and flavours. These novel creations have elicited much excitement from food critiques and diners alike. Much stands to be gained if these developments were to be matched by a better understanding of how the pleasure of food comes about in the brain. This review summarises the current state-of-the-art of the science of pleasure and specifically the brain's fundamental computational principles for eating and the pleasures evoked. It is shown how the study of food has advanced our understanding of the unitary pleasure system that is used for all pleasures. As such, these novel insights may come to serve as a guide for chefs of how to combine science and art in order to maximise pleasure—and perhaps even increase happiness.

Keywords: Dinner, Gastronomy, Brain, Pleasure cycle, Satiety, Satiation, Hedonic, Pleasure, Food, Multimodal integration, Insula, Operculum, Orbitofrontal cortex, Cingulate cortex, Wanting, Liking, Learning, Anhedonia

Introduction

The novella "Babette's Feast" by the Danish writer Karen Blixen (writing under her nom du plume of Isak Dinesen) is set in the 1870s, describing an austere religious sect, whose members "…renounced the pleasures of this world, for the earth and all that it held to them was but a kind of illusion, and the true reality was the New Jerusalem toward which they were longing" [1]. Martine and Phillipa are the unmarried daughters of the founder of the religious sect who have a French maid-of-all-work, Babette, appearing from war-torn Paris under mysterious circumstances. Upon her arrival, the pious daughters are anxious to avoid any "… French luxury and extravagance" and therefore at the time explained that they "… were poor and that to them luxurious fare was sinful. Their own food must be as plain as possible". As it happens, their worries are allayed; and for next 12 years, Babette serves them such that the whole community come to acknowledge her excellence and depend on her quiet gifts. When Babette unexpectedly wins a princely sum of money in the French lottery, they become afraid she may leave

them. Accordingly, against their better judgement, the sisters agree that Babette may cook them a special dinner celebrating the 100th anniversary of the sect's founding father. Unbeknownst to the sisters, Babette used to be a cordon bleu cook who prepares a sumptuous once-in-a-lifetime meal, leaving the guests questioning their lifelong denial of mortal pleasures.

In the novella, this cathartic meal is not described in much detail, following the vow of the devout and taciturn guests "… not to utter a word about the subject". In contrast, Danish director Gabriel Axel's Oscar-winning film adaptation tries hard to use visuals to convey the splendour of the dinner but still falls short of conveying the multisensory experience of a fine meal. Blixen is astute in using linguistic sparseness as a plot device, given that language, even that employed by great writers [2], very often fails to convey the exquisite sensory experiences of food upon which the story hinges. Blixen even feels moved to suggest that it is "… when man has not only altogether forgotten but has firmly renounced all ideas of food and drink that he eats and drinks in the right spirit". Language for all its powers is powerless when it comes to evoking the food's sensory routes to pleasure, yet the unity of pleasure is beautifully evoked: "Of what happened later in the evening nothing definite

Correspondence: Morten.Kringelbach@psych.ox.ac.uk
[1]Department of Psychiatry, University of Oxford, Warneford Hospital, Oxford OX3 7JX, England
[2]Center of Functionally Integrative Neuroscience, Aarhus University, Aarhus, Denmark

can here be stated. None of the guests later on had any clear remembrance of it. They only knew that the rooms had been filled with a heavenly light as if a number of small halos had blended into one glorious radiance. Taciturn old people received the gift of tongues; ears that for years had been almost deaf were opened to it. Time itself had merged into eternity. Long after midnight the windows of the house shone like gold, and golden song flowed out into the winter air".

Thus, Babette's feast becomes a route to intense well-being, and the pleasure is not just about the food but instead about providing unity and transcendence for the virtuous dinner guests who all leave the meal changed, suddenly awake to the potential of earthly pleasures.

For many years, such pleasures have remained mysterious and firmly within the domain of much great art. Yet, the advent of modern neuroscience has started to uncover some of the underlying mechanisms of associated brain changes.

This review describes what is known of the processing of food in scientific terms; from sensory identification of the uni- and multisensory properties of food to the associated prediction, memory and evaluation involved which may give rise to the experience of pleasure. Like all rewards, food depends on processing in interconnected and widespread brain regions to identify and characterise the different sensory properties and their multimodal integration. This processing is detailed in a multilevel model of the constituent processes involved in food intake over time. The focus here, however, is on the fundamental underlying brain mechanisms governing the initiation and termination of a meal leading to pleasure. Overall, the accumulated evidence shows that the pleasure evoked by food is remarkably similar to that of other rewards, suggesting a unitary pleasure system, whether engaging with food, sex, social or higher-order rewards. Food is thus not only highly pleasurable but also an excellent tool for discovering fundamental principles of brain function.

Brain principles of eating
While food clearly is essential to survival, it is the pleasure involved that makes eating worthwhile. While the members of the religious sect in Blixen's novella may try hard to deny the pursuit of pleasure in its many forms, their well-being is ultimately strongly enhanced as they submit to Babette's cooking, i.e. to the strong primal drive for pleasure. The evolutionary imperatives of survival and procreation are not possible without the principle of pleasure for the fundamental rewards of food, sex and conspecifics—and as such may well be evolution's boldest trick [3]. The scientific study of pleasure, *hedonia research*, is dedicated to searching for the functional neuroanatomy of hedonic processing, taking its name from the ancient Greek for pleasure (ἡδονή; transl. hédoné) derived from the word for "sweet" (ἡδύς, transl. hēdús) [4].

In the novella, the sect's initial food asceticism may stem from their religious beliefs but is guided by the basic homeostatic regulation of human eating behaviour [5], of which animal models have elucidated in great details the many subcortical circuits and molecules shared amongst mammals including humans [6-8]. Yet, as illustrated by the effects of Babette's Feast, homeostatic processes are not solely responsible for human eating. This *hedonic eating* is difficult to suppress and is even more poignantly illustrated by the current worldwide obesity pandemic [9]. There is often very little well-being linked to such over-eating, with *anhedonia*—the lack of pleasure—being a prominent feature of affective disorders. From this public health perspective, it is imperative that we better understand the fundamental pleasure systems such that we find new and more effective ways of rebalancing the system and potentially reducing obesity which is threatening to undermine public health [10].

Eating can seem simple but at its most basic, human food intake is still rather complex. The procurement of food can be surprisingly difficult in the wide variety of often hostile climates inhabited by humans. Once food is available, the preparation and eating of food are also complex processes, involving a multitude of peripheral and central processes for carefully orchestrated acts requiring significant brain processing. The necessary, sophisticated motivational, emotional and cognitive processing are likely to have been main drivers for the evolution of large primate brains [11]. The brain principles underlying eating have been investigated for a long time in many mammalian species [6,12]. Here, the focus is on the pleasure component of human eating, which over the last decade has started to transform our understanding [13,14].

To understand pleasure in the brain, it is important to consider the main challenge for the brain which is to successfully balance resource allocation for survival and procreation [15]. In order to achieve this balance, different rewards compete for resources over time. In understanding the multi-faceted nature of pleasure, it can therefore be useful to consider the typical cyclical time course shared between all rewards with distinct appetitive, consummation and satiety phases [16,17] (Figure 1). The research has demonstrated that pleasure consists of multiple brain networks and processes and involves a composite of several components: "liking" (the core reactions to hedonic impact), "wanting" (motivational processing of incentive salience) and learning (typically Pavlovian or instrumental associations and cognitive representations) [18-21]. These component processes have discriminable neural mechanisms, which wax and wane during the cycle. The neural mechanisms of *wanting*, *liking* and *learning* can occur at any time during the

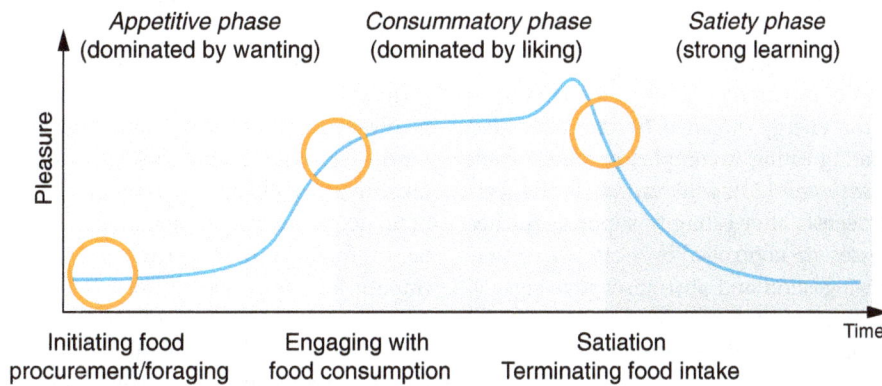

Figure 1 The pleasure cycle. The cyclical processing of rewards has classically been proposed to be associated with appetitive, consummatory and satiety phases [16,17]. Research has demonstrated that this processing is supported by multiple brain networks and processes, which crucially involves *liking* (the core reactions to hedonic impact), *wanting* (motivational processing of incentive salience) and *learning* (typically Pavlovian or instrumental associations and cognitive representations) [18-21]. These components wax and wane during the pleasure cycle and can co-occur at any time. Importantly, however, wanting processing tends to dominate the appetitive phase, while liking processing dominates the consummatory phase. In contrast, learning can happen throughout the cycle.

pleasure cycle, though wanting processes tend to dominate the appetitive phase (and are primarily associated with the neurotransmitter dopamine), while liking processes dominate the consummatory phase (and are associated with opioids) [13]. In contrast, learning can happen throughout the cycle (and is thought to be associated with synaptic plasticity). A neuroscience of pleasure seeks to map the necessary and sufficient pleasure networks

allowing potentially sparse brain resources to be allocated for survival.

This basic cyclical model of pleasure can be expanded into an elaborate multilevel model of food intake taken in account the episodic and tonic changes over time (Figure 2) [12]. The model links the pleasure cycle with the cyclical changes in hunger levels related to the initiation and termination of meals and the way food intake

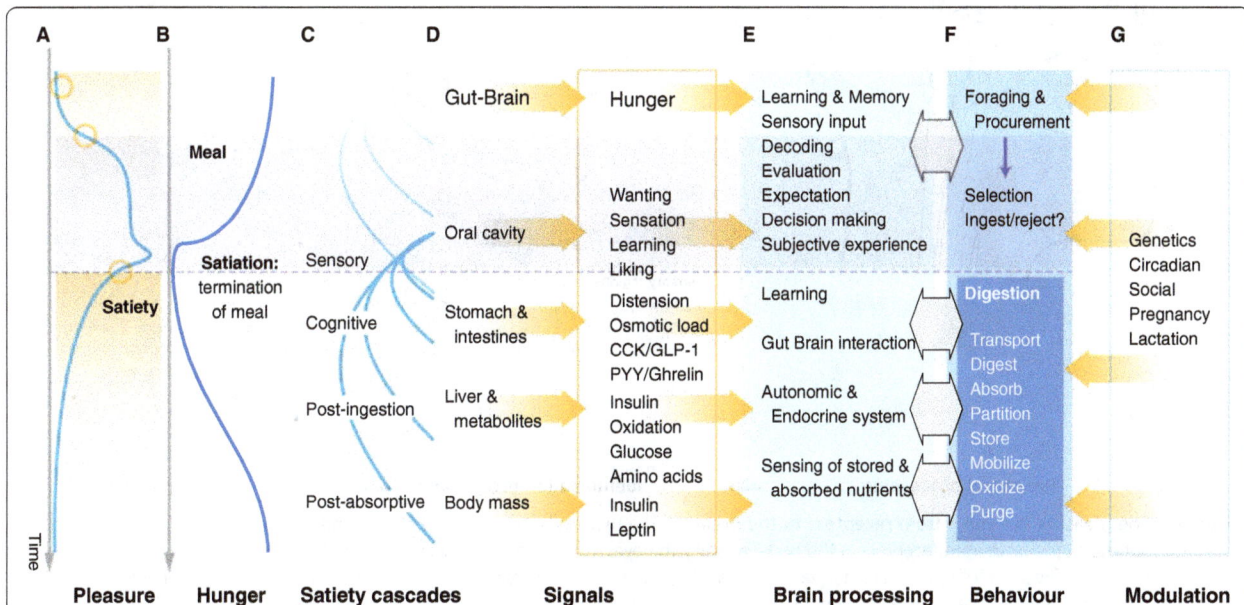

Figure 2 Multilevel model of food intake over time. The control of eating over time involves many different levels of processing as illustrated by the food. The changes at each level before, during and after meals are shown in each column which summarises the episodic and tonic changes over time (moving from top to bottom): **A)** pleasure cycle, **B)** the levels of hunger, **C)** satiation/satiety cascade (sensory, cognitive, post-ingestion and post-absorptive signals), **D)** origin of signals (gut-brain, oral cavity, stomach and intestines, liver and metabolites and body mass) and signal carriers, **E)** brain processing, **F)** behavioural changes including digestive system and **G)** general modulatory factors (see text for further information) [12].

comes about through the interaction given signals from the body, e.g. from the brain, gut-brain, oral cavity, stomach and intestines, liver and metabolites and body mass.

The dual processes of satiation and satiety are central to the model and to the energy obtained by the associated meals [22]. Terminating eating is complex process, which is encapsulated by *satiation* [23], while *satiety* is the feeling of fullness that persists after eating to suppress further eating. These processes are controlled by a cascade of sensory, cognitive, post-ingestion and post-absorptive signals, beginning with the consumption of a food in a meal and continuing as the food is digested and absorbed.

The multilevel model of food intake describes the changes over time in A) pleasure, B) the levels of hunger, C) satiation/satiety cascade signals, D) origin of signals and signal carriers, E) brain processes, F) behavioural changes including those in the digestive system and G) general modulatory factors (Figure 2). Many of these changes have been described elsewhere, e.g. the mechanisms of the changes after the termination of a meal such as the gut-

brain interactions, include signals from receptors in the digestive tract which are sensitive to calorie-rich nutrients (even in the absence of taste receptors) [24,25].

Here, however, the focus is on the processing principles involved primarily in the initiation and termination of a meal (Figure 3). The multisensory experience of food intake involves all the senses with different routes into the brain; from the distant processing of sight, sound and tactile of food to more proximal smell, taste and tactile (mouth-feel) processing. Smell is the most important determinant of the flavour of food and comes to the brain via orthonasal and retronasal pathways, experienced as we breathe in and out, respectively [26]. As demonstrated by the case with coffee, the subjective olfactory experience can feel very different from smelling the coffee in the cup to tasting the coffee in the mouth, which also relies on pure tastants (such as bitter) and mouth feel factors (such as the smoothness of the crema) (Figure 3A).

This sensory information about food is coming from receptors in the body, typically the eyes, ears, nose and

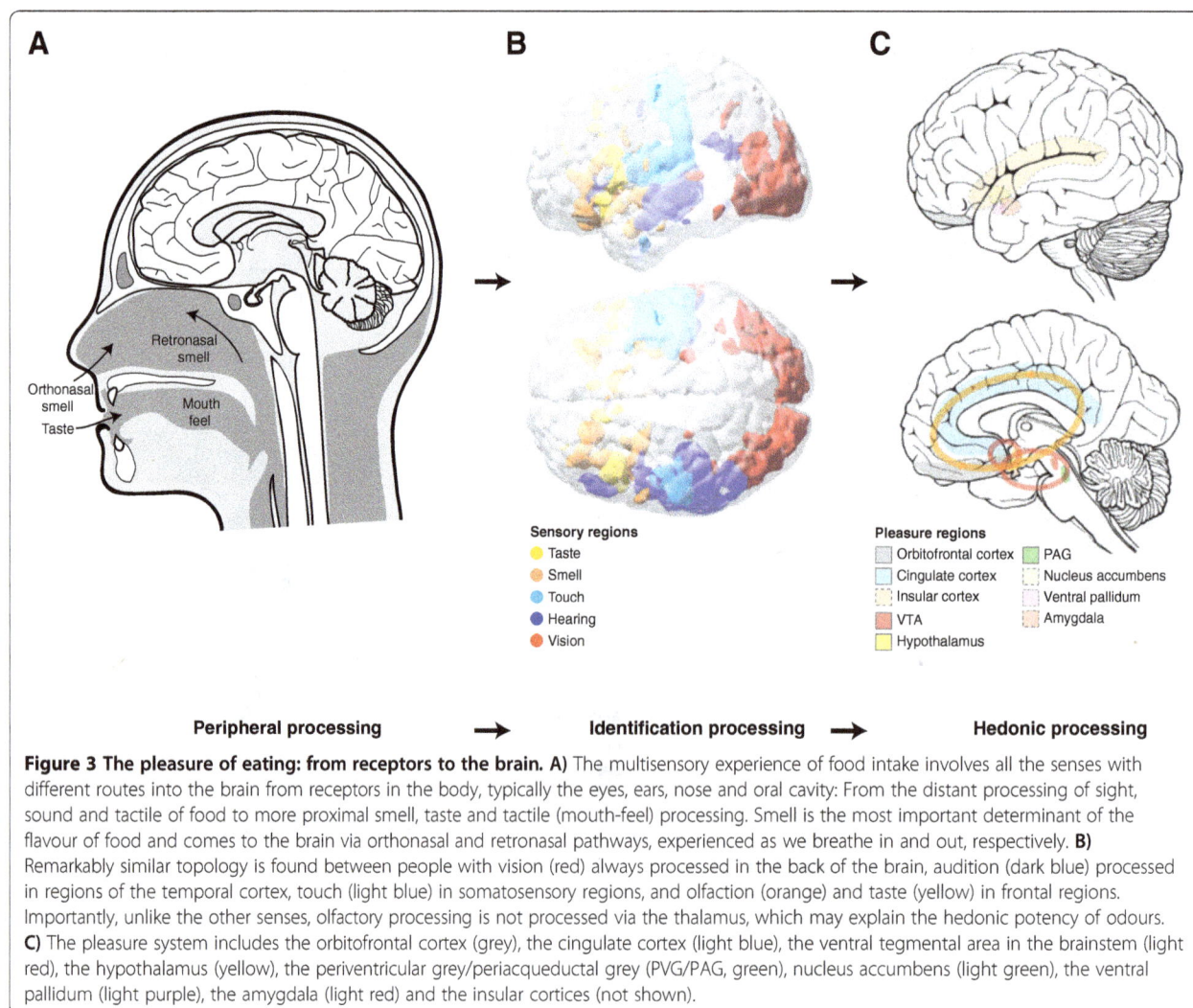

Peripheral processing → Identification processing → Hedonic processing

Figure 3 The pleasure of eating: from receptors to the brain. A) The multisensory experience of food intake involves all the senses with different routes into the brain from receptors in the body, typically the eyes, ears, nose and oral cavity: From the distant processing of sight, sound and tactile of food to more proximal smell, taste and tactile (mouth-feel) processing. Smell is the most important determinant of the flavour of food and comes to the brain via orthonasal and retronasal pathways, experienced as we breathe in and out, respectively. **B)** Remarkably similar topology is found between people with vision (red) always processed in the back of the brain, audition (dark blue) processed in regions of the temporal cortex, touch (light blue) in somatosensory regions, and olfaction (orange) and taste (yellow) in frontal regions. Importantly, unlike the other senses, olfactory processing is not processed via the thalamus, which may explain the hedonic potency of odours. **C)** The pleasure system includes the orbitofrontal cortex (grey), the cingulate cortex (light blue), the ventral tegmental area in the brainstem (light red), the hypothalamus (yellow), the periventricular grey/periacqueductal grey (PVG/PAG, green), nucleus accumbens (light green), the ventral pallidum (light purple), the amygdala (light red) and the insular cortices (not shown).

oral cavity and gets processed in the primary sensory cortices of the brain. The topology of these regions are remarkably similar between people with vision (red) always processed in the back of the brain, audition (dark blue) processed in regions of the temporal cortex, touch (light blue) in somatosensory regions and olfaction (orange) and taste (yellow) in frontal regions (Figure 3B). Importantly, unlike the other senses, olfactory processing is not processed via the thalamus which may explain the hedonic potency of odours [27]. Note that it is important that we are able to identify a food stimulus independently of whether we are hungry or sated, and accordingly, sensory information in primary sensory cortices is remarkably stable and not modulated by motivational state.

The sensory information is further integrated in multisensory areas before it is evaluated for reward value in the pleasure system. Here, the processing depends on prior memories, expectations and state and may give rise to brain activity which gives rise involuntary pleasure-evoked behaviour (such as licking of lips or soft moaning) and, at least in humans, subjective pleasure (Figure 3C).

Neuroscience has started to map the pleasure system in many species. This has been shown to include a number of important regions such as pleasure hotspot regions in subcortical areas of the brain such as the nucleus accumbens and ventral pallidum [28,29]. Manipulations of these regions with opioids have been shown to causally change pleasure-elicited reactions [13]. Other regions involved in pleasure have been found using human neuroimaging in the orbitofrontal, cingulate, medial prefrontal and insular cortices [30-37]. The pleasure system does not act in splendid isolation but is of course embedded within much larger brain networks. We are beginning to understand the metastable nature as well as the topological and functional features of these networks using advances in network science and graph theory together with advanced whole-brain computational models [38,39].

Computational processing principles for eating
Overall, eating has been demonstrated to rely on at least five fundamental processing principles: 1) hunger and attentional processing; 2) motivation-independent discriminative processing of identity and intensity; 3) learning-dependent multisensory representations; 4) reward representations of valence and 5) representations of hedonic experience [12,40]. In the following, these are briefly described.

Hunger and other attentional processing
Typically, changes in ongoing brain activity are driven by changes in the internal or external environment, signalling that the brain needs to start to reallocate resources and change behaviour. This motivational drive for change is strong for food intake, where hunger is a major attentional signal that along with other homeostatic signalling

can influence the brain to initiate food-seeking behaviours, typically following the satiety phase from the previous meal. The hunger information comes primarily from gut-brain interactions signalling if the nutrients eaten in the previous meal have yielded the expected amount of energy but a large part is also played by habit (such as regular meal times) and learning, including social interactions which may lead to overeating due to diminished attention towards the food [41,42]. Signals from receptors in the gut and in the circulatory system are vital in initiating eating through conveying messages for the need of nutrients or energy uptake [6,43].

The healthy system is balanced through careful monitoring and learning throughout life. In the presence of sufficient nutrients, healthy adults are able to maintain a stable body weight by careful management of nutrient uptake, energy needs and the balance with energy expenditure [44]. In animal models, this homeostatic component has been shown to relate to activity in hypothalamic circuits including the arcuate nucleus [6,43]. Hedonic influences beyond homeostasis can lead to malfunction to this control of energy balance, e.g. leading to obesity, potentially through a mismatch between the expected pleasure compared to the actual energy uptake from food intake [11,45].

Motivation-independent processing of identity and intensity
It is vital that reliable sensory food information is provided for the brain to guide ingestion decision-making. Eating has to be controlled very carefully since erroneous evaluation of the sensory properties of foods can potentially be fatal if ingesting toxins, microorganisms or non-food objects. Mammals have been shown to have brainstem reflexes (stereotypical for each basic taste) that are based on rudimentary analyses of the chemical composition, and which are not altered, even by the loss of all neural tissue above the level of the midbrain [46]. Eating-related behaviours in humans and other animals can usefully be described as a strategy to maintain a balance between conservative risk-minimising and life-preserving strategies (exploitation) with occasional novelty seeking (exploration) in the hope of discovering new, valuable sources of nutrients [47].

The sensory information about the identity and intensity of a food—sometimes called a *flavour object*—reaching the primary sensory cortices appears to be motivation-independent [48]. This principle has been demonstrated by neurophysiological and neuroimaging experiments using five basic pure tastes of salt, bitter, sour, sweet and umami to locate the primary taste area in humans in the bilateral anterior insula/frontal operculum [49-53] (Figure 4). Please note that one study has reported changes in activity in the primary taste cortex by expectancy [54]; but unfortunately, the authors did not publish the

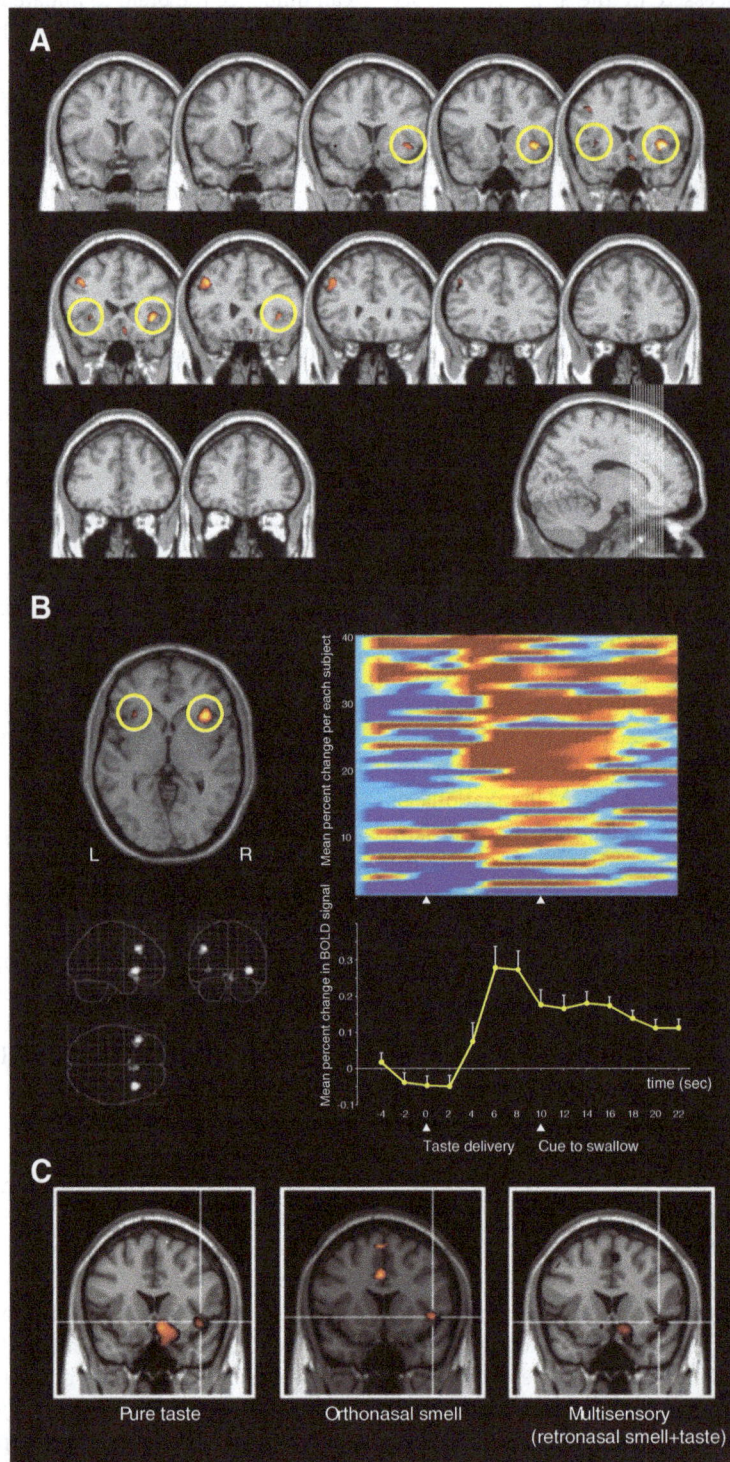

Figure 4 (See legend on next page.)

(See figure on previous page.)
Figure 4 Motivation-independent representations of food in primary sensory cortices. Pure taste is the archetypical reinforcer associated with food. **A)** Consistent with findings in non-human primates, neuroimaging has located the primary human taste cortex in bilateral anterior insular/frontal opercular cortices (yellow circles) with peak MNI coordinates of [x, y, z: 38,20,–4] and [x, y, z: –32,22,0] [53]. **B)** This data is based on 40 datasets from four experiments using eight unimodal and six multimodal taste stimuli ranging from pleasant to unpleasant. Each small aliquot of 0.75 mL taste stimulus was delivered via polythene tubes to the mouth of the participant who was asked to move it around before being cued to swallow after typically 10 s. To properly control and rinse out the effects of each stimulus, the taste stimulus was followed by a tasteless solution with the main ionic components of saliva. The time course of blood oxygen-level detection (BOLD) activity in right primary taste cortex is shown for all 40 subjects (top) and averaged across all (bottom) (for taste minus tasteless solution). **C)** Multisensory sensory integration was found in a region of the anterior insular cortex which responded to pure taste, orthonasal smell and flavour (retronasal smell and taste) [63].

exact coordinates of their putative primary taste cortex. It is thus difficult to trust this finding which is further undermined by visual inspection of the published figure, which clearly shows that the authors' purported primary taste cortex is located significantly posterior in the medial insular cortex, in contrast to the anterior insular primary taste region reported above and in all other careful neuroimaging taste studies.

Learning-dependent multisensory representations
Food-related decision-making depends on the integration of multisensory information about the food which includes information about temperature, viscosity, texture, fat contents, pungency and irritation mediated by a large variety of neural systems [25]. Neuroimaging this learning-dependent multisensory integration has found that the human orbitofrontal cortex integrates information from auditory [55], gustatory [51], olfactory [56], somatosensory [57] and visual [58] inputs, as well as information from the visceral sensory system [59]. The role of expectation and motivational control of appetite has also been investigated using restaurant menus which also found engagement of the orbitofrontal cortex [60] [61].

These human findings are consistent with neurophysiological recordings showing that the non-human primate orbitofrontal cortex receives input from all of the five senses [62]. These sensory inputs enter the orbitofrontal cortex primarily through its posterior parts and are integrated in more anterior areas [34]. The interaction between taste and smell revealed by neuroimaging is found in the orbitofrontal cortex and nearby agranular insula (Figure 4C) [33,50,63].

Reward representations of sensory stimuli
Subsequent to establishing motivation-independent representations and multisensory representations of information about a food, affective valence is assigned, helping to guide prediction and decision-making. Again, pure taste serves as a good example with a neuroimaging study finding a dissociation between the brain regions responding to the intensity of the taste and its affective valence [64]. Another study found that subjective ratings of taste pleasantness correlated with activity in the medial orbitofrontal cortex (medial OFC) and in the anterior cingulate

cortex [65] but, importantly, not with activity in the primary taste region, which was motivation-independent. Further evidence comes from experiments using orthonasal olfaction to show dissociable encoding of the intensity and pleasantness of olfactory stimuli, with the intensity encoded in the amygdala and nearby regions, and the pleasantness correlated with activity in the medial OFC (Figure 5A) and anterior cingulate cortex [66-68].

These reward-related findings in the medial OFC cohere with neuroimaging studies using other rewards. One study found a correlation between activity in the medial OFC with the amount of monetary wins and losses [69] (Figure 5B). Similarly, the subjective experience of methamphetamine over minutes was found to correlate with activity in the medial OFC [70] (Figure 5C). Even studies on the much shorter timescales of milliseconds have found activity in the medial OFC related to the reward of images of cute babies [71] (Figure 5D). These results point to the unity of reward-related activity in the pleasure system across many different rewards, which in turn suggest a system with a common currency of reward. Such a system would make it easier to decide and choose between different rewards.

Representations of hedonic experience
Finally, the evidence suggests that the subjective hedonic experience of food is encoded in activity in the pleasure system. In humans, the mid-anterior orbitofrontal cortex (mid-OFC) appears to be a key region as demonstrated by a selective-satiety neuroimaging study where activity in this region shows not only a selective decrease in the reward value to the food eaten to satiety (and not to the food not eaten) but also a correlation with pleasantness ratings (Figure 5E) [33]. This result indicates that the reward value of the taste, olfactory and somatosensory components of a food are represented in the orbitofrontal cortex and, therefore, that the subjective pleasantness of food might be represented in this region. Other studies have supported this finding, including an experiment investigating true taste synergism, where the intensity of a taste is dramatically enhanced by adding minute doses of another taste. The strong subjective enhancement of the pleasantness of umami taste that occurs when 0.005 M inosine 5′-monophosphate is added to

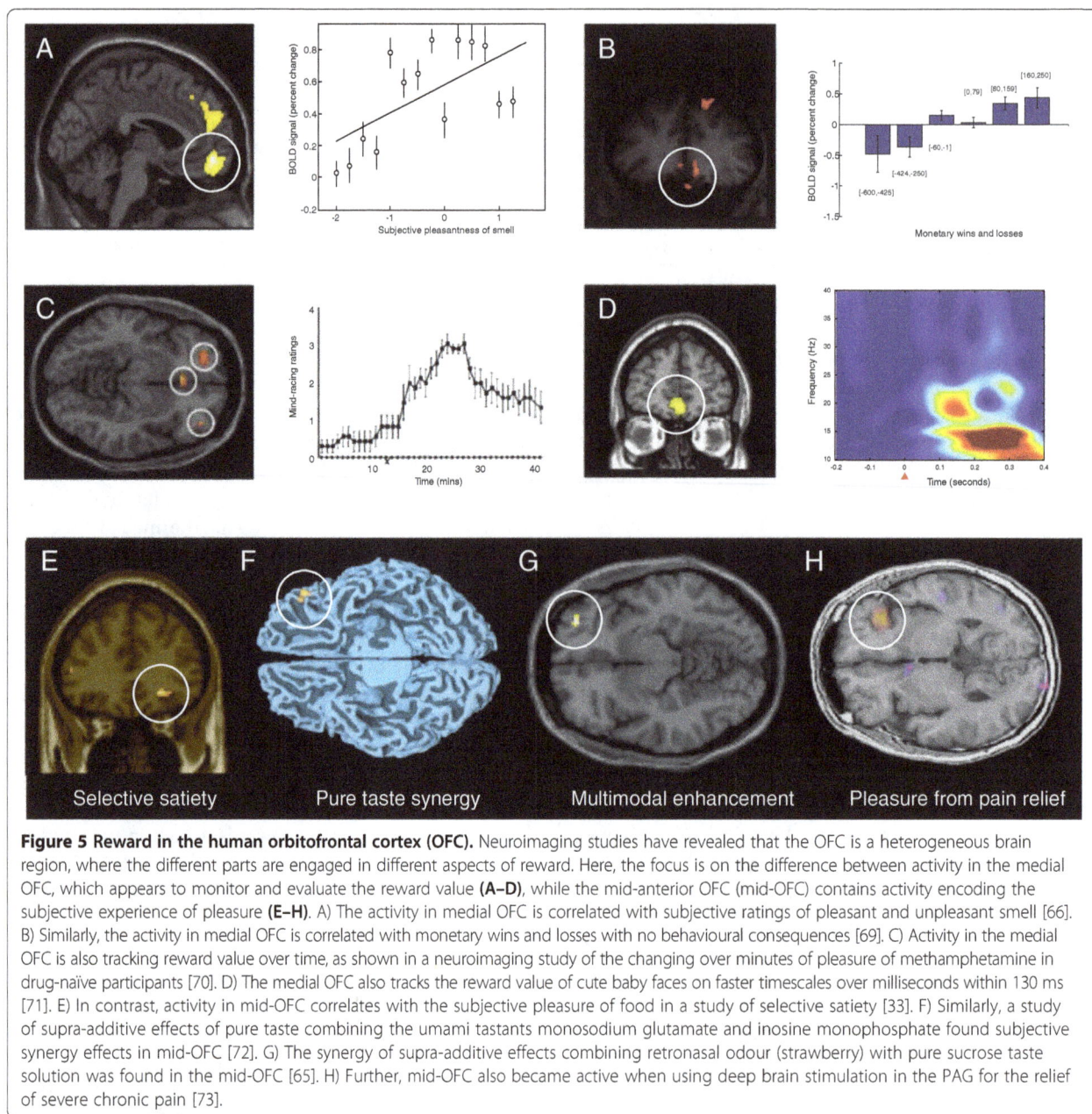

Figure 5 Reward in the human orbitofrontal cortex (OFC). Neuroimaging studies have revealed that the OFC is a heterogeneous brain region, where the different parts are engaged in different aspects of reward. Here, the focus is on the difference between activity in the medial OFC, which appears to monitor and evaluate the reward value (**A–D**), while the mid-anterior OFC (mid-OFC) contains activity encoding the subjective experience of pleasure (**E–H**). A) The activity in medial OFC is correlated with subjective ratings of pleasant and unpleasant smell [66]. B) Similarly, the activity in medial OFC is correlated with monetary wins and losses with no behavioural consequences [69]. C) Activity in the medial OFC is also tracking reward value over time, as shown in a neuroimaging study of the changing over minutes of pleasure of methamphetamine in drug-naïve participants [70]. D) The medial OFC also tracks the reward value of cute baby faces on faster timescales over milliseconds within 130 ms [71]. E) In contrast, activity in mid-OFC correlates with the subjective pleasure of food in a study of selective satiety [33]. F) Similarly, a study of supra-additive effects of pure taste combining the umami tastants monosodium glutamate and inosine monophosphate found subjective synergy effects in mid-OFC [72]. G) The synergy of supra-additive effects combining retronasal odour (strawberry) with pure sucrose taste solution was found in the mid-OFC [65]. H) Further, mid-OFC also became active when using deep brain stimulation in the PAG for the relief of severe chronic pain [73].

0.5 M monosodium glutamate (compared to both delivered separately) correlated with increased activity in mid-OFC (Figure 5 F) [72]. Similarly, investigations of the synergistic enhancement of a matched taste and retronasal smell found significant activity in the same mid-OFC region (Figure 5G) [63]. These food-related hedonic findings fit well with evidence coming from the study of other pleasures, including the finding of significant activity in mid-OFC in a study using magnetoencephalography (MEG) with deep brain stimulation to investigate the pleasurable relief from severe chronic pain (Figure 5H) [73].

Conclusions

As demonstrated poignantly by *Babette's Feast*, food is not only an important part of a balanced diet; it is also one of our main routes to pleasure. The novella opens many interesting question with regard to well-being and the good life and in particular shows that to allow oneself to be open to the possibility of pleasure of food is also allowing for the deep experiences of the multitude of pleasures. This is in sharp contrast to the denial of the pleasure of food leading to anhedonia, the lack of pleasure, which is a key constituent component of affective disorders.

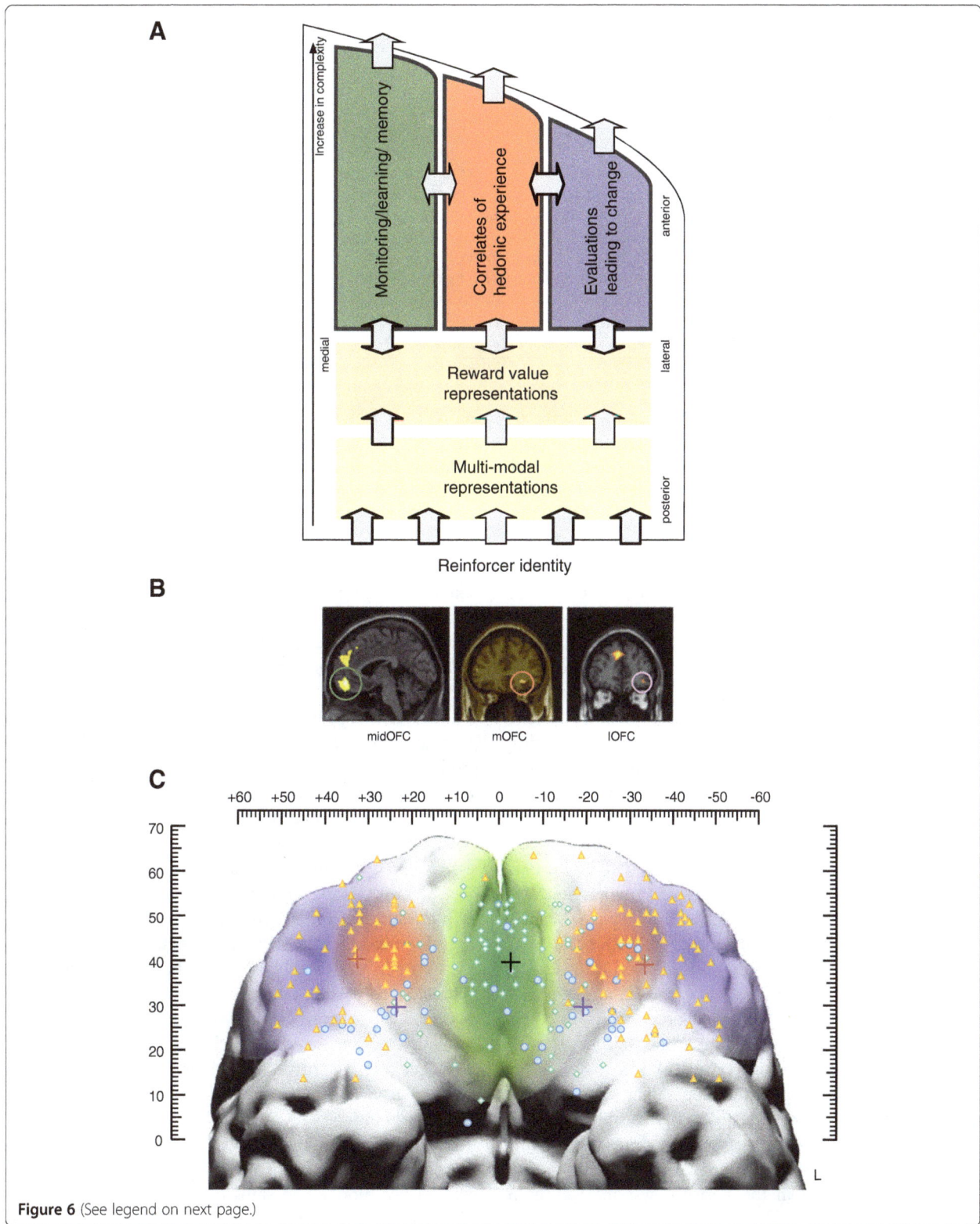

Figure 6 (See legend on next page.)

(See figure on previous page.)
Figure 6 Model of information flow in the orbitofrontal cortex (OFC). The spatial heterogeneity of the human OFC has been revealed with neuroimaging. **(A-C)** The OFC is involved in most of the phases of the pleasure cycle, including evaluation, expectation, experience as well as decision-making and selection. Sensory information comes to the OFC where it is available for pattern association between primary (e.g. taste) and secondary (e.g. visual) reinforcers. Sensory information is combined in multisensory representations in the posterior OFC with processing increasing in complexity towards more anterior areas. The reward value of reinforcers is assigned in more anterior regions. This information is stored for valence monitoring/learning/memory (in medial OFC, green) and made available for subjective hedonic experience (in mid-OFC, orange) and used to influence subsequent behaviour (in lateral OFC with links to regions of anterior cingulate cortex, blue). The OFC participates in multiple modulatory brain-loops with other important structures in the pleasure system such as the nucleus accumbens, ventral pallidum, amygdala and hypothalamus, as well as modulation with autonomic input from the gut. [34]. B) Examples of monitoring reward value in medial OFC (green) was found in a study of orthonasal smell where the activity correlated with subjective ratings of pleasant and unpleasant smell [66]. Activity in mid-OFC (orange) correlates with the subjective pleasure of food in a study of selective-satiety [33]. In contrast, the activity in lateral OFC (shown in red) was found when changing behaviour in a rapid context-dependent reversal task of simple social interactions [84]. C) A large meta-analysis of neuroimaging studies confirmed the differential functional roles of these regions [34]. Future avenues of research include describing temporal unfolding of activity, similar to early involvement of the medial OFC (<130 ms) in processing rewards such as cute babies and guide attentional resources [71].

The science of pleasure has made great strides in recent years [4], due not in small parts to using food as a pleasure-eliciting stimulus. As demonstrated in this review, the research has uncovered many of the fundamental brain mechanisms governing eating and pleasure in general. It has helped understand the brain's complex resource allocation problems with food competing with other rewards for time and resources. In particular, the brain must make important decisions of how best to balance exploration and exploitation to ensure survival. These decisions involve deciding when to pursue a reward, and whether to initiate, sustain and terminate the wanting, liking and learning processes involved in the different phases of the pleasure cycle (Figure 1). Eating is a complex process that involves many different factors over time as described in a multilevel model (Figure 2). The model demonstrates the cyclical changes in hunger levels related to the initiation and termination of meals, as they relate to signals from the brain, gut-brain, oral cavity, stomach and intestines, liver and metabolites and body mass.

Here, the focus has been on the computational principles for the multisensory processing of food information that initiates and terminates a meal, as well as the pleasure involved (Figure 3). Five main processing principles were discussed: 1) hunger and attentional processing; 2) motivation-independent processing of identity and intensity (Figure 4); 3) learning-dependent multisensory representations; 4) reward representations and 5) representations of hedonic experience. These principles are implemented within the orbitofrontal cortex that is a key, heterogeneous region in the pleasure system (Figures 5 and 6).

Furthermore, pleasure research has shown that food, sex and social interactions are fundamental to our survival and these basic stimuli take priority in resource allocation. It has also shown the unity of pleasure processing of different rewards, with food, sex, social and higher-order stimuli (such as music and money) in a unified pleasure system [12,13,74-76,84].

Much remains to be done, but finally science has gained a toehold in understanding how pleasure can come to transform lives. Understanding the pleasure of food has played a major part in hedonia research and may even offer some insights into well-being. We have previously taken a lead from Aristotle's distinction between *hedonia* and *eudaimonia* (a life well-lived) to show how the study of pleasure may offer some insights into well-being [77].

Gastronomy offers the potential to expand on these findings and create exciting experiences and great pleasure. The rise of molecular gastronomy and gastrophysics have afforded chefs with unprecedented control over the production of novel flavours and textures of food [78,79]. These experiences are by their very nature multisensory and like all experiences highly dependent on expectation and prior experiences [80]. Using scientific tools and insights allows playful chefs to create unique and highly pleasurable dining experiences, e.g. using touch and sound as interesting extras in their gastronomical palette [81]. Yet, all foods are ultimately dependent on the state of the diner's brain and body [82], and the emergence of the neuroscience of the pleasure of gastronomy could help guide further progress [11,83]. Both the science and art of cooking stand to benefit much from future collaborations between scientists and chefs, especially in so far this research can help increase the pleasure of eating and well-being.

Babette's Feast shows how a sumptuous dinner can bring about much pleasure and transform lives. Babette uses all her money and skills on creating the once-in-a-lifetime dinner, yet at the end she tells the sisters: "A great artist, Mesdames, is never poor. We have something, Mesdames, of which other people know nothing". While it is true that creating great art takes skills and years of practice, it is also important to remember that every moment and every bite of food carries within it the possibility of pleasure. The brain is built for pleasure and it is through learning to appreciate the extraordinary in ordinary experiences, through pursuing the variety of pleasures rather than the relentless single-minded pursuit (hedonism) or denial of pleasure (asceticism) that a life well-lived can be constructed.

Competing interests
The author declares that he has no competing interests.

Acknowledgements
This research is supported by the TrygFonden Charitable Foundation.

References
1. Dinesen I. Babette's Feast. In: Anecdotes of destiny. London: Penguin; 1958.
2. Kringelbach G. Forfatterne går til bords. Copenhagen: Erichsen; 1971.
3. Kringelbach ML. The orbitofrontal cortex: linking reward to hedonic experience. Nature Reviews Neuroscience. 2005;6:691–702.
4. Kringelbach ML, Berridge KC. Pleasures of the brain. New York: Oxford University Press; 2010.
5. Batterham RL, Ffytche DH, Rosenthal JM, Zelaya FO, Barker GJ, Withers DJ, Williams SC: PYY modulation of cortical and hypothalamic brain areas predicts feeding behaviour in humans. Nature 2007, 450:106–109.
6. Berthoud HR, Morrison C. The brain, appetite, and obesity. Annu Rev Psychol. 2008;59:55–92.
7. Saper CB, Chou TC, Elmquist JK. The need to feed: homeostatic and hedonic control of eating. Neuron. 2002;36:199–211.
8. Woods SC. The control of food intake: behavioral versus molecular perspectives. Cell Metab. 2009;9:489–98.
9. Kohn M, Booth M. The worldwide epidemic of obesity in adolescents. Adolesc Med. 2003;14:1–9.
10. Kringelbach ML. Balancing consumption: brain insights from the cyclical nature of pleasure. In The interdisciplinary science of consumption. Edited by Preston S, Kringelbach ML, Knutson B. Cambridge, Mass.: MIT Press; 2014
11. Kringelbach ML. Food for thought: hedonic experience beyond homeostasis in the human brain. Neuroscience. 2004;126:807–19.
12. Kringelbach ML, Stein A, van Hartevelt TJ. The functional human neuroanatomy of food pleasure cycles. Physiol Behav. 2012;106:307–16.
13. Berridge KC, Kringelbach ML. Neuroscience of affect: brain mechanisms of pleasure and displeasure. Curr Opin Neurobiol. 2013;23:294–303.
14. Kringelbach ML, Berridge KC. A joyful mind. Scientific American. 2012;307:40–5.
15. Lou HC, Joensson M, Kringelbach ML. Yoga lessons for consciousness research: a paralimbic network balancing brain resource allocation. Frontiers in Psychology. 2011;2:366.
16. Craig W. Appetites and aversions as constituents of instincts. Biological Bulletin of Woods Hole. 1918;34:91–107.
17. Sherrington CS. The integrative action of the nervous system. New York: C. Scribner's sons; 1906.
18. Robinson TE, Berridge KC. The neural basis of drug craving: an incentive-sensitization theory of addiction. Brain Res Brain Res Rev. 1993;18:247–91.
19. Robinson TE, Berridge KC. Addiction. Annu Rev Psychol. 2003;54:25–53.
20. Berridge KC, Kringelbach ML. Affective neuroscience of pleasure: Reward in humans and animals. Psychopharmacology 2008;199:457-480.
21. Finlayson G, King N, Blundell JE. Liking vs. wanting food: importance for human appetite control and weight regulation. Neurosci Biobehav Rev. 2007;31:987–1002.
22. Blundell JE, Burley VJ. Satiation, satiety and the action of fibre on food intake. Int J Obes. 1987;11 Suppl 1:9–25.
23. De Graaf C, De Jong LS, Lambers AC. Palatability affects satiation but not satiety. Physiol Behav. 1999;66:681–8.
24. De Araujo IE, Oliveira-Maia AJ, Sotnikova TD, Gainetdinov RR, Caron MG, Nicolelis MAL, et al. Food reward in the absence of taste receptor signaling. Neuron. 2008;57:930–41.
25. De Araujo IE, Simon SA. The gustatory cortex and multisensory integration. Int J Obes. 2009;33:S34–43.
26. Shepherd GM. Smell images and the flavour system in the human brain. Nature. 2006;444:316–21.
27. Van Hartevelt TJ, Kringelbach ML. The olfactory system. In The human nervous system 3rd Ed. Edited by Mai J, Paxinos G: San Diego: Academic Press; 2011: 1219–1238
28. Peciña S, Berridge KC. Hedonic hot spot in nucleus accumbens shell: where do mu-opioids cause increased hedonic impact of sweetness? J Neurosci. 2005;25:11777–86.
29. Smith KS, Berridge KC. Opioid limbic circuit for reward: interaction between hedonic hotspots of nucleus accumbens and ventral pallidum. J Neurosci. 2007;27:1594–605.
30. Cardinal RN, Parkinson JA, Hall J, Everitt BJ. Emotion and motivation: the role of the amygdala, ventral striatum, and prefrontal cortex. Neurosci Biobehav Rev. 2002;26:321–52.
31. Everitt BJ, Robbins TW. Neural systems of reinforcement for drug addiction: from actions to habits to compulsion. Nat Neurosci. 2005;8:1481–9.
32. Kringelbach ML. The hedonic brain: A functional neuroanatomy of human pleasure. In pleasures of the brain. Edited by Kringelbach ML, Berridge KC. Oxford, U.K.: Oxford University Press; 2010: 202–221
33. Kringelbach ML, O'Doherty J, Rolls ET, Andrews C. Activation of the human orbitofrontal cortex to a liquid food stimulus is correlated with its subjective pleasantness. Cerebral Cortex. 2003;13:1064–71.
34. Kringelbach ML, Rolls ET. The functional neuroanatomy of the human orbitofrontal cortex: evidence from neuroimaging and neuropsychology. Progress in Neurobiology. 2004;72:341–72.
35. Berridge KC. Food reward: brain substrates of wanting and liking. Neuroscience and Biobehavioral Reviews. 1996;20:1–25.
36. Watson KK, Shepherd SV, Platt ML. Neuroethology of pleasure. In pleasures of the brain. Edited by Kringelbach ML, Berridge KC. New York: Oxford University Press; 2010: 85–95
37. Amodio DM, Frith CD. Meeting of minds: the medial frontal cortex and social cognition. Nat Rev Neurosci. 2006;7:268–77.
38. Cabral J, Kringelbach ML, Deco G. Exploring the network dynamics underlying brain activity during rest. Prog Neurobiol. 2014;114:102–31.
39. Deco G, Kringelbach ML. Great expectations: Using Whole-Brain Computational Connectomics for Understanding Neuropsychiatric Disorders. Neuron 2014;84:892-905.
40. Kringelbach ML. Cortical systems involved in appetite and food consumption. In Appetite and body weight: integrative systems and the development of anti-obesity drugs. Edited by Cooper SJ, Kirkham TC. London: Elsevier; 2006: 5–26
41. Hetherington MM, Anderson AS, Norton GNM, Newson L. Situational effects on meal intake: a comparison of eating alone and eating with others. Physiology and Behavior. 2006;88:498–505.
42. de Graaf C, Kok FJ. Slow food, fast food and the control of food intake. Nat Rev Endocrinol. 2010;6:290–3.
43. Lenard NR, Berthoud HR. Central and peripheral regulation of food intake and physical activity: pathways and genes. Obesity (Silver Spring). 2008;16 Suppl 3:S11–22.
44. Shin AC, Zheng H, Berthoud HR. An expanded view of energy homeostasis: neural integration of metabolic, cognitive, and emotional drives to eat. Physiol Behav. 2009;97:572–80.
45. Zheng H, Lenard NR, Shin AC, Berthoud HR. Appetite control and energy balance regulation in the modern world: reward-driven brain overrides repletion signals. Int J Obes (Lond). 2009;33 Suppl 2:S8–13.
46. Grill HJ, Norgren R. The taste reactivity test. II. Mimetic responses to gustatory stimuli in chronic thalamic and chronic decerebrate rats. Brain Res. 1978;143:281–97.
47. Rozin P. Food preference. In: Smelser NJ, Baltes PB, editors. International Encyclopedia of the Social & Behavioral Sciences. Amsterdam: Elsevier; 2001.
48. Veldhuizen MG, Rudenga KJ, Small D. The pleasure of taste flavor and food. In Pleasures of the brain. Edited by Kringelbach ML, Berridge KC. Oxford, U.K.: Oxford University Press; 2010: 146–168
49. Kinomura S, Kawashima R, Yamada K, Ono S, Itoh M, Yoshioka S, et al. Functional anatomy of taste perception in the human brain studied with positron emission tomography. Brain-Res. 1994;659:263–6.
50. Small DM, Jones-Gotman M, Zatorre RJ, Petrides M, Evans AC. Flavor processing: more than the sum of its parts. Neuroreport. 1997;8:3913–7.
51. Small DM, Zald DH, Jones-Gotman M, Zatorre RJ, Pardo JV, Frey S, et al. Human cortical gustatory areas: a review of functional neuroimaging data. Neuroreport. 1999;10:7–14.

52. O'Doherty J, Rolls ET, Francis S, Bowtell R, McGlone F. Representation of pleasant and aversive taste in the human brain. J Neurophysiol. 2001;85:1315–21.

53. Kringelbach ML, de Araujo IE, Rolls ET. Taste-related activity in the human dorsolateral prefrontal cortex. Neuroimage. 2004;21:781–8.

54. Nitschke JB, Dixon GE, Sarinopoulos I, Short SJ, Cohen JD, Smith EE, et al. Altering expectancy dampens neural response to aversive taste in primary taste cortex. Nat Neurosci. 2006;9:435–42.

55. Frey S, Kostopoulos P, Petrides M. Orbitofrontal involvement in the processing of unpleasant auditory information. European Journal of Neuroscience. 2000;12:3709–12.

56. Zatorre RJ, Jones-Gotman M, Evans AC, Meyer E. Functional localization and lateralization of human olfactory cortex. Nature. 1992;360:339–40.

57. Rolls ET, O'Doherty J, Kringelbach ML, Francis S, Bowtell R, McGlone F. Representations of pleasant and painful touch in the human orbitofrontal and cingulate cortices. Cerebral Cortex. 2003;13:308–17.

58. Aharon I, Etcoff N, Ariely D, Chabris CF, O'Connor E, Breiter HC. Beautiful faces have variable reward value: fMRI and behavioral evidence. Neuron. 2001;32:537–51.

59. Critchley HD, Mathias CJ, Dolan RJ. Fear conditioning in humans: the influence of awareness and autonomic arousal on functional neuroanatomy. Neuron. 2002;33:653–63.

60. Hinton EC, Parkinson JA, Holland AJ, Arana FS, Roberts AC, Owen AM. Neural contributions to the motivational control of appetite in humans. *Eur J Neurosci*, vol. 20. pp. 1411–1418; 2004:1411–1418.

61. Arana FS, Parkinson JA, Hinton E, Holland AJ, Owen AM, Roberts AC. Dissociable contributions of the human amygdala and orbitofrontal cortex to incentive motivation and goal selection. J Neurosci. 2003;23:9632–8.

62. Rolls ET. The brain and emotion. Oxford: Oxford University Press; 1999.

63. De Araujo IET, Rolls ET, Kringelbach ML, McGlone F, Phillips N. Taste-olfactory convergence, and the representation of the pleasantness of flavour, in the human brain. European Journal of Neuroscience. 2003;18:2059–68.

64. Small DM, Gregory MD, Mak YE, Gitelman D, Mesulam MM, Parrish T. Dissociation of neural representation of intensity and affective valuation in human gustation. Neuron. 2003;39:701–11.

65. De Araujo IET, Kringelbach ML, Rolls ET, McGlone F. Human cortical responses to water in the mouth, and the effects of thirst. Journal of Neurophysiology. 2003;90:1865–76.

66. Rolls ET, Kringelbach ML, de Araujo IET. Different representations of pleasant and unpleasant odors in the human brain. European Journal of Neuroscience. 2003;18:695–703.

67. Anderson AK, Christoff K, Stappen I, Panitz D, Ghahremani DG, Glover G, et al. Dissociated neural representations of intensity and valence in human olfaction. Nature Neuroscience. 2003;6:196–202.

68. Gottfried JA, Deichmann R, Winston JS, Dolan RJ. Functional heterogeneity in human olfactory cortex: an event-related functional magnetic resonance imaging study. Journal of Neuroscience. 2002;22:10819–28.

69. O'Doherty J, Kringelbach ML, Rolls ET, Hornak J, Andrews C. Abstract reward and punishment representations in the human orbitofrontal cortex. Nature Neuroscience. 2001;4:95–102.

70. Völlm BA, de Araujo IET, Cowen PJ, Rolls ET, Kringelbach ML, Smith KA, et al. Methamphetamine activates reward circuitry in drug naïve human subjects. Neuropsychopharmacology. 2004;29:1715–22.

71. Kringelbach ML, Lehtonen A, Squire S, Harvey AG, Craske MG, Holliday IE, et al. A specific and rapid neural signature for parental instinct. *PLoS ONE* 2008, 3:e1664. doi:1610.1371/journal.pone.0001664.

72. De Araujo IET, Kringelbach ML, Rolls ET, Hobden P. The representation of umami taste in the human brain. *Journal of Neurophysiology*, vol. 90. pp. 313–319; 2003:313–319.

73. Kringelbach ML, Jenkinson N, Green AL, Owen SLF, Hansen PC, Cornelissen PL, et al. Deep brain stimulation for chronic pain investigated with magnetoencephalography. Neuroreport. 2007;18:223–8.

74. Georgiadis JR, Kringelbach ML, Pfaus JG. Sex for fun: a synthesis of human and animal neurobiology. Nat Rev Urol. 2012;9:486–98.

75. Georgiadis JR, Kringelbach ML. The human sexual response cycle: Brain imaging evidence linking sex to other pleasures. Prog Neurobiol. 2012;98:49–81.

76. Gebauer L, Kringelbach ML, Vuust P. Ever-changing cycles of musical pleasure: the role of dopamine and anticipation. Psychomusicology, Music, Mind & Brain. 2012;22:152–67.

77. Kringelbach ML, Berridge KC. Towards a functional neuroanatomy of pleasure and happiness. Trends in Cognitive Sciences. 2009;13:479–87.

78. McGee H. Q&A: Harold McGee, the curious cook. Flavour. 2013;2:13.

79. Mouritsen OG. The emerging science of gastrophysics and its application to the algal cuisine. Flavour. 2012;1:6.

80. Harrar V, Smith B, Deroy O, Spence C. Grape expectations: how the proportion of white grape in champagne affects the ratings of experts and social drinkers in a blind tasting. Flavour. 2013;3:25.

81. Spence C, Hobkinson C, Gallace A, Fiszman BP. A touch of gastronomy. Flavour. 2013;2:14.

82. Møller P. Gastrophysics in the brain and body. Flavour. 2013;2:8.

83. Shepherd GM. Neurogastronomy: how the brain creates flavor and why it matters. New York: Columbia University Press; 2011.

84. Kringelbach ML, Rolls ET. Neural correlates of rapid context-dependent reversal learning in a simple model of human social interaction. Neuroimage. 2003;20:1371–83.

10

Q&A: the science of cocktails

Tony Conigliaro[1,2*]

Abstract

Tony Conigliaro, London-based mixologist and bar owner, talks to *Flavour* about the scientific methods and equipment he has brought into cocktail making.

Opinion

Tony Conigliaro is a London-based mixologist and owner of cocktail bars *69 Colebrooke Row* and *The Zetter Town House*, both in London, UK. He also runs the Drink Factory, a bespoke lab that focuses on alcohol-development and research into liquid flavour, drinks and cocktails. Widely acknowledged as one of the UK's pioneering drinks creators, he won the award for International Bartender of the Year 2009, as well as the *Observer Food Monthly* award for Bar of the Year in 2010. His new book, *Drinks*, is released later this year.

In a Q & A with *Flavour*, Tony Conigliaro talks about the scientific methods and equipment he has brought into cocktail making and some of the new drinks he has created.

You made your name as a cocktail maker who uses scientific techniques to create great drinks. Do you have any formal training in science?

My background isn't in science. I did Fine Art and Art History for years at University and college, so I think that took care of the creative side of things. But as for the science, I'm more of a magpie, I suppose – I pick up bits and pieces as I go along, but over the years we've built a collection of people whom we can ask specific questions, so if we ever get stuck with a line of thought or there's something we don't understand, those people will jump in and help us out. Beyond that, we've never really been afraid of asking questions. If we don't have the answer, we find someone who does and actually ask, and we aren't afraid of asking until we find out the answer to the question.

I've been working in the bar cocktail industry for sixteen years now. There was, you might say, a pivotal moment where I was in the kitchen in Isola in Knightsbridge back in 1999 and I was trying to make a

fresh puree. I was kind of obsessed with fresh purees back then because all the ones that we were buying had too much water, too much sugar in them, and I wanted to add that fresh element. I think we were working on Bellinis at the time, and the pastry chef came over and just shook his head at me and said "That's not how you do it". He threw everything in the bin and said "This is how you do it". From that point it really opened up an avenue of what was possible in the kitchen and how that affected the quality of what we were using. That led at a slightly later stage to asking questions such as "How can we do this better?" and "Why does this work?", which inevitably led to talking to food scientists and people who were very well versed in preparing food. That really opened up the door; I suppose once you start asking those questions there's no turning back; there are always more questions than you can actually answer, so it's a perpetual motion of finding things out.

Is there a particular scientific project you're working on now?

For the past year and a half we've been looking at flavours that are at the fringes of what you taste, what you experience, and then looking at how those work in drinks. We've recently been focusing on terroirs in wines - how flint stone affects the flavour of wine. For example, if it does affect the flavour, how, and if it doesn't, why not? What is actually happening? So we've been doing a lot of that sort of research, and last year we started distilling powdered flint stone in our vacuum still, which actually produces a flint stone note and a flintstony, minerally flavour. So it's almost like we're adding terroirs to cocktails in a way, and we've started looking at chalks and clays and things like that, which has obviously been really interesting because we've been working with some wine connoisseurs who specialise in terroir. It's also beneficial for us to work out when certain flavours arrive and why they arrive, where they're coming

Correspondence: info@drinkfactory.com
[1]Drink Factory, 35 Britannia Row, London N1 8QH, UK
[2]69 Colebrooke Row, London N1 8AA, UK

from and why they are there, because it gives us an idea as to how well we can replicate that into a cocktail.

One part is "Where are these flavours coming from and how do they work?" and the second part is "When do they work?". We can create a progressive line of flavours. One of the drinks, The Sirocco, is based on that theory of where the flavour will appear, or register, in your mouth. It's got a grapefruit note followed by a pink peppercorn note, which is really interesting because it registers as sweet but then as incense, and then there's the flint note. You can almost read them on your tongue like a story, in that order. If you look at the size of the molecules and the way they work, they're almost progressionally small to big, and then we put a benzyl syrup underneath to smooth it out and tie it all together. That ties in with the second part of the process, working out where the flavours are arriving and how to describe the journey of those flavours. I suppose that's the romantic part that's not so scientific, but it only works by knowing the science.

When designing the drink, there's a theory which we look at, but we also consider the different elements, and then it's just testing and seeing how they work, if they work, and then making minute adjustments to all the variations of every amount you put in. So, for example, the flint: you put very little flint in, but you put in more pink peppercorn and less grapefruit. When we first made the drink it was all kind of higgledy-piggledy and everything was either too big or too small, so it's just balancing it out. Then you've got the structure of how they work together. Also, you've got flavour by itself, which reacts in one way, but flavour with alcohol can work in a completely different way, and flavour with another ingredient might do something else entirely.

How do you test your new drinks, in terms of their sensory properties?

Everyone here tries everything incessantly until all of us are happy with it. Anyone who walks in here, we usually go "Taste that". We always test things on people before we put it on the menu, sometimes randomly, and see what they think, unless we think it's amazing first off and we just have that feeling that it's going to work. But we test some of the more experimental varieties a little first to see where they register, and if we've made a huge mistake, it's quite frightening. Sometimes the drinks can take up to six months or longer - there was one drink that took two years just to get right.

Distilling a rock, like flint, requires specialist scientific equipment. How did you, as a bartender, realise that this equipment was available?

One of our main pieces of kit is the Buchi Rotavapor, which we've got two of and use practically every single day for distilling various different tinctures or other

things. I first came across it after a friend of mine had actually worked with Jordi Roca over at Roca Can Celler in Girona. When I saw it I just thought, that kind of belongs to us. In essence, it's a still that we can make our own ingredients with. So I was lucky enough that the guys at the Fat Duck had one which they let me use every now and then on Sunday afternoons. I remember it was in the back room in the lab, and I'd just sit in the back room distilling things, which was very generous of them. But then I realised how much potential it actually had for what I wanted to do, which was create flavours you wouldn't necessarily find in cocktails or spirits, so I bought one, and I suppose that was the start of the collection. Things like the vacuum machine and the centrifuge came later, but I was influenced by what I was seeing other people use.

The Rotavapor is basically a lab still; it works on the same principle as most other stills, but it works under vacuum, so it allows you to distil stuff at a very low pressure. You're boiling food pressure rather than heat, so if you're doing very delicate things like roses, for example, the heat source won't damage the chemicals in the rose, which allows you to achieve a very clean, more complete imprint of that rose on the other side of the distillate.

You use a scientific method, did this come naturally to you or was there a process where you thought you needed to be more rigorous about how you treated cocktails?

I don't really see the process as scientific, but I suppose it is in a way; it had to become more rigorous. The best example of this is when I finally got myself a vacuum still. I literally put everything I could possibly find through it just to see what would happen, and there was one particular flavour that I put through, I forget what it was now, which was absolutely magnificent, but I hadn't written anything down so I've never been able to replicate it since. I've tried millions of different variations but it just doesn't come out the same. It was from that point I realised that the process needed to be more rigorous for numerous reasons; firstly, just so we can replicate it, so that we're not repeating ourselves and we get regular results, but also in the bars it means you can then create something very accurately, very quickly and repeat that over and over again.

Do you see yourself as creating a product for your bars or do you have wider purposes, perhaps trying to change the way people think about cocktails?

Both, I think. I really like the idea that we have very bespoke ingredients at the bars because there's a uniqueness there, but it also betrays what we're trying to do. There are ingredients that aren't out there that I would like to use which would bring a whole new kind of space to cocktails; you can't, for example, go out and buy a flint vodka, whereas making a flint vodka introduces an entirely new and different way of doing things to cocktails.

Do you collaborate with academics and scientists?

Yes, we do. We talk a lot with various different field scientists and academics. Not necessarily in the field of food, far more than that; we work with perfumers, we work with perfume technicians, we even worked with an architect recently. So it's very varied and it all feeds back into what we do.

For example, there was one thing in particular that Hervé This had written about that I was very interested in, because it looked like the answer to my question but I didn't really understand how it worked. I had got back from the bar I was working in very late one night and put a draft email together of a whole load of really jumbled questions. But instead of saving it as a draft, I actually sent it to him. I was mortified. I could hardly sleep that night thinking "Oh my god, you know he'll never answer this". But by the time I'd woken up he'd actually answered it, understood it and written answers to all the questions. I think people are as enthusiastic about what we do on the other side, so it's just a question of asking them and the enthusiasm comes back.

What are your longer-term aspirations for your research?

We're going to investigate terroir because it is so fascinating. I'm very interested in the idea that minerals can register on your palate when you drink mineral water and what happens with that, because that can open up whole new avenues. A really good example of this was when we did a test where we were running electrodes through various vermouths and gins to change the flavours, which was really interesting because for a limited period the actual flavours would change in the drink. Then we started putting in silver rods and we made colloidal silver in some of the spirits. One drink we made was called the silver phantom, which was a martini with colloidal silver in. It had this shimmer that registered as a slight metallic note, almost like blood in a way, it had that kind of tingle in your mouth. But that, paired with the juniper and the gin, was fantastic because it made the juniper sing, almost like a high pine note, and that was very exciting. It had a real visual act to it too, because it had that shimmer. The whole idea of changing flavour through electricity was intriguing, but it was too limited for us to actually do in the bar. I mean we could have electrodes in the bar, but health and safety might get a little bit worried about that.

Competing interests

Tony Conigliaro is the owner of *69 Colebrooke Row* and *The Zetter Town House*, cocktail bars in London, and the Drinks Factory, a laboratory that researches liquid flavour, drinks and cocktails.

Author information

Tony Conigliaro is a mixologist, bartender and bar owner who has pioneered a scientific approach to cocktail making. He has created several new drinks including the Twinkle, Spitfire, Oh Gosh, and a re-invention of the Prairie Oyster and Bloody Mary, many of which appear on menus across the world. His cocktail bars in London, *69 Colebrooke Row* and *The Zetter Town House*, were winner and runner-up in the *Observer Food Monthly* Bar of the Year in 2010 and 2011, respectively. His new book, *Drinks*, is published in 2012.

11

Seaweeds for umami flavour in the New Nordic Cuisine

Ole G Mouritsen[1,2*], Lars Williams[2,3], Rasmus Bjerregaard[4] and Lars Duelund[1]

Abstract

Use of the term 'umami' for the fifth basic taste and for describing the sensation of deliciousness is finding its way into Western cuisine. The unique molecular mechanism behind umami sensation is now partly understood as an allosteric action of glutamate and certain 5'-ribonucleotides on the umami receptors. Chefs have started using this understanding to create dishes with delicious taste by adding old and new ingredients that enhance umami. In this paper, we take as our starting point the traditional Japanese soup broth *dashi* as the 'mother' of umami and demonstrate how *dashi* can be prepared from local, Nordic seaweeds, in particular the large brown seaweed sugar kelp (*Saccharina latissima*) and the red seaweed dulse (*Palmaria palmata*), possibly combined with bacon, chicken meat or dried mushrooms to provide synergy in the umami taste. Optimal conditions are determined for *dashi* extraction from these seaweeds, and the corresponding glutamate, aspartate and alaninate contents are determined quantitatively and compared with Japanese *dashi* extracted from the brown seaweed *konbu* (*Saccharina japonica*). Dulse and *dashi* from dulse are proposed as promising novel ingredients in the New Nordic Cuisine to infuse a range of different dishes with umami taste, such as ice cream, fresh cheese and bread.

Keywords: umami, seaweed, *dashi*, glutamate, kelp, dulse, New Nordic Cuisine

Authors' summary for chefs

Herein we review the concept of umami and deliciousness in a historical context and describe recent advances in the scientific understanding of the sensory perception of umami and the involved taste receptors. The primary stimulatory agent in umami is the chemical compound glutamate, which is found in large amounts in the Japanese seaweed *konbu*, which is used to prepare the soup broth *dashi*. We have explored the potential of local Nordic seaweeds, in particular sugar kelp and dulse, for *dashi* production and have discovered that dulse is high in free glutamate and hence a good candidate for umami flavouring. We describe methods by which to optimise the umami flavour using *sous-vide* techniques for extraction of the seaweeds, and we demonstrate how dulse *dashi* can be used in concrete recipes for ice cream, fresh cheese and sourdough bread.

Background

Although umami was suggested as a basic taste in 1908 by the Japanese chemist Kikunae Ikeda [1], umami only caught on very slowly in the Western world [2-5]. Being a verbal construction to describe the essence of delicious taste ('umai' (旨い) is delicious, and 'mi' (味) is essence, inner being or taste), the term 'umami' was coined by Ikeda to signify a unique and savoury taste sensation that should be ranked as the fifth basic taste along with the four classical basic taste modalities: sour, sweet, salty and bitter. In the past couple of decades, along with the globalisation of the Asian kitchen, and in particular the Japanese kitchen, umami is being used more commonly in a culinary context among chefs [6,7] and food scientists [8]. The term has now entered the diverse world of cooking recipes and has been the main topic of a couple of cookbooks [9,10] and most recently a popular science book [11].

Ikeda based his suggestion of umami as a specific taste on the discovery of a particular substance, monosodium glutamate (MSG), which he found in large quantities in free chemical form in one of the key ingredients that enters *dashi*, the soup stock behind all Japanese soups.

* Correspondence: ogm@memphys.sdu.dk
[1]MEMPHYS, Center for Biomembrane Physics, Department of Physics, Chemistry, and Pharmacy, University of Southern Denmark, Campusvej 55, DK-5230 Odense M, Denmark, ogm@memphys.sdu.dk
Full list of author information is available at the end of the article

Dashi is made by a warm extract of the large brown Japanese seaweed *konbu* (*Saccharina japonica*). This extract is called *konbu dashi*. To the *konbu dashi* is then added a particular, highly processed fish product, *katsuobushi*, leading to the so-called first *dashi* (*ichiban dashi*). Ikeda discovered that *konbu* contains about 2 to 3 g of free MSG per 100 g dry weight of *konbu* [1]. Whereas a lot of processed foods, in particular fermented products, can contain as much MSG as, and even more than, *konbu* [3,12-16], no other unprocessed, raw organic material is known to contain more free MSG than *konbu*. In addition to linking MSG to umami, Ikeda immediately realised the technological importance of his discovery as a means to produce artificial flavouring agents to foodstuff, leading the way to the establishment of the international company Ajinomoto (Tokyo, Japan) [17]. MSG is now used across the world as a flavouring enhancer, potentiator and additive in a wide variety of foodstuff [14]. In Europe, it has to be declared as E621 on food product labels.

There are several reasons for the slow acceptance of umami as a basic taste in the Western world. First, in contrast to Japanese cuisine, there is no single common ingredient in Western cuisines that provides as clean a sensation of umami as *dashi*, whereas Western cuisine has kitchen salt (sodium chloride) for salty, ordinary table sugar (sucrose) for sweet, quinine for bitter and acid for sour. Second, cultural differences imply fundamental differences in taste description and codability of taste [18]. Third, the taste sensation of pure MSG appears to be different in different individuals, probably because umami interacts strongly with sweet and salty [19,20]. This has led to confusion regarding MSG's being the source of a unique taste or simply a taste enhancer [21]. Fourth, a shear resistance to accept a new basic taste without a scientifically established sensational physiological basis probably also has played a role, certainly among food scientists and neurophysiologists.

A breakthrough came in 2000 when the first umami receptor was discovered [22]: the metabotropic glutamate receptor *taste*-mGluR4, which is a special dimeric G protein-coupled receptor [23] located in the membranes of the taste cells in the taste buds. *Taste*-mGluR4 is a truncated version of the well-known glutamate receptor mGluR4 in the brain, and it is selectively sensitive to L-glutamate. Since then, two other umami receptors have been found: T1R1/T1R3 [24,25] and a special mGlu receptor [26] that is related to the brain glutamate receptor mGluR1. It remains unclear whether the different umami receptors use different signalling pathways [27,28].

The T1R1/T1R3 receptor is particularly interesting for an informed, molecularly based use of umami in the kitchen. In contrast to *taste*-mGluR4, which is sensitive only to L-glutamate, T1R1/T1R3 is also strongly stimulated by certain 5'-ribonucleotides, in particular inosinate (inosine-5'-monophosphate, IMP) and guanylate (guanosine-5'-monophosphate, GMP), which in a synergistic fashion potentiate the receptor's sensitivity to glutamate. This type of synergy in umami taste has been known phenomenologically for centuries in Japanese cuisine, and the first scientific basis for it was provided in 1957 by the Japanese chemist Akira Kuninaka [29]. Building on earlier work by Shintaro Kodama [30], who had found inosinate in *katsuobushi*, Kuninaka discovered that guanylate from dried *shiitake* mushrooms or inosinate from *katsuobushi* enter a synergistic relationship with glutamate from *konbu*. These synergies underlie classic preparations of *dashi* [15]. The molecular basis for the synergy in umami sensation has recently been revealed as a cooperative, allosteric binding of glutamate and ribonucleotides on the Venus flytrap motif on the T1R1 part of the T1R1/T1R3 receptor complex ([31] and H Khandelia and OG Mouritsen, unpublished data). Umami also enters an interaction with other tastes, in particular sweet and salty but also bitter, and a particularly complex relationship has been found between the umami receptor and the receptors for sweet and bitter [32]. In addition to glutamate, L-aspartate (monosodium aspartate, MSA) has also been associated with umami, but with much less potency and a still unknown sensory mechanism [19,25].

Although in a much less pure form than in Japanese cuisine, umami also plays a key role in Western cuisine [11]. Many types of food contain large natural amounts of free MSG. The most well-known are Marmite, fish sauces, mature hard cheeses such as Parmesan cheese, blue cheeses, sun-dried tomatoes, anchovy paste, soy sauce, cured ham and so on. Similarly, synergy in umami sensation is used extensively in Western food pairing, such as tomatoes with anchovies, vegetables with meat, eggs with bacon, green peas with scallops, and so on.

The outline of the remaining part of the paper is as follows. First, we describe how *dashi* traditionally is made from *konbu*, then we move on to introduce some Nordic seaweed species that are candidates for *dashi* production. The main core of the paper that follows those sections describes improved methods of *dashi* production, including in particular production in controlled temperature conditions. Data regarding the content of free glutamate and other amino acids in different seaweed extracts is presented. Moreover, three concrete recipes are provided for dishes that take advantage of the *dashi* and umami flavours from the red seaweed species dulse that turns out to release large amounts of free glutamate. In the "Discussion" and "Conclusion" sections, we highlight the potential of using seaweed species from local Nordic waters in the New Nordic Cuisine.

Dashi from seaweeds

Classic Japanese cuisine [33] revolves around *dashi* made from *konbu* and *katsuobushi*. In the strictly vegetarian temple kitchen, *shōjin ryōri*, also known as 'the enlightened kitchen' [34], deriving from 12th-century Japan, *katsuobushi* is replaced by dried *shiitake*. *Konbu* provides glutamate and *shiitake* provides guanylate to replace inosinate from *katsuobushi*. The synergetic action in umami is even stronger in the pairing of glutamate with guanylate than with inosinate [31,35].

There are several variants of *konbu*, all of which belong to the algal order Laminariales, with *Saccharina japonica* being the most commonly used species for *dashi*. *Konbu* hence belongs to the same genus as sugar kelp (*Saccharina latissima*). The blades of *konbu* can grow to be several metres long and have a maximum width of 10 to 30 cm. Most of the *konbu* on the world market for human consumption is farmed at lines in the seas around Japan and China. After harvest, *konbu* is sun-dried and the best quality *konbu* is aged in cellars (*kuragakoi*) for from one to ten years, with the typical ageing period being two years. During ageing, the seaweed matures and obtains a milder flavour and a less strong taste and smell of the sea. The umami flavour in *dashi* made from aged and matured *konbu* appears to stand out more clearly.

Konbu contains large amounts of free amino acids, of which 80% to 90% are glutamic acid in the form of MSG [3]. Other important free amino acids are alanine and proline, which impart a sweet taste to the seaweed. *Konbu* does not contain any of the 5'-ribonucleotides that enter synergistically with glutamate in umami. Often some of the free MSG precipitates together with salt and mannitol to form a white layer on the surface of the dried and aged *konbu* blades. This layer should not be removed before the *dashi* is extracted from the *konbu*, because it dissolves readily in water and provides a combination of flavours: umami, salty and sweet. Mannitol is a sweet-tasting sugar alcohol which has about 60% of the relative sweetness of table sugar (sucrose). Mannitol is often found in the *Saccharina* family, in particularly large amounts in sugar kelp (hence its name). *Konbu* contains 2 to 3 g of free MSG per 100 g dry weight. With the exception of the pulp of mature tomatoes [36], *konbu* is possibly the kind of foodstuff that, with the least amount of processing, develops free glutamate in any appreciable amounts.

Of the many different variants of Japanese *konbu*, *ma-konbu*, *rausu-konbu* and *rishiri-konbu* are considered to be the best for extraction to *dashi* [10], and they lead to a very light *dashi* with a mild and somewhat complex taste. *Ma-konbu* is the *konbu* with the largest amount of free glutamate, 3,200 mg/100 g, whereas *rausu-konbu* has 2,200 mg/100 g and *rishiri-konbu* has 2,000 mg/100 g. The lower-quality *hidaka-konbu* has 1,300 mg/100 g [10,12]. The red seaweed laver (*Porphyra yezoensis*) used to produce *nori* has comparable amounts (1,378 mg/100 g), whereas *wakame* (*Undaria pinnatifida*) has very little (9 mg/100 g) [12].

A remarkable feature of dried and aged *konbu*, as well as a number of other seaweeds, is that the free glutamate in the tissues of the seaweed can be transferred to water by a rather mild, warm extraction process solely involving water. Although there is a substantial variation in the way different chefs prepare *konbu dashi*, the recipes used generally prescribe soaking the dry *konbu* in water at room temperature (typically using 10 g of dry *konbu* per litre of water) for about half an hour, heating it in an open pan to just below the water boiling point at 100°C and then quickly removing the *konbu* from the water before bitter-tasting compounds seep out. During this procedure, only a relatively small amount of the total free glutamate is released into the water. A typical *konbu dashi* or *ichiban dashi* prepared using the traditional Japanese recipe (K Ninomiya, unpublished data from the Umami Information Center, and [37]) contains about 20 to 30 mg of glutamate per 100 g of aqueous *dashi* extract. The extraction process can be optimised by varying conditions such as extraction temperature and possibly the quality (hardness) of the water used for the extraction. We shall address the optimisation of *dashi* preparation in this paper and show that by extracting the seaweeds at a lower temperature but for a longer time, a much more flavourful *dashi* characterised by significantly higher levels of glutamate as well as aspartate can be obtained.

Although *konbu* is the kind of seaweed that contains the largest percentage of free glutamate, other seaweeds can also provide some umami flavour despite their smaller glutamate content. Laver (*Porphyra* spp.) that is used to produce the paper-thin *nori* sheets well-known from *maki-zushi* [38] contains rather large amounts of free glutamate [12] in addition to some inosinate and guanylate, which enhances the umami flavour in sushi dishes. In fact, *nori* is the only kind of seaweed that contributes to both basic (glutamate) and synergetic (nucleotides) umami flavour.

Until now, little work has been reported on the use of seaweeds other than *konbu* for *dashi* preparations, although it is well-known that seaweeds in general impart delicious flavours to food [39,40]. Moreover, the use of seaweeds for human consumption is little developed in the Western world [41]. To investigate the potential of using local seaweeds from the waters around the Nordic countries for *dashi* preparations and umami flavouring, we have undertaken quantitative scientific and qualitative gastronomic investigations of *dashi* extracts from a brown seaweed, sugar kelp (*Saccharina latissima*), and a red seaweed, dulse (*Palmaria palmata*). Some preliminary work has also been done on the red seaweed graceful red weed (*Gracilaria verrucosa*). For comparison, we have simultaneously studied classic *dashi* prepared from various qualities of Japanese *konbu* (*Saccharina japonica*). Our investigations can be seen as an attempt to explore the gastronomic potential of local seaweeds [42] for use in the New Nordic Cuisine [43,44].

The seaweeds
Sugar kelp (Saccharina latissima)
The large, metre-long, brown seaweed (kelp) of the order Laminariales, sugar kelp *(Saccharina latissima)* is rich in iodine and minerals, in particular calcium, potassium, manganese and iron. It is very common in the sublittoral zones of the cold seas of the North Atlantic Ocean, and it has a strong flavour of the ocean. Its name derives from its distinct sweet taste caused by large amounts of the sugar alcohol mannitol. Sugar kelp can release significant amounts of extracellular polysaccharides, such as alginates, that easily seep out in water extracts, making these undesirably viscous for most gastronomical uses. The tissues of sugar kelp are tougher than those of *konbu*.

In the present work, we used sugar kelp farmed in Denmark. Both young and old specimens are harvested, and before use for *dashi* they are all subject to drying. Some samples were also stored and aged for a period of time before use. Specimens of dried, farmed sugar kelp are shown in Figure 1.

Dulse (Palmaria palmata)
Dulse *(Palmaria palmata)* is a red, intertidal seaweed well-known in the traditional cuisines of Ireland, Scotland and Iceland, as well as along the coasts of North America. It is common in the Atlantic Ocean, where it grows to a size of up to 50 cm. It has thin and delicate purple fronds with simpler polysaccharides than those found in the brown seaweeds, providing dulse with a more delicate flavour and soft texture. When dried, dulse develops hints of liquorice and smoke, and when toasted it has a nutty taste. Although dulse appears to be one of the seaweeds with the more interesting potential for gastronomical applications, it is surprisingly little used in modern cuisine.

In the present work, we used a variety of different supplies of dulse. Some were harvested in the wild in Iceland, and some were farmed in Denmark. All samples were dried before use. A specimen of dried, farmed dulse is shown in Figure 2. This particular specimen has an almost isotropic shape due to its being grown freely in a pool with constantly moving and swirling water. This is in contrast to dulse naturally grown by being anchored on a substrate at the bottom of the sea, which usually leads to a directional, treelike structure of the shape of a hand, as suggested by the Latin name *palmata*.

Graceful red weed (Gracilaria verrucosa)
This red and stringy seaweed, also called 'sea moss' because of its thread-thin fronds, has not been used to date in Western cuisine. It is traditionally used as a sea vegetable, for example, in Hawaii and Japan, where it is called *ogonori*. *Gracilaria* is a rich source of agar that is used as a thickening agent. In recent years, it has made its way into the Nordic waters, where it is considered one of the invasive species that threatens domestic marine life. In the present work, we used *Gracilaria* farmed in Denmark, and all samples were dried before use.

Figure 1 Dried sugar kelp (*Saccharina latissima*) farmed in Horsens, Denmark. (Photography: Jonas Drotner Mouritsen.).

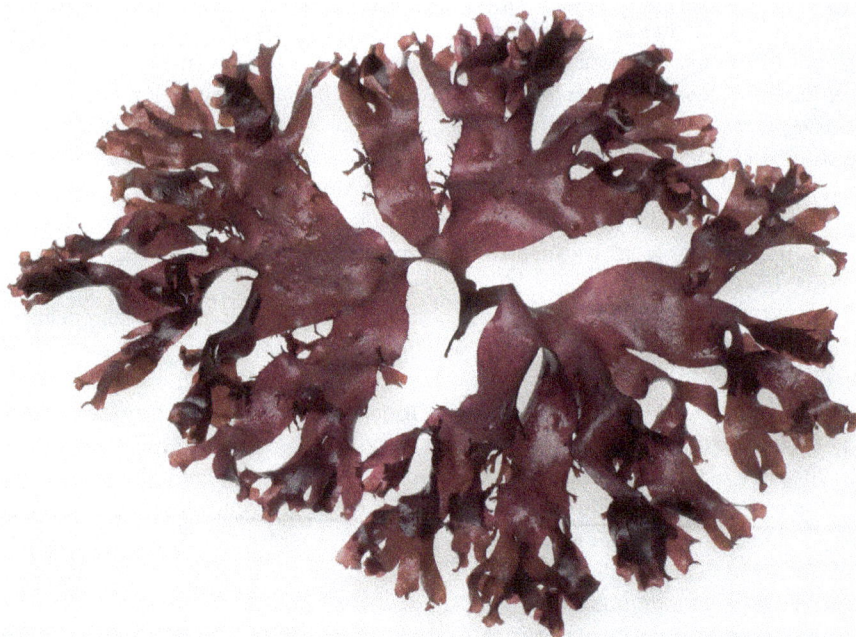

Figure 2 Dried dulse (*Palmaria palmata*) **farmed in Horsens, Denmark**. (Photography: Jonas Drotner Mouritsen.).

Konbu (Saccharina japonica)

Konbu is a large brown seaweed that is farmed in large quantities in China and Japan, amongst other countries. In these Asian countries it is a common staple in traditional cuisine as a sea vegetable and a source for *dashi*, and it is also used in a wide range of processed *konbu* products. *Konbu* is appreciated for its mild and umami flavour and has an interesting and soft texture, despite the considerable thickness of its fronds. Similarly to sugar kelp, *konbu* is high in iodine. It secretes much less polysaccharide than sugar kelp when extracted from water, and the resulting *dashi* is light in colour with a fluidity similar to that of pure water. In the present work, we used two commercial supplies of dried Japanese *konbu* of two distinct qualities: a first-quality *rausu-konbu* and a second-quality *hidaka-konbu*.

Results

The *dashi* from the seaweeds is prepared as described in 'Materials and methods'. Figure 3 shows photographic images of samples of *dashi* prepared from, respectively, *konbu* (*konbu dashi*), *konbu* and *katsuobushi* (*ichiban dashi*) and dulse (dulse *dashi*). The *konbu dashi* is very light in colour, the *ichiban dashi* is darker because of colouring from the fermented *katsuobushi* and the dulse *dashi* has a light purple colour.

The colours and flavours of the different types of *dashi* vary quite significantly. The colour and flavour of the sugar kelp *dashi* differed in relation to the age of the seaweeds and whether they were in a sorus stage.

The sorus contains the sporophylls and is the reproductive organ of the seaweed. *Dashi* prepared from sorus had far less flavour and were lighter in colour than those not in a reproductive cycle. All of the sugar kelp *dashi* is viscous. With regard to the dulse, there is a distinct difference between *dashi* from fresher seaweed and that which has been aged. This is evident in the appearance of the original material. The aged seaweed precipitates more salts (including glutamate) on the surface of the fronds in a rather conspicuous manner, and the taste is far stronger and more complex. Therefore, we can conclude that this ageing process is critical for a superior taste and a method employed for producing high-quality *konbu*.

The difference in the *dashi* prepared from the two types of *konbu* is as visually dramatic as it is in taste. The *hidaka-konbu* produces a *dashi* with a somewhat greenish hue, with a smell evocative of the term 'sea vegetable'. The taste is briny, with a slight metallic vegetable tone. In contrast, the *rausu-dashi* is golden in colour and tastes almost like the chicken bouillon it resembles, very meaty and intense. This taste was amplified when we gently reduced the *rausu-dashi* in a dehydrator at 60°C, and the stock eventually turned into crispy flakes, as shown in Figure 4. Biting in these flakes was almost as overpowering as biting into a bouillon cube. The *konbu-dashi* seemed to be the only *dashi* suitable for this process. The dehydrated *dashi* from the other types of seaweed became bitter, or unable to fully dry, and it remained sticky.

Figure 3 *Dashi* **based on three different seaweed extracts**. Left: *Konbu dashi* from *Saccharina japonica*. Middle: Traditional Japanese *ichiban-dashi* from *konbu* and *katsuobushi*. Right: Nordic *dashi* from dulse (*Palmaria palmata*). For illustration, a piece of dulse is left in the dulse *dashi*. (Photography: Jonas Drotner Mouritsen.).

The different *dashi* preparations made from the different samples of dried seaweeds were analysed for their full amino acid content and profile, with a focus on glutamate and aspartate, which provide umami flavour, and alaninate, which provides a sweet flavour. The results are shown in Table 1. Figure 5 shows the full amino acid profiles for two types of *dashi* made from *rausu-konbu* and dulse, respectively. We shall return to a discussion of these profiles in the 'Discussion' section below and to a comparison of *dashi* made from different seaweeds as opposed to the use of different extraction techniques.

The data in Table 1 refer to extractions by the techniques introduced in this paper and described in 'Materials and methods'. It should be noted that the precise amount of amino acids can be very sample-dependent and that the error bars quoted in Table 1 reflect only the estimated uncertainties in the actual amino acid analysis. The data show that *rausu-konbu* releases very large amounts of free glutamate and aspartate, whereas the levels of alaninate are rather low. The *hidaka-konbu*

provides for about half the amount of glutamate and aspartate compared to *rausu-konbu*. These differences in the umami-producing amino acids reflect the qualitative taste sensations described above and explain why *rausu-konbu* is considered to be superior to *hidaka-konbu* for *dashi* production. In contrast to the differences in umami taste compounds, the two types of *konbu* release similar amounts of alaninate.

It turns out, somewhat surprisingly, that extraction temperatures of 60°C and 100°C led to similar results with respect to the concentrations of glutamate, aspartate and alaninate in *konbu dashi*. Also, the softness of the water seems to have had little influence on the concentration of the three amino acids in the *dashi* extracts studied.

Turning to the *dashi* prepared from sugar kelp, Table 1 shows that *dashi* from sugar kelp contains very little of any of the free amino acids analysed, and the results did not depend on whether we used extraction temperatures of 60°C or 100°C. Additionally, the measured amounts are so low that we can find no discernible dependence on

Figure 4 Kelp 'crisps' that are formed as flakes when dehydrating a *dashi* produced by extraction of *rausu-konbu*. (Photography: Lars Williams.).

Table 1 Extraction of glutamate, aspartate and alaninate from dried seaweeds

Seaweed	Extracted glutamate (mg/100 g)	Extracted aspartate (mg/100 g)	Extracted alaninate (mg/100 g)
Rausu-konbu (farmed, Japanese)	145 ± 5	85 ± 15	20 ± 3
Hidaka-konbu (farmed, Japanese)	70 ± 15	40 ± 20	20 ± 10
Sugar kelp (farmed, Danish)	3 ± 3	3 ± 2	7 ± 4
Dulse (farmed, Danish)	40 ± 10	27 ± 8	25 ± 6
Dulse (wild, Icelandic)	10 ± 5	11 ± 2	12 ± 2
Graceful red weed (farmed, Danish)	6	1	4

All extractions are based on 10 g of dry seaweed in 500 ml of water extracted over a period of 45 minutes in a vacuumed, sealed plastic bag placed in a water bath at the prescribed constant extraction temperature. The extracted amounts of amino acids refer to the concentrations measured in the specific aqueous extract. The concentrations quoted refer to the amino acids in their deprotonated form. The quoted error bars reflect the variation over two to five independent measurements (except for graceful red weed, for which only one measurement was performed). It should be noted that the precise amount of amino acids measured could be very sample-dependent and that the error bars quoted reflect only the estimated uncertainties in the actual amino acid analysis.

the age of the kelp, whether it was matured or not or whether it contained the sorus or not. It is possibly that sugar kelp grown under other nutritional conditions with more nitrogen may contain more glutamic acid and free glutamate than the sugar kelp used for the present study. As noted above, *dashi* from sugar kelp displays an undesirable viscous behaviour that makes it less suitable as a soup stock.

In contrast to sugar kelp, the data for the dulse *dashi* in Table 1 show that this red seaweed has an exceptionally good potential for umami flavour. The farmed Danish dulse releases significantly more glutamate and aspartate than the wild Icelandic dulse. In the case of the Icelandic dulse, we measured both young dulse and aged dulse, but found no significant differences in the amount of released amino acids. For none of the dulse samples studied are we able discern any variation in amino acid concentrations for the two different extraction temperatures. Table 1 shows that *Gracilaria* is a poor source of umami flavours, similar to farmed Danish sugar kelp.

Examples of dishes flavoured by dulse

In this section, we provide some specific recipes using dulse for flavouring dishes developed for the New

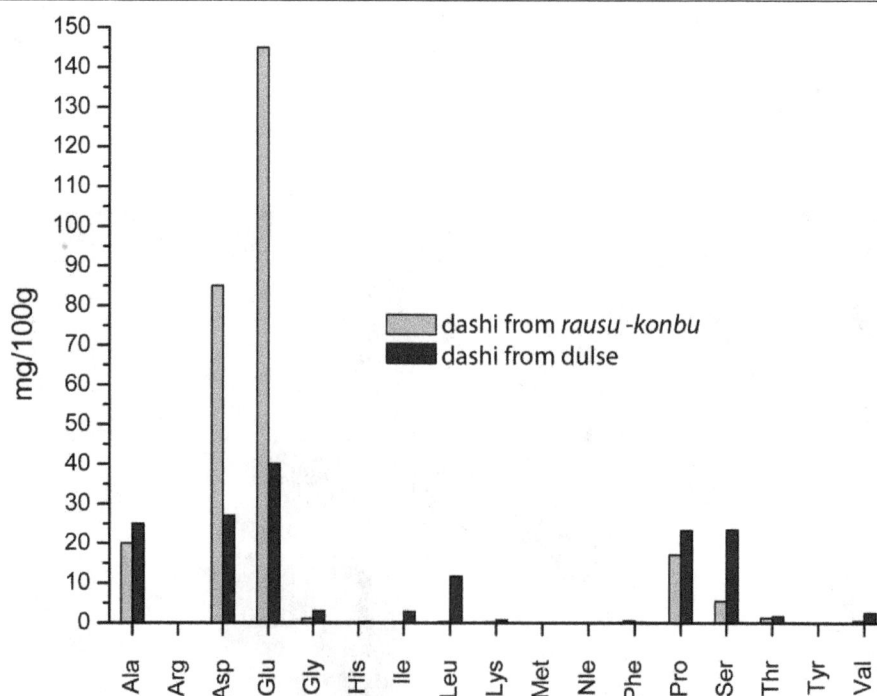

Figure 5 Amino acid profiles for two types of *dashi* made from *rausu-konbu* and dulse, respectively. Both types of *dashi* are based on 10 g of dry seaweed in 500 ml of water extracted over a period of 45 minutes in a vacuumed, sealed plastic bag placed in a water bath at the prescribed constant extraction temperature. The extracted amounts of amino acids refer to the concentrations measured in the specific aqueous extract. The concentrations quoted refer to the amino acids in their deprotonated form.

Nordic Cuisine. The recipes represent a range of novel experiments conducted at the Nordic Food Lab. Photographs of the resulting dishes are shown in Figure 6.

Ice cream with dulse

> 600 g of dulse-infused milk (infuse at 20 g of dulse/ litre of milk)
> 100 g of cream
> 80 g of trimoline (inverted sugar syrup)
> 35 g of sugar
> 24 g of ColdSwell cornstarch (KMC, Brande, Denmark)

Place the dulse and milk in a plastic bag under vacuum and seal, leaving it in a refrigerator overnight to cold-infuse. Strain the dulse and blend it into a fine purée, and preserve to be added later. Dissolve the sugar and trimoline in a small amount of warmed milk. When cooled, add the milk to the rest of the components, mix thoroughly and freeze the mixture in Pacojet containers. Just before serving, the ice cream is prepared in the Pacojet by high-speed precision spinning and thin-layer shaving to produce a creamy consistency of the ice cream.

The dulse ice cream was conceived to demonstrate the culinary versatility of seaweed in an often unexpected fashion. We chose a low-fat base of almost all milk used, allowing the flavour to emerge. Although there was initial reluctance among some tasters to the idea of seaweed ice cream, the vast majority responded with satisfaction upon actually consuming the ice cream.

The colour of the dulse ice cream is a very pleasing light mauve. The flavour is delicate, light and floral. Some tasters have compared the dulse ice cream with Japanese green tea ice cream, which is indicative of a nuanced, acceptable flavour profile.

We also observed an improvement in texture of the dulse-infused ice cream, which is creamier and smoother than the same ice recipe without dulse. This change in texture is likely caused by the polysaccharides released from the dulse.

Fresh cheese with dulse

> 1,250 g of dulse-infused milk (infuse at 20 g of dulse/litre of milk)
> 60 g of cream
> 25 g of buttermilk
> 5 g of rennet

Place dulse and milk in a plastic bag under vacuum and seal, leaving it in a refrigerator overnight to cold-infuse. Strain the milk, heat it to 33°C and add the remaining ingredients, including the rennet. Pour the mixture into plastic containers, cover tightly and cook in an oven at 36°C for 45 minutes.

The result is a slightly acidic, fresh cheese with a light tofulike consistency. There is a bit more brininess and more of a rounder seaweed flavour than with the ice cream. Again, there is a desirable improvement in texture and a slightly more elastic, though very pleasant, mouthfeel. This more viscous texture is likely due to carrageenan released from the dulse, which also reduces the cooking time by half.

Bread with dulse

> 2,500 g of ølandshvede flour, a speltlike wheat species that is high in gluten and protein (13.5%)
> 500 g of spelt flour
> 2,200 g of dulse *dashi*, strained, dulse-minced and reserved
> 200 g of sourdough starter
> 50 g of fresh baker's yeast
> 60 g of salt

Whisk *dashi*, starter, yeast and salt together. Add flour and mix at low speed for 5 minutes. Incorporate the minced dulse. Oil a suitable vessel, and proof in a 5°C

Figure 6 Examples of dishes flavoured with dulse. Left: Ice cream without and with dulse infusion. Middle: Fresh cheese infused with dulse. Right: Bread made from a sourdough infused with dulse. (Photography: Lars Williams.).

refrigerator for 24 hours, folding every 8 hours. Shape and allow it to come to room temperature. Bake in an oven at 225°C for 45 minutes.

The dulse sourdough was considered to be a great success. The liquorice, almost fruity tones came forward and added a supportive savoury flavour. The tea tones steamed from a fresh piece of warm bream when torn apart. It was especially excellent with cheese, almost addictive according to many reports from tasters. The crumb seemed quite moist and had excellent texture. The mixing of the bread was quite different from normal, and some care had to be taken not to overmix the dough.

Discussion

The results presented herein provide some quantitative data for amino acid profiles in *dashi* prepared from Nordic seaweeds, which reveal their potential for providing umami taste. Clearly, the results show that among the studied species, dulse is a much more interesting candidate than sugar kelp and *Gracilaria* for umami flavouring and *dashi*. Moreover, the three recipes presented demonstrate in a concrete setting that dulse *dashi* has a versatile use in the kitchen.

To assess the relative potential of dulse in umami flavouring, we compared the amino acid composition of *dashi* prepared from dulse with classical *dashi* on the basis of Japanese *konbu*. This comparison is provided in Table 2, together with data for two types of chicken soup stock [45,46]. Caution should be exercised when comparing data for the same seaweed species from different sources, because the sample material can be vastly different and the methods of amino acid analysis used and reported in the literature can differ as well. The latter can be particularly troublesome for analysis of extracts of glutamate from brown seaweeds, because it is known that their alginate and salt contents can interfere with the derivation of the amino acids when using classic high-performance liquid chromatography (HPLC) methods [47]. Moreover, different workers have used different

amounts of seaweed for their *dashi* preparations. In the present work, we used about twice the amount of dry seaweed per litre of water compared with many classic Japanese *dashi* recipes. Still, upon normalising to the same weight ratios, we generally found that the use of the extraction techniques described in the present paper released at least twice as much glutamate and aspartate and up to almost ten times as much alaninate compared to values reported in the literature for *konbu* (Table 2).

Notwithstanding the above-mentioned difficulties in comparing, on a quantitative basis, the amino acid contents in *dashi* quoted in reports from different workers, we can compare the contents in *dashi* which we have prepared in the present work using the same weight ratios for seaweed and water and applying the same preparation techniques. As seen in Table 2, we found that *dashi* prepared by our extraction technique has substantially more umami capacity than *dashi* prepared by the classic Japanese recipe, which typically involves cold soaking of the *konbu* for 30 to 60 minutes, followed by warm extraction in a pan and increasing the temperature to just below boiling, then immediately straining the extract. It is well-known that keeping the *konbu* in the boiling *dashi* leads to an unpleasantly bitter flavour. More recently, the classic Japanese recipe has been optimised to provide a better flavour and a clearer *dashi* by heating the solution to only 60°C (as in the present work), but still in an open pan [37]. The many different recipes for preparing classical Japanese *dashi* probably reflect different preferences among the chefs with respect to the balance between bitter, sweet and umami notes of the extract. The comparison of data in Table 2 between *dashi* based on various seaweeds and traditional soup stock prepared from chicken meat and vegetables shows that seaweeds may be superior to these meat-vegetable soups with respect to providing compounds that induce umami.

The taste of the dulse *dashi* is found to be more sweet and complex than traditional *konbu dashi*. Part of the explanation may be found in the differences in the full amino acid profiles shown in Figure 5. Compared to

Table 2 Comparison of amino acid contents in *dashi* and various soup stocks

Dashi or soup stock	Glutamic acid (mg/100 g)	Aspartic acid (mg/100 g)	Alanine (mg/100 g)
Rishiri-konbu dashi[a]	22	16	1
Rausu-konbu dashi (traditional)[b]	100	60	7
Rausu-konbu dashi (sous-vide)[b]	145	85	20
Ichiban dashi[c]	25	18	4
Dulse *dashi*[b]	40	27	25
Western chicken stock[d]	18	6	11
Chinese tang (chicken base)[e]	14	4	8

[a]Recipe: 1.8 L of water and 30 g of *rishiri-konbu*. The *konbu* is soaked in the water, placed in a pan and heated to 60°C for 1 hour (K Ninomiya (personal communication) and [37]). [b]This work. [c]K Ninomiya (unpublished data from the Umami Information Center). [d]From [45]. [e]From [46]. Note that the *rausu dashi* and the dulse *dashi* are prepared by using about twice as much seaweed per litre of water as that used for the *rishiri dashi* and *ichiban dashi*.

konbu, dulse releases more of the sweet amino acids alanine, proline, glycine and serine. The dulse extracts also contain small amounts of the bitter-tasting amino acids such as isoleucine, leucine and valine, which may account for the more complex flavour of the dulse *dashi*. It is noteworthy that the farmed Danish dulse is a serious competitor with *hidaka-konbu* for umami flavour. One reason may be that the much thinner and delicate fronds of dulse are more susceptible to releasing their contents of free amino acids during extraction.

Often it is stated that the softness of water is important for *dashi* preparation [10]. This may be true for the total flavour of the extract, but we cannot discern any significant dependence on water hardness in terms of the actual amounts of free amino acids. Also, the amino acid contents and profiles did not depend on whether we used extraction temperatures of 60°C or 100°C. Hence it is not bitter amino acids that cause the bitter taste often found in a *dashi* which has been prepared close to boiling.

We found that sugar kelp with and without sorus did not lead to a different *dashi* in terms of amino acid composition. Still, it is well-known that the sorus of some seaweeds can be more flavourful than the other parts of the fronds. This is recognised particularly in the case of the brown seaweed *wakame* (*Undaria pinnatifida*), in which the reproductive organs, the sporophylls (*mekabu*), are appreciated for their mouth-filling flavour, which might be caused by their higher fatty acid contents compared to the fronds.

Traditional Japanese *dashi* recipes prescribe addition of the prepared fish product *katsuobushi* to the seaweed extract to provide for synergy compounds, specifically inosinate. In an attempt to find a suitable alternative to *katsubushi* to finish a *dashi* from dulse, we considered typical cured products that Scandinavians consume on a normal basis. Bacon was a delicious-sounding first thought. Pork bacon contains high levels of inosinate and glutamate [12] and would be an ideal starting point. The smokiness of the bacon, combined with the tealike dulse *dashi*, accounted for a surprisingly complex flavour profile. There was an obvious meatier taste, but it was very well balanced with a floral sweetness and slight mineral notes. We declared it a consummate success. We have also experimented with other sources to provide synergetic compounds for umami in Nordic *dashi*. Specifically, we have found that the inosinate contents in dried and salted chicken meat enhance the umami of the dulse *dashi* to some degree, whereas dried local mushrooms such as champignon seemingly contain too little guanylate to furnish anything interesting to pursue.

Conclusion

In a first attempt to explore the gastronomic potential of seaweeds from local waters to provide umami flavouring in the New Nordic Cuisine, we have undertaken a systematic study of a small selection of brown and red seaweeds and compared their umami flavouring amino acid contents with those of the traditional Japanese soup broth *dashi* prepared from the large, brown Japanese seaweed *konbu*.

Although there is a documented, historical tradition of using seaweeds in the diet of some of the Nordic countries, in particular Iceland and Greenland, seaweeds are practically absent in traditional and modern Nordic cuisine [41]. With the rise of the New Nordic Cuisine and the efforts to use local foodstuff ingredients for a New Nordic Diet, many of them almost forgotten, we have focused in particular on two seaweed species: the red seaweed dulse (*Palmaria palmata*) and the brown seaweed sugar kelp (*Saccharina latissima*). Both of these species arc available in large amounts in the wild in Nordic waters and can be farmed under controlled conditions. The wild resources can be harvested in a sustainable fashion, and, because they grow in the cold, pristine Nordic waters, the seaweeds are very clean and suitable for human consumption.

We have investigated the gastronomical potential of these seaweed species by focusing on the flavouring potential of simple water extracts, similarly to the Japanese *dashi* that is the classic source of umami. *Dashi* owes its umami taste to the sodium salts of glutamic acid and aspartic acid, and any sweetness is predominantly due to free alanine. We have measured the concentration of these three amino acids in various extracts prepared under different well-controlled conditions, such as extraction temperature.

The main finding of the present work is that the use of well-controlled extraction techniques may release much more glutamate, aspartate and alaninate than the use of classic recipes involving cold-water soaking and subsequent heating in an open pan. Hence techniques involving extraction in a sealed plastic bag under vacuum pressure appear to have improved the extraction efficiency for free amino acids from seaweeds without compromising the flavour. Whereas the extraction temperature has a definite influence on the overall taste of the *dashi*, in particular the bitter notes developed at the higher temperatures, the amounts of glutamate, aspartate and alaninate appear to be little sensitive to whether the extraction temperature is 60°C or 100°C. Similarly, we could not detect any significant dependence of water hardness on the amount of released umami-flavouring free amino acids.

We believe that the findings of the present study may be of use for improving recipes for making *dashi*, not least from *konbu*. Specifically, we found that whereas sugar kelp is a poor umami source, dulse is an excellent source with similar amounts of umami agents compared to Japanese *dashi* prepared from *konbu* in the classic way. *Dashi* from dulse also contains a significant amount of alaninate, which probably contributes to its mild, sweet taste.

The flavouring abilities of Nordic dulse *dashi* by its infusion in various dishes have been demonstrated in the case of three specific examples: ice cream with dulse infusion, fresh cheese infused with dulse and bread made from a sourdough infused with dulse. Subjective tasting experiments suggest that dulse is indeed an attractive flavouring agent and holds great promise for novel uses, not only in the New Nordic Cuisine but in general.

Materials and methods
Seaweed cultivation and harvesting
Sugar kelp (*Saccharina latissima*) was cultivated in the open coastal waters of Kattegat in Denmark. The sporophytes of sugar kelp are sawn on smaller ropes in a hatchery, and they attach to the rope with their holdfasts. The seeded and sprouted kelp ropes are fixed on cultivation longlines, cultivated for about 18 months and harvested when they reach a length of about 1.5 m. The controlled cultivation produces high quality with nearly no fouling. The harvested kelp is sun-dried immediately after harvest.

Dulse (*Palmaria palmata)* and graceful red weed (*Gracilaria verrucosa*) were grown in open tanks (pools) fed with seawater and by using air turbulence to move the seaweeds, to provide nutrition and to facilitate photosynthesis. The dulse grows by making new proliferations and then building new main tissues. The controlled cultivation in pools not only enables a fouling-free quality but also facilitates a highly red pigmentation and large protein content. The harvested dulse and *Gracilaria* are sun-dried immediately after harvest.

Commercial seaweed supplies
Commercially available dried *konbu* was purchased in two different qualities: first-quality *rausu-konbu* from Sunaga village, Rausu District (Japan Fooding Ltd, London, UK) and second-quality *hidaka-konbu* from Hokkaido, Japan (Wakou Corp, Shiga, Japan). Commercially available dried dulse was purchased from Íslensk hollusta ehf (Reykjavik, Iceland). The dulse is hand-harvested from wild Icelandic resources and subsequently dried.

Sous-vide water extracts from seaweeds
Two types of water were used for seaweed extracts: ordinary tap water (Copenhagen, Denmark; water hardness = 20°dH) and filtered, demineralised soft water. All extractions were based on 10 g of dry seaweed in 500 ml of water placed in a plastic bag sealed under vacuum pressure (*sous-vide*) at 98.5 kPa in a Komet Plus Vac 20 (KOMET Maschinenfabrik GmbH, Plochingen,

Germany) and immersed over a period of 45 minutes in a water bath at the prescribed constant extraction temperature.

Sensory perception
Because the present paper is not intended to be a quantitative study of the sensory perception of umami flavour in the seaweed extracts and the dishes flavoured by the extracts, we have not used a formal panel of professional tasters but employed a subjective and qualitative measure of taste sensation by integrating statements from experimenters and colleague chefs who are very experienced in evaluating and describing taste. The subjective analysis was carried out by a minimum of five qualified chefs who are considered trained tasters. In the case of the dulse ice cream, the tasting was part of a master's degree thesis on the complexity in food (Faculty of Life Sciences, University of Copenhagen, Copenhagen, Denmark) that was favourably received by a tasting panel of 60 persons. In addition, we registered responses from a large number of individual tasters on different occasions when the dulse-infused dishes were presented and tasted.

Amino acid analysis
All chemicals used were from Sigma-Aldrich (Copenhagen, Denmark) and of HPLC quality or better. Amino acid analysis was performed using the Biochrom 31+ Protein Hydrolysate System amino acid analyser (Biochrom, Cambridge, UK). Prior to analysis, proteins were precipitated by addition of trichloroacetic acid, and lipids were extracted with hexane. The amino acids were identified and quantified by comparison with pure amino acid standards with a major focus on glutamic acid, aspartic acid and alanine in their deprotonated states.

Abbreviations
GMP: guanosine-5'-monophosphate (guanylate); IMP: inosine-5'-monophosphate (inosinate); MSA: monosodium aspartate (aspartate); MSG: monosodium glutamate (glutamate).

Acknowledgements
Dr Kumiko Ninomiya is thanked for useful correspondence regarding *dashi* preparations, for information on umami flavour of soup broths, and for making available to us unpublished data on glutamate content in *ichiban dashi*. Prof Stefan Vogel is thanked for providing access to an HPLC installation and Prof Peter Højrup for help with the amino acid analysis. Masami Suenaga of Japan Fooding Ltd is gratefully acknowledged for help in purchasing *rausu-konbu*. MEMPHYS was supported as a centre of excellence by the Danish National Research Foundation for the period from 2001 to 2011. Nordic Food Lab is an independent institution fuelled by finances from external funds, private companies and foundations, including NordeaFonden, as well as from government sources. This work was supported by a grant (J.nr. 3414-09-02518) from the Danish Food Industry Agency.

Author details
[1]MEMPHYS, Center for Biomembrane Physics, Department of Physics, Chemistry, and Pharmacy, University of Southern Denmark, Campusvej 55, DK-5230 Odense M, Denmark, ogm@memphys.sdu.dk. [2]Nordic Food Lab, 93

Strandgade, DK-1401 Copenhagen K, Denmark, lw@nordicfoodlab.org. ³Restaurant Noma, 93 Strandgade, DK-1401 Copenhagen K, Denmark, marikultur@hotmail.com. ⁴Blue Food ApS, 2 Nordre Kaj, DK-8700 Horsens, Denmark, lad@memphys.sdu.dk.

Authors' contributions

OGM designed the study, suggested some of the pairing of ingredients for *dashi* production, took part in some of the *dashi* preparations, researched the literature on umami and composed and wrote the paper. LW designed procedures for optimal seaweed extraction to optimise umami flavour, designed the paring of ingredients for umami flavour, prepared *dashi*, exercised qualitative sensory evaluation of *dashi* and was a coinventor of dishes flavoured with dulse-based *dashi*. RB designed and implemented production facilities for seaweed farming in Denmark and harvested and processed the sugar kelp, dulse and *Gracilaria* used in this paper. LD set up and performed the amino acid analyses and processed the data.

Competing interests

The authors declare that they have no competing interests.

References

1. Ikeda I: **New seasonings.** *Chem Senses* 2002, **27**:847-849.
2. Yamaguchi S, Ninomiya K: **What is umami?** *Food Rev Int* 1998, **14**:123-138.
3. Ninomiya K: **Umami: a universal taste.** *Food Rev Int* 2002, **18**:23-38.
4. Kawamura Y, Kare M: *Umami: A Basic Taste: Physiology, Biochemistry, Nutrition, Food Science* New York: Marcel Dekker; 1986.
5. Proceedings of the 100th Anniversary Symposium of Umami Discovery: the roles of glutamate in taste, gastrointestinal function, metabolism, and physiology. Tokyo, Japan. September 11-13, 2008. In *Am J Clin Nutr* Edited by: Fernstrom JD 2009, **90(suppl)**:705S-885S.
6. Marcus JB: **Culinary applications of umami.** *Food Technol* 2005, **59**:24-30.
7. Blumenthal H: *The Fat Duck Cookbook* London: Bloomsbury; 2008.
8. Barham P, Skibsted LH, Bredie WL, Frøst MB, Møller P, Risbo J, Snitkjaer P, Mortensen LM: **Molecular gastronomy: a new emerging scientific discipline.** *Chem Rev* 2010, **110**:2313-2365.
9. Kasabian A, Kasabian D: *The Fifth Taste: Cooking with Umami* New York: Universe Publishing; 2005.
10. Blumenthal H, Barbot P, Matsushisa N, Mikuni K: *Dashi and Umami: The Heart of Japanese Cuisine* London: Eat-Japan, Cross Media Ltd; 2009.
11. Mouritsen OG, Styrbæk K: *Umami: Gourmetaben og den femte smag* Copenhagen: Nyt Nordisk Forlag Arnold Busck; 2011.
12. Ninomiya K: **Natural occurrence.** *Food Rev Int* 1998, **14**:177-211.
13. Giacometti T: **Free and bound glutamate in natural products.** In *Glutamic Acid: Advances in Biochemistry and Physiology*. Edited by: Filer LJ Jr, Garattini S, Kare MR, Reynolds AW, Wurtmann RJ. New York: Raven; 1979:25-34.
14. Maga JA: **Flavor potentiators.** *Crit Rev Food Sci Nutr* 1993, **18**:231-312.
15. Yamaguchi S, Ninomiya K: **Umami and food palatability.** *J Nutr* 2000, **130**:921S-926S.
16. Yoshida Y: **Umami taste and traditional seasonings.** *Food Rev Int* 1998, **14**:213-246.
17. Sano C: **History of glutamate production.** *Am J Clin Nutr* 2009, **90**:728S-732S.
18. O'Mahony M, Ishii R: **A comparison of English and Japanese taste languages: Taste descriptive methodology, codability and the umami taste.** *Br J Psychol* 1986, **77**:161-174.
19. Chandrashekar J, Hoon MA, Ryba NJ, Zucker CA: **The receptors and cells for mammalian taste.** *Nature* 2006, **444**:288-294.
20. Fuke S, Ueda Y: **Interactions between umami and other flavor characteristics.** *Trends Food Sci Technol* 1996, **7**:407-411.
21. Bachmanov A: **Umami: Fifth taste? Flavor enhancer?** *Perfum Flavor* 2010, **35**:52-57.
22. Chaudhari N, Landin AM, Roper SD: **A novel metabotropic glutamate receptor functions as a taste receptor.** *Nature Neurosci* 2000, **3**:113-119.
23. Kunishima N, Shimada Y, Tsuji Y, Sato T, Yamamoto M, Kumasaka T, Nakanishi S, Jingami H, Morikawa K: **Structural basis of glutamate recognition by a dimeric metabotropic glutamate receptor.** *Nature* 2000, **407**:971-977.
24. Nelson G, Chandrashekar J, Moon MA, Feng L, Zhao G, Ryba NJ, Zucker CS: **An amino-acid taste receptor.** *Nature* 2002, **416**:199-202.
25. Li X, Staszewski L, Xu H, Durick K, Zoller M, Adler E: **Human receptors for sweet and umami taste.** *Proc Natl Acad Sci USA* 2002, **99**:4692-4696.
26. San Gabriel A, Uneyama H, Yoshie Y, Torii K: **Cloning and characterization of a novel mGluR1 variant from vallate papillae that functions as a receptor for L-glutamate stimuli.** *Chem Senses* 2005, **30**:i25-i26.
27. Yasuo T, Kusuhara Y, Yasumatsu K, Ninomiya Y: **Multiple receptor systems for glutamate detection in the taste organ.** *Biol Pharm Bull* 2008, **31**:1833-1837.
28. Jyotaki M, Shigemura N, Ninomiya Y: **Multiple umami receptors and their variants in human and mice.** *J Health Sci* 2009, **55**:647-681.
29. Kuninaka A: **Studies on taste of ribonucleic acid derivatives.** *J Agric Chem Soc Jpn* 1960, **34**:487-492.
30. Kodama S: **On a procedure for separating inosinic acid.** *J Tokyo Chem Soc* 1913, **34**:751-755.
31. Zhang FB, Klebansky B, Fine RM, Xu H, Pronin A, Liu H, Tachdjian C, Li X: **Molecular mechanism for the umami taste synergism.** *Proc Natl Acad Sci USA* 2008, **105**:20930-20934.
32. Temussi PA: **Sweet, bitter and umami receptors: a complex relationship.** *Trends Biochem Sci* 2009, **34**:296-302.
33. Tsuji A: *Japanese Cooking: A Simple Art* Tokyo: Kodansha; 1980.
34. Fujii M: *The Enlightened Kitchen: Fresh Vegetable Dishes from the Temples of Japan* Tokyo: Kodanska; 2005.
35. Yamaguchi S: **Basic properties of umami and its effects on food flavor.** *Food Rev Int* 1998, **14**:139-176.
36. Oruna-Concha MJ, Methven L, Blumenthal H, Young C, Mottram DS: **Differences in glutamic acid and 5'-ribonucleotide contents between flesh and pulp of tomatoes and the relationship with umami taste.** *J Agric Food Chem* 2007, **55**:5776-5780.
37. Kurihara K: **Glutamate: from discovery as a food flavor to role as a basic taste.** *Am J Clin Nutr* 2009, **90**:719S-722S.
38. Mouritsen OG: *Sushi: Food for the Eye, the Body & the Soul* New York: Springer; 2009.
39. Mouritsen OG: *Tang: Grøntsager fra havet* Copenhagen: Nyt Nordisk Forlag Arnold Busck; 2009.
40. Rhatigan P: *The Irish Seaweed Kitchen* Co Down, Ireland: Booklink; 2010.
41. Mouritsen OG: **The emerging science of gastrophysics and its application to the algal cuisine.** *Flavour* 2012, **1**:6.
42. Mouritsen OG, Vildgaard T, Westh S, Williams L: **Nordisk dashi: Smagsdommerne på Nordic Food Lab.** *Gastro* 2010, **51**:72-75.
43. Redzepi R, Meyer C: *Noma Nordic Cuisine* Copenhagen: Politiken; 2006.
44. Redzepi R: *Noma: Time and Place in Nordic Cuisine* New York: Phaidon; 2010.
45. Ozawa S, Miyano H, Kawai M, Sawa A, Ninomiya K, Mawatari K, Kuroda M: **Changes of free amino acids during cooking process of chicken consommé.** 2004, 322, (in Japanese) presented at the 58th Congress of Japanese Society for Nutrition and Food Science, Sendai, May 21-23.
46. Ozawa S, Miyano H, Kawai M, Sawa A, Ninomiya K, Mawatari K, Kuroda M: **Changes of free amino acids during cooking process of Chinese chicken bouillon.** 2005, 322, (in Japanese) presented at 59th Congress of Japanese Society for Nutrition and Food Science, Tokyo, May 13-15.
47. Bergeron E, Jolivet P: **Quantitative determination of glutamate in Rhodophyceae (*Chondrus crispus*) and four Phaeophyceae (*Fucus vesiculosos, Fucus serratus, Cystoseira elegans, Cystoseira barbata*).** *J Appl Phycol* 1991, **33**:115-120.

12

On the psychological impact of food colour

Charles Spence

Abstract

Colour is the single most important product-intrinsic sensory cue when it comes to setting people's expectations regarding the likely taste and flavour of food and drink. To date, a large body of laboratory research has demonstrated that changing the hue or intensity/saturation of the colour of food and beverage items can exert a sometimes dramatic impact on the expectations, and hence on the subsequent experiences, of consumers (or participants in the lab). However, should the colour not match the taste, then the result may well be a negatively valenced disconfirmation of expectation. Food colours can have rather different meanings and hence give rise to differing expectations, in different age groups, not to mention in different cultures. Genetic differences, such as in a person's taster status, can also modulate the psychological impact of food colour on flavour perception. By gaining a better understanding of the sensory and hedonic expectations elicited by food colour in different groups of individuals, researchers are coming to understand more about why it is that what we see modulates the multisensory perception of flavour, as well as our appetitive and avoidance-related food behaviours.

Keywords: Flavour, Taste, Expectations, Disconfirmed expectations, Sensory, Hedonic, Multisensory

Review

Under most everyday conditions (excepting perhaps the dine-in-the-dark restaurant; see [1]), consumers have the opportunity to inspect food and drink visually before deciding on whether or not to buy or taste it [2]. Indeed, it has long been recognized that colour constitutes one of the most salient of visual cues concerning the likely sensory properties (for example, taste/flavour) of that which we are about to eat or drink (for example, [3-10]). There is a very long history of colouring being added to food and drink [11-13]. Furthermore, although little studied, those colours that we take to suggest that a food may have gone off can exert a particularly powerful effect on our food avoidance behaviours [14,15]. As such, food colour can be considered as perhaps the single most important product-intrinsic sensory cue governing the sensory and hedonic expectations that the consumer holds concerning the foods and drinks that they search for, purchase, and which they may subsequently consume.[a]

At the outset, though, it is important to distinguish clearly between taste and flavour, two terms that are used more or less interchangeably in everyday language [16,17]. The reason being that colour cues appear to have a somewhat different effect on taste *versus* flavour perception (see [18], for a review). Strictly speaking, 'taste' refers to the perception of sweet, sour, bitter, salty, and the other basic tastes, which are detected by the gustatory receptors found primarily in the oral cavity. By contrast, 'flavour' refers to those experiences that also involve a retronasal olfactory component, such as meaty, burnt, floral, fruity, citrusy, and so on (see [17]). However, confusing matters somewhat, in everyday language, people typically use the term taste to describe their overall experience of food and drink. Here, the terms are used with their more precise scientific meaning.

A growing body of scientific research now suggests that our experience of taste and flavour is determined to a large degree by the expectations that we generate (often automatically) prior to tasting [19-21]. Such expectations can result from branding, labelling, packaging, and other contextual effects (that is, from a host of product-extrinsic cues) but also from a variety of product-intrinsic cues as well. The smell and aroma of food and drink are clearly important here, as are, on occasion, the sounds of food preparation (see [22], for a review). That said, olfactory cues can often be obscured by product packaging, and the products on the supermarket shelf rarely make any sound when inspected visually. Hence, it is vision, and most often colour, that is the cue

Correspondence: charles.spence@psy.ox.ac.uk
Crossmodal Research Laboratory, Department of Experimental Psychology, University of Oxford, 9 South Parks Road, Oxford OX1 3UD, UK

used by the brain in order to help identify sources of food and make predictions about their likely taste and flavour [20,23]. Or, as the spokesperson for the Institute of Food Technologists put it a few years ago: 'Color creates a psychological expectation for a certain flavor that is often impossible to dislodge.' [24].

The focus in this article is on the psychological effect, or better said, effects, that food colour exerts over the mind and behaviour of the consumer. The review starts by looking at the effect of food colouring on sensory expectations and hence on people's judgments of taste/flavour intensity and flavour identity (see [18], for a review). The literature on off-colours in foods and drinks is reviewed briefly, and popular concerns regarding artificial food colouring highlighted. Attention will be drawn to research showing the important individual differences in terms of the meaning, and hence psychological influence, of colour in food. Along the way, some of the problems associated with the interpretation of much of the laboratory research that has been conducted to date will be highlighted.

Although falling beyond the scope of the present review, it is worth noting that colour is but one aspect of vision's influence over taste and flavour perception. Researchers have, for instance, reported that people tend to judge the freshness of fish, in part, based on the luminance distribution (that is, the glossiness) of fish eyes [25]. The luminance distribution also appears to be an important cue for judging the freshness of certain fruit and vegetables as well [26]. The influence of visual food texture on people's sensory perception and consumption behaviour has also been studied by researchers [27-29]. However, given space constraints, the focus here will be squarely on colour and its psychological impact on the perception/behaviour of the consumer.

Psychological effects of food colour: setting sensory expectations

Taste/flavour intensity

It would seem reasonable to assume that wherever in the world one finds oneself, more intensely coloured foods are likely to be more intensely flavoured. What also seems likely is that consumers will have picked up on this statistical regularity in the environment and hence will tend to expect that more intensely coloured foods and beverages (not to mention the packaging in which such products come) will have a more intense taste/flavour. Should those expectations not be met, then a negatively valenced disconfirmation of expectation response may well ensue (for example, [30-32]).[b] Over the last 50 years or so, a large body of laboratory research has demonstrated that adding more colouring to a food, or more often, to a beverage (see [33], for a review), can lead the participants in laboratory research to rate the taste and/or flavour as more intense (for example, [18,34-38]).

The addition of food colouring influences sensory thresholds for certain of the basic tastes. In one classic study, Maga [39] demonstrated that adding food colouring (red, green, or yellow) to an otherwise clear solution exerted a significant effect on thresholds for the detection of certain of the basic tastes when presented in solution. Adding green food colouring decreased people's detection threshold for sourness, while at the same time increasing the threshold for the detection of sweetness. The addition of yellow colouring reduced the detection threshold for both sourness and sweetness, while the addition of red colouring reduced the threshold for the detection of bitterness.[c] Intriguingly, the threshold for the detection of salt was unaffected by the addition of food colouring. Maga's [39] suggestion at the time was that this null effect resulted from the fact that salty foods are associated with foods of many different colours and hence that salt is not associated with a particular colour. As Maga himself put it: 'numerous foods of varying color can be characterized as tasting salty, examples would be pretzels (brown), potato chips (yellow), popcorn (white), olives (green, black), and pickles (green).' ([39], p. 118). That said, more recent research has clearly demonstrated that most people do tend to associate salt with the colour white (see [40]). Perhaps, then, had Maga tested a different range of colours, he might have come to a somewhat different conclusion.

Perhaps the most convincing evidence published to date concerning the influence of food colouring on ratings of taste intensity comes from research published by Clydesdale et al. [41]. These researchers conducted a number of psychophysical studies showing that the addition of food colouring can deliver as much as 10% perceived sweetness. Indeed, such results have led some to wonder whether food colouring could be used as an effective means of reducing the sugar content of foods. While this is certainly a theoretical possibility, it is worth bearing in mind that the majority of the studies that have been published to date have involved fairly short-term exposure to particular combinations of colour-taste/flavour. While demonstrating that food colouring has an impact on sweetness perception in the short term is one thing, it is quite another to convincingly demonstrate that it will necessarily have psychological effects that last over the long term (cf. [42]). Hence, longer-term follow-ups are most definitely in order. What is more, as the studies discussed below make only too clear, psychological effects of food colouring on the perception of taste and flavour intensity have not always been demonstrated.

One null result in this area was reported by Norton and Johnson [38]. These researchers manipulated the intensity of four typical drink colours. They were unable to find any meaningful relationship between the intensity of the colour and flavour ratings on either a sweet-sour scale or on a distinct-indistinct flavour scale in the 18 participants whom they tested. Meanwhile, Lavin and Lawless [43] investigated the influence of varying the intensity of food colouring on ratings of sweetness intensity. The participants were given two pairs of strawberry-flavoured drinks to compare and to rate in terms of their sweetness, using nine-point scales. One pair of drinks was light and dark red, whereas the other pair was light and dark green. The drinks were equally physically sweet, varying only in terms of their appearance properties (that is, colour). Those adults who took part in this study rated the dark-red and light-green drinks as tasting sweeter than the light-red and dark-green samples, respectively. By contrast, colour intensity had no effect on the responses of 5- and 14-year-old children. Elsewhere, Alley and Alley [44] similarly failed to demonstrate any effect of the addition of colour (red, blue, yellow, or green) to an otherwise colourless base (either liquid or solid) on the perceived sweetness of sugar solutions in a group of 11 to 13 year olds.

In a study by Philipsen et al. [45], a group of young adults (aged 18 to 22 years) and a group of older participants (aged 60 to 75 years) rated a number of attributes (for example, sweetness, flavour intensity, flavour quality, flavour identification, and so on) of 15 samples of an artificially flavoured cherry drink that varied in terms of its sucrose content, flavour, and colour. Interestingly, variations in colour intensity had no effect on sweetness ratings in either age group but did impact on flavour intensity ratings in the older participants.

In another study, Chan and Kane-Martinelli [46] examined the effect of food colouring on perceived flavour intensity and acceptability ratings in samples of chicken bouillon and chocolate pudding. These foods were presented with no colour added, with the normal (that is, commercial) level of food colouring, or with twice the normal level of colour added. The participants tasted and evaluated the three samples of either food, using visual analogue scales. Younger adults (20 to 35 years of age) were found to be more affected by the presence of food colouring than were the older adults (60 to 90 years of age). Interestingly, the younger group's judgment of the overall flavour intensity of the chicken bouillon was influenced by the amount of colouring that had been added to the sample.

Zampini et al. [47] conducted a study in which a group of adults had to try and identify the flavour of a variety of drinks and rate perceived flavour intensity using a labelled magnitude scale. The drinks were flavourless, or else had an orange, lime, or strawberry flavour, and could be presented as colourless solutions, or else artificially coloured red, green, or orange. The food colouring was added at either a standard or double concentration. However, variations in the intensity of the food colouring (no matter whether that colour was appropriate or inappropriate to the flavour of the drink) had no effect on the perceived

flavour intensity. That said, the addition of inappropriate food colouring significantly impaired the accuracy of participants' flavour identification responses (see Figure 1), thus suggesting that the participants were unable to ignore the colour of the drinks completely, as they had been encouraged to do by the experimenter.

To date, the majority of the research on the psychological influence of colour on judgments of taste/flavour intensity (not to mention flavour identity, see below) has been conducted with beverages. This is presumably because it is easier to manipulate the level of colour in solutions [33]. That said, intriguing research by Shermer and Levitan [48] has recently demonstrated that people also expect more intensely red-coloured salsas to be spicier (that is, more piquant). In fact, over the last 80 years or so, researchers have looked at the psychological impact of food colour on everything from noodles [49] through vegetables [50] and from cheese [51] through to yoghurt [34,52], not to mention cake [53], jams, jellies, chocolates, and sherbets [7,54,55].

One final point to note here is that it may be important to pay careful attention to the methodological details from the various studies of the psychological effect of colour on flavour intensity. The reason being that one study has obtained differing effects of food colour on orthonasal and retronasal judgments of a commercial fruit-flavoured water drink [56]. In particular, colouring a tangerine-pineapple-guava-flavoured solution red led to odour enhancement in those participants who sniffed the odour orthonasally, while giving rise to a reduction in perceived odour intensity when the same olfactory stimulus was presented retronasally instead.

Koza et al. [56] attempted to account for this surprising pattern of results by suggesting that it may be more important for us to correctly evaluate foods once they have entered our mouths, since that is when they pose a greater risk of poisoning us. By contrast, the threat of poisoning from foodstuffs located outside the mouth is obviously going to be less severe. Whatever the explanation for Koza et al.'s results turns out to be, the main point that these results highlight is that one cannot simply assume that colour's effect on orthonasal olfactory judgments of a food or drink's flavour will necessarily be the same when people come to actually taste it.

Interim summary

As the results reviewed in this section have made only too clear, the psychological effects of either adding or changing the intensity of food colouring on the *intensity* of taste/flavour perception are not altogether clear. Null results have been obtained by some researchers (for example, [44,57]). And even those who have obtained significant effects of colour on taste/flavour intensity ratings/perception have tended to do so only under a subset of experimental

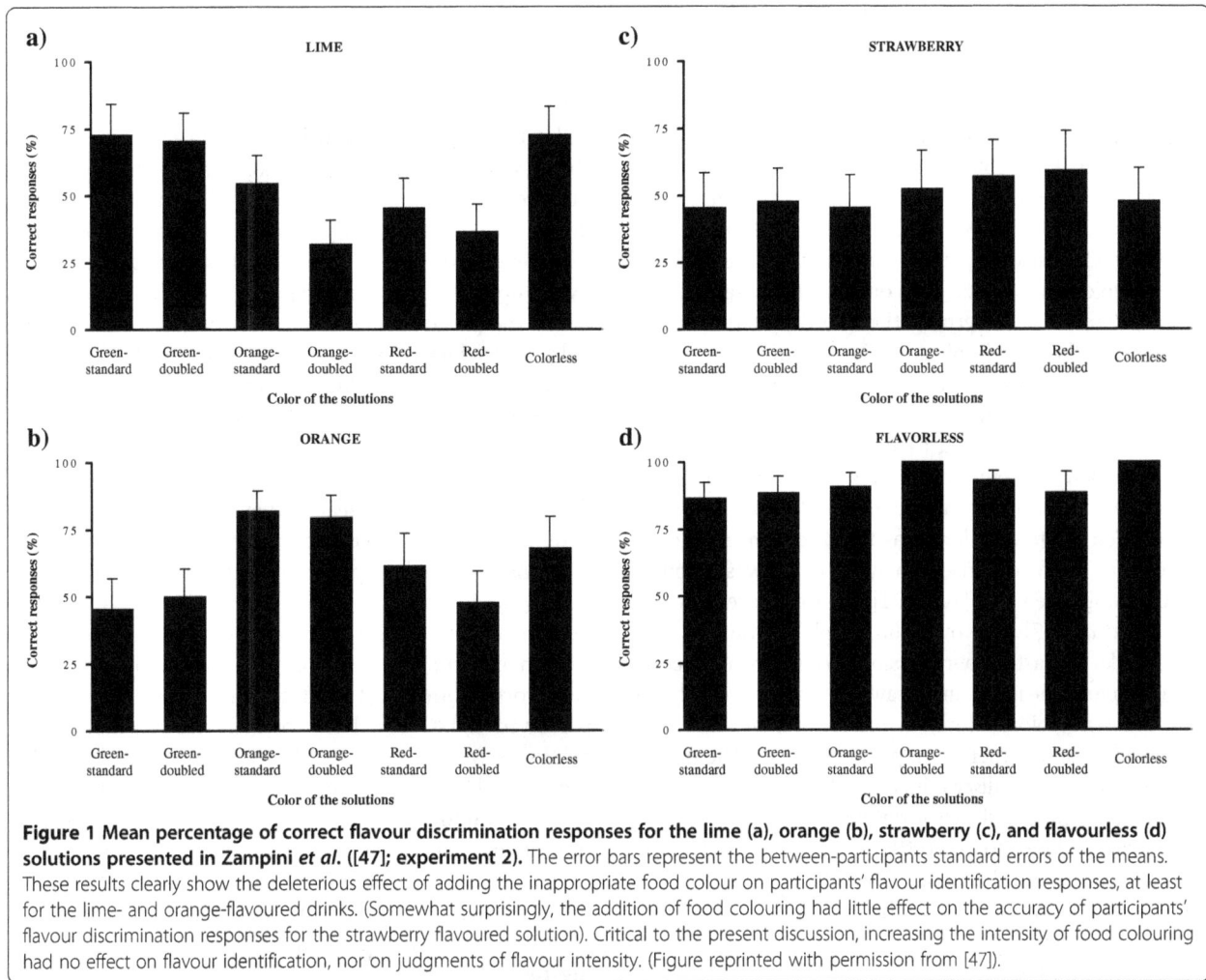

Figure 1 Mean percentage of correct flavour discrimination responses for the lime (a), orange (b), strawberry (c), and flavourless (d) solutions presented in Zampini et al. ([47]; experiment 2). The error bars represent the between-participants standard errors of the means. These results clearly show the deleterious effect of adding the inappropriate food colour on participants' flavour identification responses, at least for the lime- and orange-flavoured drinks. (Somewhat surprisingly, the addition of food colouring had little effect on the accuracy of participants' flavour discrimination responses for the strawberry flavoured solution). Critical to the present discussion, increasing the intensity of food colouring had no effect on flavour identification, nor on judgments of flavour intensity. (Figure reprinted with permission from [47]).

conditions or else in a subset of those individuals whom they have tested (for example, see [39,43,45-47,53,58-66]). As such, it is difficult to draw any overarching conclusions from the range of results that have been published to date as to when exactly the addition of food colouring will influence ratings of taste/flavour intensity. Clearly, the addition of food colouring can influence thresholds and ratings of stimulus intensity. However, when exactly such crossmodal effects will be observed is harder to predict with any confidence.[d] Indeed, one question left unresolved by much of the research that has been published to date in this area concerns why it is that these seemingly inconsistent results might have been obtained in the first place. According to Koza et al.'s [56] findings, part of the answer might relate to methodological details concerning whether olfactory stimuli are presented orthonasally or retronasally.

However, perhaps, one also needs to take a step back and consider what happens if the sensory expectations set by the intensity of food colouring fail to match up with the experience when a food or beverage item is actually tasted by the participant or consumer. In the real world, this might be expected to give rise to a negatively valenced disconfirmation of expectation response [21,31,32]. However, in much of the laboratory research published to date, there is a question as to whether the participants actually believed that the colours of the foods or drinks that they were tasting had any meaning - that is, to what extent did they really believe that the food colouring they saw was linked to the actual taste/flavour of the drinks that they were tasting? (One might also be tempted to wonder whether or not participants noticed any discrepancy between what they saw and what they tasted [67]). While such an assumption is presumably likely out there in the real world, it is not so clearly the case for those participants taking part in laboratory research where they may have been exposed to a whole series of inappropriately coloured samples to taste and evaluate over the course of their experimental session. What is more, the research varies between those studies in which the researchers have been very explicit about the fact that the colour cues were designed to be misleading [47,68], through to those who have done their utmost to hide the purpose of their study (and the potentially misleading nature of the colours) from their participants [69].

Flavour identity

Perhaps the most robustly demonstrated effect of adding (or changing) food colouring has been on people's identification of the flavour of food or, more commonly, drink (see [33], for a review). Classic research by DuBose and his colleagues [53] demonstrated that the addition of food colouring (green, red, or orange) biased participants' judgments concerning the identity of the flavour of a cherry-flavoured solution. So, for instance, nearly 20% of the participants in this study reported that the drink tasted of orange when the cherry-flavoured solution was coloured orange as compared to no such responses when the same drink was coloured red, green, or remained colourless. Meanwhile, colouring the same drink green led to 26% lime-flavoured responses as compared to no such responses when the drink was coloured red or orange (see also [10,70,71], for similar results).

Oram *et al.* [67] gave over 300 people (of various ages) four drinks to taste. Four possible drink flavours (chocolate, orange, pineapple, and strawberry) were presented in four different colours (brown, orange, yellow, and red), thus giving rise to a total of 16 possible drinks. The participants had to try and discriminate the flavour of the drinks. The results highlighted a clear developmental trend toward an increased ability to correctly report the flavour of the drinks, regardless of the colour in which the drink was presented (see Figure 2). That is, the crossmodal modulation of flavour perception by vision apparently decreased with age (from 2 years of age up).

Importantly, in this and the majority of the other studies that have been reported so far, the participants were given no information about the possibility that the colour of the solutions might have been misleading (a point to which we will return later). Research by Zampini *et al.* [47,68] has shown that adults can easily be confused by the addition of inappropriate colour to a range of fruit-flavoured soft drinks (see Figure 3). Importantly, the crossmodal effects of beverage colour on flavour identification demonstrated by Zampini *et al.* were obtained under those conditions where the participants had been told to ignore the potentially misleading colouring of the beverages that they had been presented with. Such results therefore hint at the automaticity of such crossmodal effects.

Interim summary

The majority of the research that has been published to date has convincingly demonstrated that food colour affects the ability of people to correctly identify the flavour of food and drink (see Spence *et al.* [18], for a comprehensive review of the literature on this question). Although beyond the scope of this review, it is perhaps also worth noting that food colouring can influence the perceived thirst-quenching (or refreshing) properties of drinks as well [41,72-74]. That said, I would argue that there is a danger that one can get a biased impression of just how important colour is to the consumer's perception of, and response to, food and drink. This is because in the majority of laboratory studies, the colour of the foodstuff was pretty much the only cue, sensory or otherwise, that the participants had to go on when making their decisions as to the taste/flavour of that which

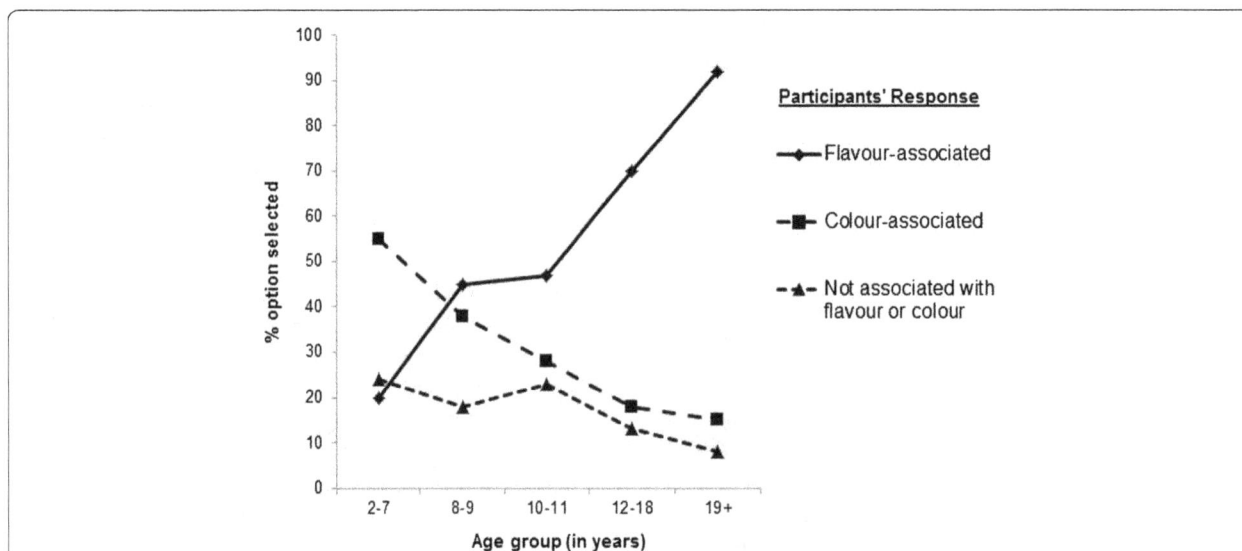

Figure 2 Graph highlighting the percentage of trials in which the participants' flavour discrimination response matched the colour of the drink, the actual flavour of the drink, or matched neither the colour or flavour of the drink as a function of the age of the participants in a developmental study of the psychological impact of colour on people's flavour discrimination responses reported by **Oram *et al.* [67].** [Reprinted from [67], with permission].

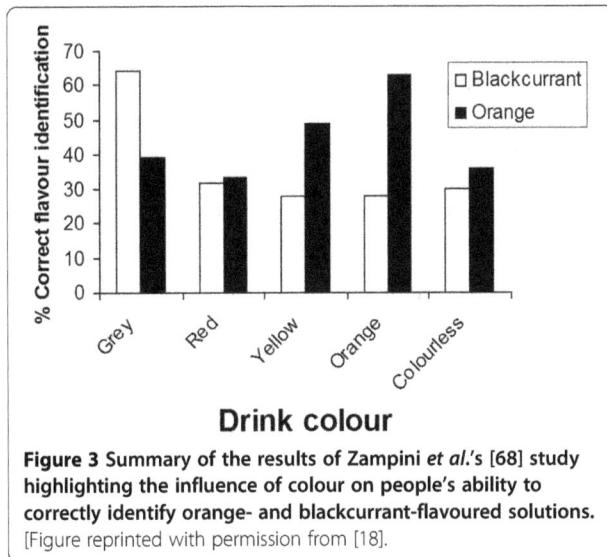

Figure 3 Summary of the results of Zampini et al.'s [68] study highlighting the influence of colour on people's ability to correctly identify orange- and blackcurrant-flavoured solutions. [Figure reprinted with permission from [18].

they were tasting. In the real world (see below), the consumer normally has a number of other cues to utilize when trying to judge the likely sensory and hedonic qualities of food and drink. What is more, there is always a danger that being confronted with a whole range of drinks, say, similar in flavour and differing most noticeably in terms of their colouring may have drawn, or focused, the participants' attention on colour as the most salient dimension (that is, in a way that may not be representative of everyday life).

One other thing to note here is that food colours are not necessarily associated with just one taste/flavour. As shown by Zampini et al. [47], for example, a red-coloured drink may be most strongly associated with the flavour of strawberry, but also, to a lesser degree with the flavour of raspberry and cherry. Hence, if one really wants to understand/predict its effect on multisensory flavour perception, it is important to bear in mind that a given beverage colour may actually prime a number of different possible flavours [75-78]. What is more, similar food colours may give rise to qualitatively different flavour expectations depending on the category of product under consideration (for example, soft drinks, cake, noodles, curry, and so on) and possibly also the brand (cf. [79]). As such, there is clearly a need for more research addressing the influence of food colour across different kinds of food product (and as a function of branding) in order to get a more complete, not to mention market-relevant, understanding of the psychological effect of food colour.

Names, brands, and colours

Given the ambiguity in the meaning of colour in foods and beverages, it can sometimes be important that the name and description of a food or beverage set the right sensory or hedonic expectations or else help to

disambiguate between the different possible meanings that may be associated with a given colour. The classic example here comes from the work of Yeomans and colleagues [80]. These researchers demonstrated that when the meaning of food colouring is misinterpreted (that is, when it sets the wrong sensory expectations), then this can have an adverse effect on people's subsequent taste ratings. The participants in this study were given a bright pink ice cream to taste. One group of participants was given no information about the dish, another group was informed that the food was called 'Food 386', and a third group was told that what they were about to eat was a frozen savoury mousse. Those participants who had not been given any information about the dish and hence who were led by their eyes into expecting that they would taste a strawberry-flavoured ice cream (which has the same pinkish-red colour) did not like the dish when they tried it. Specifically, they rated the frozen savoury smoked salmon ice cream as tasting too salty. By contrast, those participants in the other two groups rated the seasoning of the dish as being just right, and, what is more, liked the savoury ice far more as well (see [81-83], for related research). These results therefore demonstrate that the meaning of colour in food and drink can be altered simply by the description that is given to a product or dish [1]. Generally speaking (that is, in all environments excepting perhaps the modernist restaurant), it is important to avoid disconfirmed expectation [1,84].

Indeed, the typical laboratory situation can be contrasted that with that of everyday consumption episodes where a food or drink will most likely be encountered in the context of branding/packaging information, or may well have been described by whoever has prepared, or is serving the food or drink. In other words, it can be argued that the situation that is typically studied in the laboratory setting is quite unlike that of everyday life (see also [85-87]). Hence, one concern here is that the results of much of the research that has been conducted in the laboratory may actually end up giving a biased view of the importance of colour in multisensory flavour perception. Note that in the laboratory situation, colour is often the only cue that participants have to go on when making their judgments of expected flavour. By contrast, in the majority of real-world consumption situations, colour is but one of many cues (including branding, pricing, labelling, and so on) that the consumer can use.

One other product-extrinsic cue that can modulate the meaning of colour in beverages is the nature of the glass or receptacle in which that drink happens to be presented [88,89]. The same colour drink may have a very different meaning if shown in a plastic bathroom cup than in a cocktail glass, say. In the former case, a blue-coloured drink is likely to be interpreted as connoting mouthwash and hence associated with a mint flavour, whereas when exactly the same colour is seen in a cocktail glass, it may be interpreted as signifying the orange flavour of blue cura-çao instead [90].

Psychological effects of food colour on behaviour

It is important to realize that the psychological effects of food colouring are not restricted to the sensory-discriminative domain. It has often been suggested that food colouring can modulate certain of our food-related behaviours as well [91,92]. Certainly, getting the colour right can play an important role in food acceptance, liking, and hence, ultimately, food intake [24,93-97]. Though, as pointed out by Garber et al. [85], while it is often claimed that colour influences food preferences, good, marketing-relevant insights tend to be a little harder to come by in this area.

Colour can play an important role in modulating a consumer's affective expectations [32,98]. And just as there can be a sensory disconfirmation of expectation (as outlined above), there can also be a hedonic discon-firmation of expectation - that is, when a consumer real-izes that they do not like a food or beverage as much as they were expecting that they would.

In other research, it has been shown that people will consume more candy if it comes in a variety of colours than if presented in just a single colour [99], even if that colour happens to be the consumer's favourite one. Whether sensory-specific satiety or boredom is the most appropriate explanation for such results is still being de-liberated by researchers (see [92], for a review). Interest-ingly, while the use of colour (specifically increasing colour variety) is usually portrayed as a means by which the big food companies can get their consumers to con-sume more (think only of the multicoloured packs of Smarties, M&Ms, or Jelly Beans), there is some evidence to suggest that colour cues can also be used to modulate intake downward, by providing an effective cue to por-tion control ([100] see also [101,102]). So, for example, Geier et al. [100] reported that people ended up eating fewer potato chips if every seventh chip in a tube hap-pened to be coloured red.

Off-colouring in food

Researchers have been interested in the response of consumers to food colouring that they associate with products that have been in some way spoiled. That such off-colours can have a profound effect on people's food behaviours was suggested by the response of consumers to a batch of Tropicana grapefruit juice that was donated to a food bank some decades ago. According to Crumpacker ([14], p. 6), nobody wanted to drink the juice because of its abnormal brown colour. This despite the fact that those who tried it reported it to taste perfectly acceptable; see also [59,81,103], on the preferred colour of this staple of the breakfast table.

Meanwhile, the dinner party guests in Wheatley's [15] classic study were invited to dine on a meal of steak, chips, and peas. The only thing that may have struck any of the diners as odd was how dim the lighting was. However, this aspect of the atmosphere was actually de-signed to help hide the food's true colour. Part-way through the meal, the lighting was returned to normal, revealing that the steak had been artificially coloured blue, the chips looked green, and the peas had been coloured red. A number of Wheatley's guests suddenly felt ill when the lighting was turned to normal levels, with several of them apparently heading straight for the bathroom (cf. [54]).[e]

It is noticeable how the majority of the research on the psychological impact of off-colour in food is rather anecdotal in nature (presumably because it can be diffi-cult to get ethical approval to present food to partici-pants and have them believe that the colour indicates that it has gone off). Nevertheless, the evidence that has been published to date does seem to highlight the strong avoidance responses that such food colouring can in-duce, especially in the case of meats and fish that look off.[f]

Artificial/natural

Over the years, there has been ongoing concerns expressed about the negative health and well-being con-sequences that are apparently associated with the con-sumption of certain artificial food colourings, this despite their being rated as being safe and tasteless [24,104-116]. This had led some consumers to search out those foods that are free from all colouring. How-ever, such products generally do not taste that good. As Harris pointed out in an article that appeared in The New York Times [24], many commercial foods are disap-pointingly lacking in taste/flavour if served in a colourless (that is, clear or white) format.

A less extreme reaction to concerns over artificial food colourings has been to search out natural colourings that better match the sensory properties desired by the food producers: This includes everything from trying to deliver a wide enough range of natural colours [117], through to improving the stability of natural colourings, at least for those products that are likely to have a long shelf life [118-120]. Of course, that food colouring is nat-ural does not in-and-of-itself necessarily make it appeal-ing to the consumer. Here, one only needs to think of the red colouring of, for example, Smarties (the candy-covered chocolate; http://www.nestle.co.uk/brands/cho-colate_and_confectionery/chocolate/smarties) that used to be made from carminic acid extracted from scaly in-sects. Unappealing to most consumers, one imagines. Nowadays, though, the red colouring comes from red cabbage instead [116].

And what, exactly, constitutes natural is not obvious. The vibrant orange-coloured carrots that we are all familiar with nowadays, for example, are actually the result of extensive breeding. Once upon a time, the majority of carrots were naturally purple. According to some, the selective breeding was designed to deliver the orange colour of the Dutch royal family in the seventeenth century [121-123]. Although another, perhaps more plausible, explanation for why the orange variety may have been preferred over the original purple variety was because the latter would colour the soups, stews, and so on into which they were placed.

A number of the modernist chefs we have been fortunate enough to work with here at the Crossmodal Research Laboratory at Oxford University over the years have been particularly interested in surprising their diners by presenting foods that have one colour (and hence set a particular taste/flavour expectation) while actually delivering another unexpected flavour instead.[g] However, the chefs typically do not want to achieve such results by means of artificial food colourings for fear of their diners' reaction.

One elegant example of the use of natural colouring to create surprise and delight in the mind of the diner comes from the beetroot and orange jelly dish that used to be served as one of the opening courses on the menu at The Fat Duck restaurant in Bray (http://www.thefatduck.co.uk/). This dish would be presented as two blocks of jelly, one bright orange, the other a dark purple, placed side-by-side on the plate. And where the modernist chefs lead, the market sometimes follows. Pine berries, for example, which look for all-the-world like white strawberries provide an intriguing example of an otherworldly, at least to Western eyes, but entirely naturally coloured food.[h] Such unusually coloured food products have apparently been selling well in the supermarkets in recent years (see also [123]). More generally, there would appear to be renewed interest in surprisingly coloured foods in the mass market as well. For example, a few years ago, one well-known burger chain launched a pitch black bamboo and squid ink burger in Japan, that was seasoned with black squid ink ketchup, and served in a black bun [124]. As a group, children seem to be particularly fond of such miscoloured foods (think confused Skittles; http://www.wrigley.com/uk/brands/skittles.aspx) and beverages [125-128].

Marketing colour

Adding colour to food or else changing the colour of a food or beverage (or its packaging) has long been used as a marketing tool (for example, [129-133]; see also http://www.ddwcolor.com/hue/why-color/). In fact, according to an informal store audit reported by Garber et al. [85], 97% of all food brands displayed (in all categories) used food colour to indicate flavour. Food colour is used in marketing

for a number of reasons: Everything from increasing shelf stand-out through to blurring the distinction between different products. Indeed, going back three quarters of a century now, there was quite a fight by the butter lobby in order to try and prevent the makers of margarine from adding a golden yellow hue to their product in order to give it the appearance of its better established rival (for example, see [134]).

More recently, the potential role of adding food colouring in marketing was amply demonstrated by the dramatic rise in sales of tomato ketchup when Heinz decided to add a tiny amount of food colouring and turn this staple of the dining table green [135]. Other large drink brands that have, in recent years, launched drinks in unusual colours include an amber-coloured cola, called Pepsi Gold, in India [133]. However, not every attempt by marketers to use colour to boost sales has been successful. Clear cola drinks, for example, have generally failed in the marketplace [136]. And while there are a number of theories out there in the marketing literature about what went wrong in such cases, one suggestion is that when such drinks were tasted away from their packaging then the likely disconfirmation of expectation that results from experiencing a cola flavour when the sight of the drink led the consumer to expect lemonade or soda water may have been especially problematic.[i]

Individual differences in the psychological effects of colour

One thing that is noticeable about much of the early research on the psychological effects of food colouring is how little attention was paid by researchers to the profiles of the participants themselves. This turns out to be an important caveat since the latest research now shows that exactly the same food colour can elicit qualitatively different expectations concerning the likely taste/flavour of food and drink in different groups of consumers.

Cross-cultural differences

Exactly the same colour (for example, in a beverage) has been shown to set up qualitatively different expectations in the minds of different (groups of) consumers. Just take the two drinks shown in Figure 4: When they were shown to young adults in Taiwan and the UK, the former expected them to taste of cranberry and mint (mouthwash?), respectively, whereas the latter expected cherry/strawberry and raspberry, instead [75]. Wan et al. [88,89] have recently been conducting a number of internet-based studies designed to assess which food colours have a similar meaning in terms of expected flavour across culture and which differ markedly in terms of the expectations that they set. Food marketers working in the global marketplace obviously need to be aware of any cultural differences in the meaning of food colour [133]. Here, though, one potential

Figure 4 Two of the six coloured drinks shown to the participants from the UK and Taiwan in a study by Shankar *et al.* [75]. The results of this cross-cultural study demonstrated that exactly the physically same food colour can elicit qualitatively sensory different expectations as far as the likely flavour of a drink might be in consumers from different countries. The most frequently expected flavours for drinks of these colours are shown at the bottom.

limitation with the internet-based testing of consumers' colour expectations ought to be noted: Namely, it is difficult to precisely control the appearance properties of the visual stimuli on an individual participant's monitor. By contrast, in Shankar *et al.*'s [75] study, the participants actually viewed the drinks.

Developmental differences

Developmental differences in the meaning, and influence, of food colour have been reported by researchers. As noted earlier, young children seem to be more drawn to brightly (some would say artificially) coloured foods than are adults (though see [90]). In terms of changes in the psychological influence of food colouring across the lifespan, on the basis of the evidence that has been published to date [45,67,137], it would appear that, if anything, visual cues exert a somewhat greater influence on flavour identification early in development (see Figure 2), and in old age, than in adulthood (see [138], for a review). One reason as to why children might show more visual dominance (that is, simply relying on what they see) is because they have not yet learned to integrate their senses in an adult-like manner (cf. [139]). Thus far, published studies have assessed the responses of children from 2 years of age upward.

At the other end of the spectrum, the well-documented decline of taste and smell sensitivity in old age may mean that the residual senses (especially those where prostheses, such as glasses or hearing aids, are available) take on a

more important role in terms of determining the final taste/flavour experience [3,137]. However, it has to be said that the evidence that has been published on this topic to date is rather mixed (see [138], for a review). While some researchers have been able to demonstrate more pronounced psychological effects of food colouring in, say, older adults [45,103], such differences have certainly not always been found.

Here, it is perhaps also worth bearing in mind that there may be changes in the meaning and acceptability of colour over time. One only needs to remember, for example, that blue foods were traditionally considered unacceptable to a majority of consumers [140,141]. Nowadays, many foods are blue [133], although in this case, note that they are primarily marketed at the younger consumer [85]. Over a much longer timescale, one could even think of how the flavour of carrots may have switched its colour association from purple to orange (see above).

Expertise and the psychological effects of food colouring

Expertise has been shown to modulate the psychological impact of food colouring on flavour perception. Some of the most impressive studies have come from the world of wine (see [142], for early research; and [143], for a review). In one oft-cited experiment, Morrot and his colleagues [144] reported that a group of students on a university wine course in Bordeaux, France, had been fooled into choosing red wine aroma descriptors when given a white wine to evaluate that had been artificially coloured red with odourless food dye. Meanwhile, Parr *et al.* [145] conducted a follow-up in New Zealand in which they tested both experts (including professional wine taster and wine makers) and 'social' drinkers. The descriptions of the aroma of a Chardonnay wine given by the experts when it had been artificially coloured red were more accurate when the wine was served in an opaque glass than when served in a clear glass. Interestingly, this colour-induced biasing of flavour judgments occurred despite the fact that the experts had been explicitly instructed to rate each of the wines that they had been given to taste while ignoring any colour cues. Such results therefore suggest that the crossmodal effect of vision is not under cognitive control. Ironically, the social drinkers in Parr *et al.*'s study turned out to be so bad at reliably identifying the aromas present in the wine that it was difficult to discern any pattern in the data when an inappropriate wine colour was added.

Taken together, therefore, the evidence that has been published to date is consistent with the view that expert wine tasters differ from social drinkers (that is, non-experts) in the degree to which visual (colour) cues influence their orthonasal perception of flavour [145] and their perception of the taste of sweetness ([142]; see also [84]). That said, it is worth noting that not all food/flavour experts necessarily exhibit the same increased responsiveness to colour cues when evaluating the taste and flavour of food and drink. Shankar *et al.* [78], for example, reported that the flavour experts working on a descriptive panel at an international flavour house (who all had more than 3 years of experience flavour profiling food and drink products) exhibited just as much visual capture (or assimilation) of their orthonasal olfactory flavour judgments as did non-experts. Thus, based on the research that has been published to date, the most appropriate conclusion regarding flavour experts would appear to be that while some (specifically those with an expertise in wine) show an enhanced susceptibility to the crossmodal influence of colour on judgments within their area of expertise [142,145], this pattern of results does not necessarily extend to other groups of flavour experts [78].

Genetic differences in the effect of colour

Although surprisingly little studied to date, various genetic differences might also modulate the psychological effect of food colouring. Here, for example, one might think both of those individuals who are born colour blind (primarily males and constituting approximately 6% of the population; [146]). Presumably such differences in colour perception ought to have some impact of the psychological effect of food colour, though it is hard to find any published research on the topic (see http://www.colourblindawareness.org/colour-blindness/

living-with-colour-vision-deficiency/food/). Here, it is also worth noting that there are several discrete kinds of colour blindness, each likely affecting the perception of food and beverage colour in a slightly different way.

However, just as important as any deficits in colour perception, may be an individual's taster status. It turns out that genetic differences here may play an important role in determining just how much of a role colour plays in flavour perception. Some people have far more taste buds than others (the former are known as supertasters, the latter, non-tasters; with 25% of the population falling into each category). The remaining half of consumers fall into an intermediate group, known as medium tasters [147]. To give some idea of the differences in receptor density that might be involved here, it has been estimated that some individuals may have up to 14 times more taste buds than others [148]. Zampini *et al.* [68] have reported that supertasters are significantly less affected by the colour of a drink than medium tasters, who, in turn, are less affected than non-tasters (see Figure 5). It is somewhat surprising to find that this is the only study of the psychological impact of food colour to have assessed the taster status of their participants.[j] One can, perhaps, frame this result in terms of the literature on sensory dominance. That is, those individuals with a greater number of taste buds presumably exhibit lower variance in terms of their unisensory gustatory judgments. According to the maximum likelihood account of multisensory integration [149], the perceptual estimate with the lower variance will likely be weighted more heavily when it comes to estimating a given stimulus attribute.

Interim summary

One cannot hope to attain a comprehensive understanding of the psychological impact of food colour without taking into account the individual differences. The relevant differences include genetic differences in terms of taster status and colour perception, as well as cross-cultural and age-related differences. Although beyond the scope of this article, there may be racial differences in terms of colour preferences as well [150]. What is clear from the research that has been published to date is that these individual differences can influence both the meaning of colour and its influence on the consumer. Having established the importance of such individual differences (of both genetic and experiential origin), the question becomes one of how to assess the psychological impact of food colour experimentally. One solution here has been proposed in the work of Shankar and her colleagues [76-78].

According to Shankar *et al.* [77], assessing the 'degree of discrepancy' between the expected flavour set by colour and the flavour when eventually experienced by

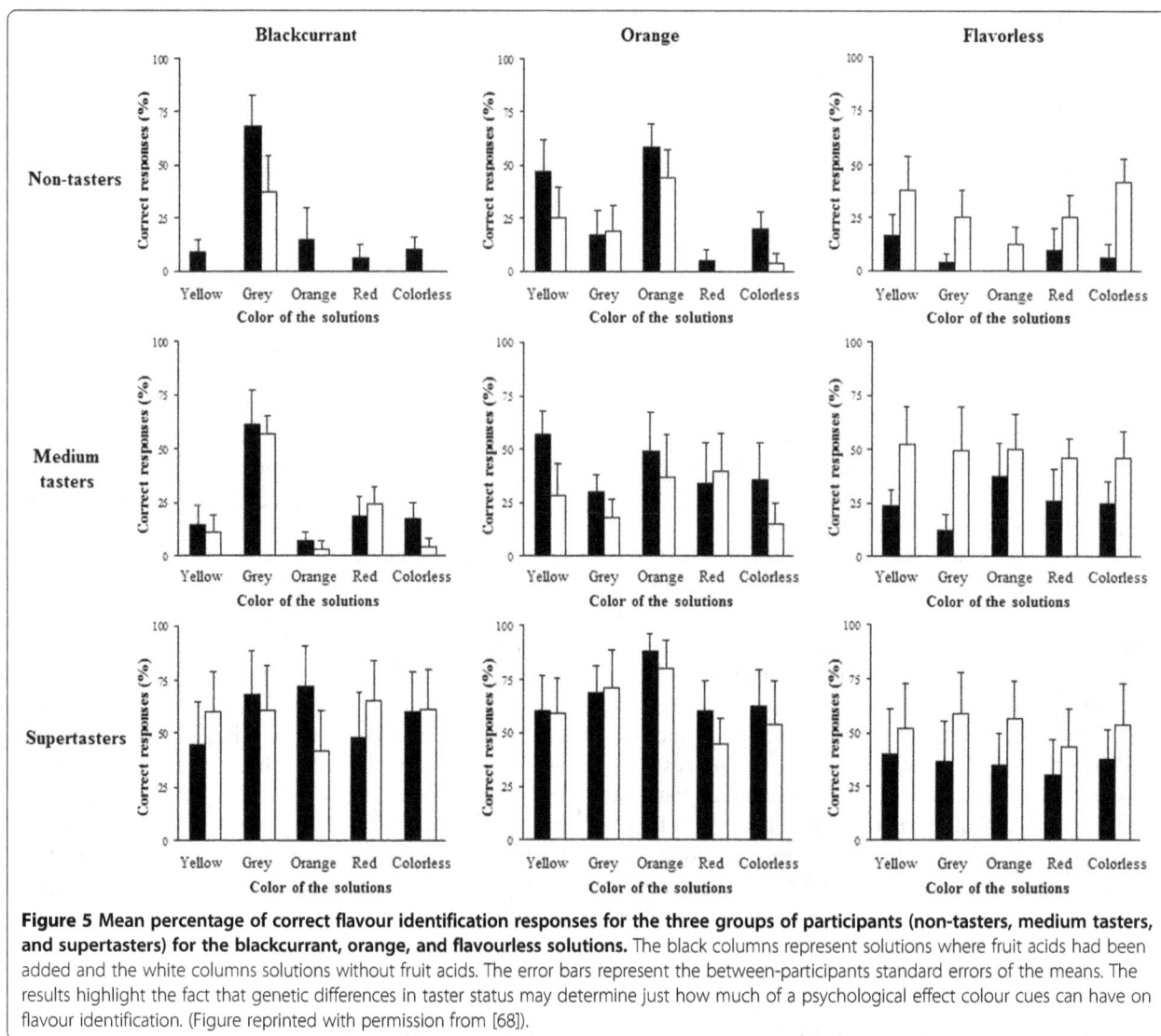

Figure 5 Mean percentage of correct flavour identification responses for the three groups of participants (non-tasters, medium tasters, and supertasters) for the blackcurrant, orange, and flavourless solutions. The black columns represent solutions where fruit acids had been added and the white columns solutions without fruit acids. The error bars represent the between-participants standard errors of the means. The results highlight the fact that genetic differences in taster status may determine just how much of a psychological effect colour cues can have on flavour identification. (Figure reprinted with permission from [68]).

the participant (or consumer) is key to understanding when colour influences flavour perception. Shankar *et al.* argued that under conditions of low discrepancy, the perceived disparity between the expected and actual flavour of a drink (or food) is small. Low discrepancy colour-flavour combinations might, for example, consist of cranberry- or blueberry-flavoured drinks coloured purple (purple being associated with grape flavour), whereas high discrepancy combinations might include banana- or vanilla-flavoured drinks that have been coloured purple. Across several experiments, when a particular colour - identified by participants as one that generated a strong flavour expectation - was added to the drinks that the participants were given to sniff (as compared with when no such colour was added), a significantly greater proportion of their identification responses were consistent with this expectation.[k] By contrast, under conditions of high discrepancy, adding the

same colours to the drinks no longer affected participants' identification responses in the same way (see Figure 6). That is, there was a significant difference in the proportion of responses that were consistent with participants' colour-based expectations in conditions of low as compared with high discrepancy. Shankar *et al.*'s results therefore demonstrate that the degree of discrepancy between an individual's expectation concerning the flavour (derived visually) and their actual experience on tasting the drink modulates the crossmodal influence of colour cues on judgments of flavour identity.

One thing to bear in mind about Shankar *et al.*'s [76-78] studies, though, is that the participants never got to taste the flavoured drinks that they were asked to judge. That is, all their judgments/ratings were made on the basis of nothing more that orthonasal olfactory cues. Of course, this should not matter all that much, given the extensive literature showing that colour cues can modulate

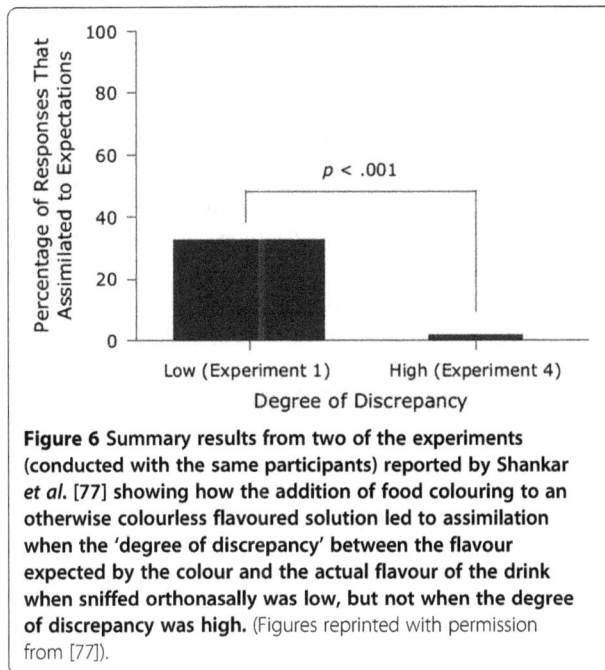

Figure 6 Summary results from two of the experiments (conducted with the same participants) reported by Shankar *et al.* [77] showing how the addition of food colouring to an otherwise colourless flavoured solution led to assimilation when the 'degree of discrepancy' between the flavour expected by the colour and the actual flavour of the drink when sniffed orthonasally was low, but not when the degree of discrepancy was high. (Figures reprinted with permission from [77]).

orthonasal olfactory discrimination/identification responses across a wide range of experimental conditions ([151-158]; see [159], for a review).[1]

Bottom-up or top-down influences of colour

Now, one further question that can, and probably should, be asked before closing concerns whether colour should be considered as exerting its psychological influence over flavour perception in more of a 'bottom-up' or more of a 'top-down' manner. On the one hand, it is clear that when people know that the colour they see is inappropriate (misleading) and so should be ignored, it nevertheless still influences their perception in a seemingly automatic manner [47,68,145]. Such results support a bottom-up account of at least part of colour's crossmodal influence over taste and flavour perception. Of course, the existence of such bottom-up effects should not be taken as evidence to deny the fact that top-down influences are also important.

Indeed, colour certainly also influence people's flavour perception in more of a top-down manner as well. Here, it is relevant to note that researchers have demonstrated that labelling, branding, and other descriptive information can all modify the meaning of a given food colour and by so doing influence the perceived taste of a food or beverage [1]. So, for instance, Shankar *et al.* [83] reported that, even when blindfolded, telling a participant that a sugar-coated chocolate candy has a particular colour (or that it is light or dark chocolate) influenced the pattern of responding that was observed. In a similar manner, a variety of non-sensory (labelling) cues have also been shown to bias the way in which a normally sighted observer interprets the meaning of a given colour (as in Yeomans *et al.*'s, [80],

study). Taken together, then, there is good evidence that colour's psychological influence on taste and flavour perception occurs not only in a bottom-up but also in more of a top-down manner as well. Studying the interaction between these influences on flavour perception is an area of growing interest from both a theoretical and more marketing-inspired perspective [42,76,82]. A related challenge comes from trying to integrate the growing literature demonstrating the influence of everything from the colour of the product packaging [160] through to the colour of the lighting [161,162] in which food and drink are consumed on the multisensory flavour experience [1].

Conclusions

Since the first reports that changing the colour of a food could change the taste/flavour were published [54,55], somewhere in the region of 150 papers have investigated the impact of food colouring on the perception and behaviour of participants/consumers. While the majority of those studies have tended to focus on colour's effect on taste/flavour identification (see [18], for a review), it is important to note that colour cues influence our food and drink-related behaviour in a number of different ways [1,92,125,163]. Food colouring undoubtedly plays an important role in driving liking and the consumer acceptability of a variety of food and beverage products. And while increasing colour variety in food can lead to enhanced consumption [92], what we see can also lead to a suppression of our appetitive behaviours when associated with off-colours (or coloration that is interpreted by the consumer as such).

Finally, given the practical difficulties associated with delivering flavours while a participant lies in the brain scanner [1], it is perhaps understandable that there has not been a great deal of neuroimaging research that has looked at the influence of colour on flavour perception as yet ([164]; see also [165]). Whether or not as the result of further neuroimaging, it is clear that additional research is most definitely needed in order to develop a better understanding of the psychological mechanisms underlying the various effects of colour on our perception of, and behaviours toward, food [166].

Certainly, the expectations, both sensory and hedonic, that are set by food colouring play an important role in determining the final flavour experience and how much it is liked. Furthermore, the degree of discrepancy between the sensory and hedonic expectations and the subsequent experience appears crucial to the question of whether assimilation or contrast will be observed. Here, recent research has increasingly demonstrated the differing meanings associated with food colour in different consumers. Identifying consistent colour-flavour mappings and training the consumer to internalize other new associations is one of the important challenges facing the food marketer interested in launching new products, or brand extensions, in a marketplace that is more colourful than ever.

Endnotes

[a]While the term 'product-intrinsic' is widely used in the literature when talking about the colour of a food or beverage, the appropriateness of this notion can be questioned from the perspective of (holistic) perception. Strictly speaking, colour is not a property of a (food) material but rather a percept in an observer that originates from an interaction with a material, under the influence of many other cues that are external to the coloured surface, but certainly internal to, the observer (for example, [167], p. 5).

[b]Here, it is perhaps worth noting that intense food colouring, while seemingly attractive to children (see [138], for a review), may lead some consumers to consider a food or beverage product as being 'artificial' and hence less liked (for example, [45,46]).

[c]Note that the participants in this study only ever had to report whether or not the solution had a taste. That is, they never had to identify the tastant. In fact, somewhat surprisingly, the question of whether colour influences the ability of people to *identify/discriminate* the basic tastes has not, as far as I am aware, been studied to date (see [18], for a review). This despite the fact that extensive evidence has been collected concerning the colours that people in different cultures associate with each of the basic tastes (see [40], for a review and cross-cultural evidence).

[d]Note that a lack of precise colour measurement has hampered comparison of the results of many of the studies that have been published to date (cf. [4,168]).

[e]No mention is made of whether ethical approval was obtained for this particular study!

[f]Though note that olfactory cues are at least as important in people's judgment of whether a food has gone off ([169]; see also [170] on the consumer evaluation of the sensory properties of fish).

[g]Note that while under the majority of everyday conditions, people prefer foods and beverages that taste as they expect them to taste (that is, people do not like surprises, especially when it comes to the stimuli that enter the mouth, and hence have the potential to poison them), there are occasions, such as at the tables of the modernist restaurant where many diners seem to positively relish having their expectations played with [1,125].

[h]These 'white strawberries' are the result of cross-breeding the South American strawberry *Fragaria chiloensis*, which grows wild in some parts of Chile, and the North American strawberry *Fragaria virginiana*.

[i]Of course, here, it needs to be remembered that changing the colour of a drink can change its flavour

perceptually. However, one has to imagine that any such crossmodal perceptual effects would have been picked up in consumer tests before the product was launched.

[j]Indeed, given the relatively small sample size and the *post hoc* nature of Zampini *et al.*'s [68] discovery, replication in a larger sample would undoubtedly be desirable to check on the generalizability of this potentially important result.

[k]Here, it is worth pointing out that when flavour experts were tested, their results were similar to those of normal participants [78].

[l]That said, Koza *et al.*'s [56] results concerning the differing effect of colour on orthonasal and retronasal olfactory intensity judgments needs to be borne in mind here.

Competing interests

The author declares that he has no competing interests.

Author's contributions

CS wrote all parts of this review. The author read and approved the final manuscript.

Acknowledgement

CS would like to acknowledge the AHRC Rethinking the Senses grant (AH/L007053/1).

References

1. Spence C, Piqueras-Fiszman B. The perfect meal: the multisensory science of food and dining. Oxford: Wiley-Blackwell; 2014.
2. Cardello AV. The role of the human senses in food acceptance. In: Meiselman HL, MacFie HJH, editors. Food choice, acceptance and consumption. New York, NY: Blackie Academic and Professional; 1996. p. 1–82.
3. Clydesdale FM. The influence of colour on sensory perception and food choices. In: Walford J, editor. Developments in food colours–2. London, UK: Elsevier Applied Science; 1984. p. 75–112.
4. Clydesdale FM. Color perception and food quality. J Food Qual. 1991;14:61–74.
5. Clydesdale FM. Color as a factor in food choice. Crit Rev Food Sci Nutrit. 1993;33:83–101.
6. Delwiche JF. You eat with your eyes first. Physiol Behav. 2012;107:502–4.
7. Hall RL. Flavor study approaches at McCormick and Company, Inc. In: In AD Little, Inc, editor. Flavor research and food acceptance: A survey of the scope of flavor and associated research, compiled from papers presented in a series of symposia given in 1956–1957. New York, NY: Reinhold; 1958. p. 224–240).
8. Kanig JL. Mental impact of colors in foods studied. Food Field Reporter. 1955;23:57.
9. Kostyla AS, Clydesdale FM. The psychophysical relationships between color and flavor. CRC Critic Rev Food Sci Nutrit. 1978;10:303–19.
10. Watson E: We eat with our eyes: flavor perception strongly influenced by food color, says DDW. Downloaded from http://www.foodnavigator-usa.com/Science/We-eat-with-our-eyes-Flavor-perception-strongly-influenced-by-food-color-says-DDW on 19/12/2014.
11. Downham A, Collins P. Colouring our foods in the last and next millennium. Int J Food Sci Technol. 2000;35:5–22.
12. Tannahill R. Food in history. New York, NY: Stein and Day; 1973.
13. Walford J. Historical development of food coloration. In: Walford J, editor. Developments in food colours. London: Applied Science; 1980.
14. Crumpacker B. The sex life of food: when body and soul meet to eat. New York, NY: Thomas Dunne Books; 2006.

15. Wheatley J: Putting colour into marketing. Marketing 1973, October:24–29, 67.

16. Rozin P. "Taste-smell confusions" and the duality of the olfactory sense. Percept Psychophys. 1982;31:397–401.

17. Spence C, Smith B, Auvray M. Confusing tastes and flavours. In: Stokes D, Matthen M, Biggs S, editors. Perception and its modalities. Oxford: Oxford University Press; 2015. p. 247–74.

18. Spence C, Levitan C, Shankar MU, Zampini M. Does food color influence taste and flavor perception in humans? Chemosens Percept. 2010;3:68–84.

19. Deliza R, MacFie HJH. The generation of sensory expectation by external cues and its effect on sensory perception and hedonic ratings: a review. J Sens Stud. 1997;2:103–28.

20. Hutchings JB. Expectations and the food industry: the impact of color and appearance. New York, NY: Plenum Publishers; 2003.

21. Piqueras-Fizman B, Spence C. Sensory expectations based on product-extrinsic food cues: an interdisciplinary review of the empirical evidence and theoretical accounts. Food Qual Prefer. 2015;40:165–79.

22. Spence C. Eating with our ears: assessing the importance of the sounds of consumption to our perception and enjoyment of multisensory flavour experiences. Flavour. 2015;4:3.

23. Cardello AV. Consumer expectations and their role in food acceptance. In: MacFie HJH, Thomson DMH, editors. Measurement of food preferences. London, UK: Blackie Academic & Professional; 1994. p. 253–97.

24. Harris G: Colorless food? We blanch. The New York Times 2011, April 3:3. Downloaded from http://www.nytimes.com/2011/04/03/weekinreview/03harris.html?_r=0 on 21/12/2014.

25. Murakoshi T, Masuda T, Utsumi K, Tsubota K, Wada Y. Glossiness and perishable food quality: visual freshness judgment of fish eyes based on luminance distribution. PLoS One. 2013;8(3):e58994.

26. Péneau S, Brockhoff PB, Escher F, Nuessli J. A comprehensive approach to evaluate the freshness of strawberries and carrots. Postharvest Biol Technol. 2007;45:20–9.

27. Prinz JF, & de Wijk RA: Effects of flavor and visual texture on ingested volume. Poster presented at the 5th Meeting of the International Multisensory Research Forum. 2-5th June, Sitges, Spain; 2004.

28. Okajima K, & Spence C: Effects of visual food texture on taste perception. i-Perception 2011, 2(8), http://i-perception.perceptionweb.com/journal/I/article/ic966.

29. Lawless HT, Klein BP. Sensory science theory and applications in foods. New York, NY: Marcel Dekker; 1991.

30. Carlsmith JM, Aronson E. Some hedonic consequences of the confirmation and disconfirmation of expectancies. J Abnormal Social Psychol. 1963;66:151–6.

31. Schifferstein HNJ. Effects of product beliefs on product perception and liking. In: Frewer L, Risvik E, Schifferstein H, editors. Food, people and society: A European perspective of consumers' food choices. Berlin: Springer Verlag; 2001. p. 73.

32. Zellner D, Strickhouser D, Tornow C. Disconfirmed hedonic expectations produce perceptual contrast, not assimilation. Am J Psychol. 2004;117:363–87.

33. Spence C: Visual contributions to taste and flavour perception. In M. Scotter (Ed.), Colour additives for food and beverages. Cambridge, UK: Woodhead Publishing; in press.

34. Calvo C, Salvador A, Fiszman S. Influence of colour intensity on the perception of colour and sweetness in various fruit-flavoured yoghurts. European Food Res Technol. 2001;213:99–103.

35. Johnson J, Clydesdale FM. Perceived sweetness and redness in colored sucrose solutions. J Food Sci. 1982;47:747–52.

36. Johnson JL, Dzendolet E, Damon R, Sawyer M, Clydesdale FM. Psychophysical relationships between perceived sweetness and color in cherry-flavored beverages. J Food Protect. 1982;45:601–6.

37. Johnson JL, Dzendolet E, Clydesdale FM. Psychophysical relationships between perceived sweetness and redness in strawberry-flavored beverages. J Food Protect. 1983;46:21–5. 28.

38. Norton WE, Johnson FN. The influence of intensity of colour on perceived flavour characteristics. Medical Sci Res. 1987;15:329–30.

39. Maga JA. Influence of color on taste thresholds. Chem Senses Flavor. 1974;1:115–9.

40. Wan X, Woods AT, van den Bosch J, Mckenzie KJ, Velasco C, Spence C. Cross-cultural differences in crossmodal correspondences between tastes and visual features. Frontiers Psychol: Cognit. 2014;5:1365.

41. Clydesdale FM, Gover R, Philipsen DH, Fugardi C. The effect of color on thirst quenching, sweetness, acceptability and flavor intensity in fruit punch flavored beverages. J Food Qual. 1992;15:19–38.

42. Levitan C, Zampini M, Li R, Spence C. Assessing the role of color cues and people's beliefs about color-flavor associations on the discrimination of the flavor of sugar-coated chocolates. Chem Senses. 2008;33:415–23.

43. Lavin JG, Lawless HT. Effects of color and odor on judgments of sweetness among children and adults. Food Qual Prefer. 1998;9:283–9.

44. Alley RL, Alley TR. The influence of physical state and color on perceived sweetness. J Psychol. 1998;132:561–8.

45. Philipsen DH, Clydesdale FM, Griffin RW, Stern P. Consumer age affects response to sensory characteristics of a cherry flavored beverage. J Food Sci. 1995;60:364–8.

46. Chan MM, Kane-Martinelli C. The effect of color on perceived flavor intensity and acceptance of foods by young adults and elderly adults. J Am Dietetic Assoc. 1997;97:657–9.

47. Zampini M, Sanabria D, Phillips N, Spence C. The multisensory perception of flavor: assessing the influence of color cues on flavor discrimination responses. Food Qual Prefer. 2007;18:975–84.

48. Shermer DZ, Levitan CA. Red hot: the crossmodal effect of color intensity on piquancy. Multisensory Res. 2014;27:207–23.

49. Zhou X, Wan X, Mu B, Du D, & Spence C: Examining colour-receptacle-flavour interactions for Asian noodles. Food Qual Prefer in press.

50. Urbányi G. Investigation into the interaction of different properties in the course of sensory evaluation. I. The effect of colour upon the evaluation of taste in fruit and vegetable products. Acta Aliment. 1982;11:233–43.

51. Wadhwani R, McMahon DJ. Color of low-fat cheese influences flavor perception and consumer liking. J Dairy Science. 2012;95:2336–46.

52. Dolnick E: Fish or foul? The New York Times 2008, September 2, downloaded from http://www.nytimes.com/2008/09/02/opinion/02dolnick.html?_r=1&scp=1&sq=chocolate%20strawberry%20yogurt&st=cse on 26/12/14.

53. DuBose CN, Cardello AV, Maller O. Effects of colorants and flavorants on identification, perceived flavor intensity, and hedonic quality of fruit-flavored beverages and cake. J Food Sci. 1980;45:1393–9. 1415.

54. Moir HC. Some observations on the appreciation of flavour in foodstuffs. J Soc Chemical Ind: Chem Ind Rev. 1936;14:145–8.

55. Duncker K. The influence of past experience upon perceptual properties. Am J Psychol. 1939;52:255–65.

56. Koza BJ, Cilmi A, Dolese M, Zellner DA. Color enhances orthonasal olfactory intensity and reduces retronasal olfactory intensity. Chem Senses. 2005;30:643–9.

57. Frank RA, Ducheny K, Mize SJS. Strawberry odor, but not red color, enhances the sweetness of sucrose solutions. Chem Senses. 1989;14:371–7.

58. Bayarri S, Calvo C, Costell E, Duran L. Influence of color on perception of sweetness and fruit flavor of fruit drinks. Food Sci Technol Int. 2001;7:399–404.

59. Fernández-Vázquez R, Hewson L, Fisk I, Vila D, Mira F, Vicario IM, et al. Colour influences sensory perception and liking of orange juice. Flavour. 2014;3:1.

60. Gifford SR, Clydesdale FM. The psychophysical relationship between color and sodium chloride concentrations in model systems. J Food Protect. 1986;49:977–82.

61. Gifford SR, Clydesdale FM, Damon Jr RA. The psychophysical relationship between color and salt concentrations in chicken flavored broths. J Sensory Stud. 1987;2:137–47.

62. McCullough JM, Martinsen CS, Moinpour R. Application of multidimensional scaling to the analysis of sensory evaluations of stimuli with known attribute structures. J Applied Psychol. 1978;65:103–9.

63. Pangborn RM. Influence of color on the discrimination of sweetness. Am J Psychol. 1960;73:229–38.

64. Pangborn RM, Hansen B. The influence of color on discrimination of sweetness and sourness in pear-nectar. Am J Psychol. 1963;76:315–7.

65. Roth HA, Radle LJ, Gifford SR, Clydesdale FM. Psychophysical relationships between perceived sweetness and color in lemon- and lime-flavored drinks. J Food Sci. 1988;53:1116–9. 1162.

66. Strugnell C. Colour and its role in sweetness perception. Appetite. 1997;28:85.

67. Oram N, Laing DG, Hutchinson I, Owen J, Rose G, Freeman M, et al. The influence of flavor and color on drink identification by children and adults. Develop Psychobiol. 1995;28:239–46.

68. Zampini M, Wantling E, Phillips N, Spence C. Multisensory flavor perception: assessing the influence of fruit acids and color cues on the perception of fruit-flavored beverages. Food Qual Prefer. 2008;19:335–43.

69. Garber Jr LL, Hyatt EM, Starr Jr RG. The effects of food color on perceived flavor. J Market Theory Practice. 2000;8(4):59–72.

70. Hyman A. The influence of color on the taste perception of carbonated water preparations. Bull Psychon Soc. 1983;21:145–8.

71. Stillman J. Color influences flavor identification in fruit-flavored beverages. J Food Sci. 1993;58:810–2.

72. Guinard JX, Souchard A, Picot M, Rogeaux M, Siefferman JM. Sensory determinants of the thirst-quenching character of beer. Appetite. 1998;31:101–15.

73. Zellner DA, Durlach P. What is refreshing? An investigation of the color and other sensory attributes of refreshing foods and beverages. Appetite. 2002;39:185–6.

74. Zellner DA, Durlach P. Effect of color on expected and experienced refreshment, intensity, and liking of beverages. Am J Psychol. 2003;116:633–47.

75. Shankar MU, Levitan C, Spence C. Grape expectations: the role of cognitive influences in color-flavor interactions. Conscious Cognit. 2010;19:380–90.

76. Shankar M, Simons C, Levitan C, Shiv B, McClure S, Spence C. An expectations-based approach to explaining the crossmodal influence of color on odor identification: the influence of temporal and spatial factors. J Sensory Stud. 2010;25:791–803.

77. Shankar M, Simons C, Shiv B, Levitan C, McClure S, Spence C. An expectations-based approach to explaining the influence of color on odor identification: the influence of degree of discrepancy. Attent Percept Psychophys. 2010;72:1981–93.

78. Shankar M, Simons C, Shiv B, McClure S, Spence C. An expectation-based approach to explaining the crossmodal influence of color on odor identification: the influence of expertise. Chemosens Percept. 2010;3:167–73.

79. Piqueras-Fiszman B, Spence C. Crossmodal correspondences in product packaging: assessing color-flavor correspondences for potato chips (crisps). Appetite. 2011;57:753–7.

80. Yeomans M, Chambers L, Blumenthal H, Blake A. The role of expectancy in sensory and hedonic evaluation: the case of smoked salmon ice-cream. Food Qual Prefer. 2008;19:565–73.

81. Hoegg J, Alba JW. Taste perception: more than meets the tongue. J Consumer Res. 2007;33:490–8.

82. Miller EG, Kahn BE. Shades of meaning: the effect of color and flavor names on consumer choice. J Consumer Res. 2005;32:86–92.

83. Shankar MU, Levitan CA, Prescott J, Spence C. The influence of color and label information on flavor perception. Chemosens Percept. 2009;2:53–8.

84. Lelièvre M, Chollet S, Abdi H, Valentin D. Beer-trained and untrained assessors rely more on vision than on taste when they categorize beers. Chemosens Percept. 2009;2:143–53.

85. Garber Jr LL, Hyatt EM, Starr Jr RG. Placing food color experimentation into a valid consumer context. J Food Products Market. 2001;7(3):3–24.

86. Garber Jr LL, Hyatt EM, Starr Jr RG. Measuring consumer response to food products. Food Qual Prefer. 2003;14:3–15.

87. Garber Jr LL, Hyatt EM, Starr Jr RG. Reply to commentaries on: "Placing food color experimentation into a valid consumer context". Food Qual Prefer. 2003;14:41–3.

88. Wan X, Velasco C, Michel C, Mu B, Woods AT, Spence C. Does the shape of the glass influence the crossmodal association between colour and flavour? A cross-cultural comparison. Flavour. 2014;3:3.

89. Wan X, Woods AT, Seoul KH, Butcher N, Spence C. When the shape of the glass influences the flavour associated with a coloured beverage: evidence from consumers in three countries. Food Qual Prefer. 2015;39:109–16.

90. Spence C. Drinking in colour. Cocktail Lovers. 2014;13:28–9.

91. Birren F. Color and human appetite. Food Technol. 1963;17(May):45–7.

92. Piqueras-Fiszman B, Spence C. Colour, pleasantness, and consumption behaviour within a meal. Appetite. 2014;75:165–72.

93. de Wijk RA, Polet IA, Engelen L, van Doorn RM, Prinz JF. Amount of ingested custard dessert as affected by its color, odor, and texture. Physiol Behav. 2004;82:397–403.

94. Gossinger M, Mayer F, Radochan N, Höfler M, Boner A, Grolle E, et al. Consumer's color acceptance of strawberry nectars from puree. J Sensory Stud. 2009;24:78–92.

95. Imram N. The role of visual cues in consumer perception and acceptance of a food product. Nutrit Food Sci. 1999;99:224–30.

96. Schutz HG. Color in relation to food preference. In: Farrell KT, Wagner JR, Peterson MS, MacKinney G, editors. Color in foods: A symposium sponsored by the Quartermaster Food and Container Institute for the Armed Forces

Quartermaster Research and Development Command U. S. Army Quartermaster Corps. Washington: National Academy of Sciences – National Research Council; 1954. p. 16–23.

97. Wei ST, Ou LC, Luo MR, Hutchings JB. Optimization of food expectations using product colour and appearance. Food Qual Prefer. 2012;23:49–62.

98. Wilson T, Klaaren K. Expectation whirls me round: the role of affective expectations on affective experiences. In: Clear MS, editor. Review of personality and social psychology: Emotion and social behavior. Newbury Park: Sage; 1992. p. 1–31.

99. Rolls BJ, Rowe EA, Rolls ET. How sensory properties of foods affect human feeding behaviour. Physiol Behav. 1982;29:409–17.

100. Geier A, Wansink B, Rozin P. Red potato chips: segmentation cues can substantially decrease food intake. Health Psychol. 2012;31:398–401.

101. Kahn BE, Wansink B. The influence of assortment structure on perceived variety and consumption quantities. J Consumer Res. 2004;30:519–33.

102. Redden JP, Hoch SJ. The presence of variety reduces perceived quantity. J Consumer Res. 2009;36:406–17.

103. Tepper BJ. Effects of a slight color variation on consumer acceptance of orange juice. J Sens Stud. 1993;8:145–54.

104. Accum F: A treatise on adulteration of food and culinary poisons. Cited in Anon. (1980); 1820.

105. Anon. Colourings - an interim review. Int Flavours Food Additives. 1979;10(3):96–7.

106. Anon. Additive use triggers consumer food concerns. Food Product Develop. 1979;13(8):8.

107. Anon. Food colors. A scientific status summary by the Institute of Food Technologists' Expert Panel on Food Safety & Nutrition and the Committee on Public Information. Food Technol. 1980;34(7):77–84.

108. Goldenberg N. Colours - do we need them? In British Nutrition Foundation (Ed.), Why food additives? The safety of foods (pp. 22–24). London, UK: Forbes Publications; 1977.

109. Kramer A. Benefits and risks of color additives. Food Technol. 1978;32(8):65–7.

110. Lucas CD, Hallagan JB, Taylor SL. The role of natural color additives in food allergy. Advances Food Nutrit Res. 2001;43:195–216.

111. Meggos H. Food colours: an international perspective. Manufacturing Confectioner. 1995;75:59–65.

112. Stevens LJ, Kuczek T, Burgess JR, Stochelski MA, Eugene Arnold L, Galland L. Mechanisms of behavioral, atopic, and other reactions to artificial food colors in children. Nutrit Rev. 2013;71:268–81.

113. Tuorila-Ollikainen H. Pleasantness of colourless and coloured soft drinks and consumer attitudes to artificial food colours. Appetite. 1982;3:369–76.

114. Weiss B, Williams JH, Margen S, Abrams B, Caan B, Citron LJ, et al. Behavioral responses to artificial food colors. Science. 1980;207:1487–9.

115. Whitehill I. Human idiosyncratic responses to food colours. Food Flavour Ingred Packag Process. 1980;1(7):23–7. 37.

116. Wilson B. Swindled: from poison sweets to counterfeit coffee - the dark history of the food cheats. London: John Murray; 2009.

117. Patel A. Going green: tuneable colloidal colour blends from natural colourants. New Food Magazine. 2014;17(2):7–9.

118. Bridle P, Timberlake CF. Anthocyanins as natural food colours–selected aspects. Food Chem. 1997;58:103–9.

119. Tolliday S. Nestlé confectionary: journey with colours. New Food Magazine. 2012;13(6):27–31.

120. Wissgott U, Bortlik K. Prospects for new natural food colorants. Trends Food Sci Technol. 1996;7:298–302.

121. Dalby A. Food in the Ancient World from A to Z. London, UK: Routledge; 2003.

122. Greene W. Vegetable gardening the Colonial Williamsburg Way: 18th century methods for today's organic gardeners. Rodale: New York, NY; 2012.

123. Macrae F: What's for dinner? Rainbow coloured carrots and super broccoli that's healthier and sweeter. DailyMail Online 2011, 15 October. Available at http://www.dailymail.co.uk/health/article-2044695/Purple-carrots-sale-Tescosupermarket-Orange-year.html (accessed January 2014).

124. Cook W: Would you eat a 'gourmet' burger made with charred bamboo and squid ink? Daily Mail Online 2012, 25th September. Downloaded from: http://www.dailymail.co.uk/news/article-2208321/Burger-King-black-burger-Japan-bamboo-charcoal-squid-ink.html.

125. Anon: 'Anything' and 'Whatever' beverages promise a surprise, every time. Press release, 17th May; 2007.

126. Garber Jr LL, Hyatt EM, Boya UO. The mediating effects of the appearance of nondurable consumer goods and their packaging on consumer behavior. In: Schifferstein HNJ, Hekkert P, editors. Product experience. London, UK: Elsevier; 2008. p. 581–602.

127. Walsh LM, Toma RB, Tuveson RV, Sondhi L. Color preference and food choice among children. J Psychol. 1990;124:645–53.

128. Piqueras-Fiszman B, Spence C. Sensory incongruity in the food and beverage sector: art, science, and commercialization. Petits Propos Culinaires. 2012;95:74–118.

129. Favre JP, November A. Colour and communication. Zurich: ABC-Verlag; 1979.

130. Gimba JG. Color in marketing: shades of meaning. Marketing News. 1998;32(6):16.

131. Hicks D. Benefits of added colourings in food and drinks. Int Flavours Food Additives. 1979;10(1):31–2.

132. Singh S. Impact of color on marketing. Manag Decis. 2006;44:783–9.

133. Garber LL, Hyatt EM, & Nafees L: The effects of food color on perceived flavor: a factorial investigation in India. J Food Products Market 2015 (in press).

134. Masurovsky BI. How to obtain the right food color. Food Industries. 1939;11 (13):55–6.

135. Farrell G: What's green. Easy to squirt? Ketchup! USA Today 2000, Monday July 10:2b.

136. Triplett T. Consumers show little taste for clear beverages. Market News. 1994;28(11):2. 11.

137. Christensen C. Effect of color on judgments of food aroma and food intensity in young and elderly adults. Percept. 1985;14:755–62.

138. Spence C. The development and decline of multisensory flavour perception. In: Bremner AJ, Lewkowicz D, Spence C, editors. Multisensory development. Oxford, UK: Oxford University Press; 2012. p. 63–87.

139. Gori M, Del Viva M, Sandini G, Burr DC. Young children do not integrate visual and haptic information. Curr Biol. 2008;18:694–8.

140. Cheskin L. How to predict what people will buy. New York, NY: Liveright; 1957.

141. Hine T. The total package: the secret history and hidden meanings of boxes, bottles, cans, and other persuasive containers. New York, NY: Little Brown; 1995.

142. Pangborn RM, Berg HW, Hansen B. The influence of color on discrimination of sweetness in dry table-wine. Am J Psychol. 1963;76:492–5.

143. Spence C. The color of wine - part 1. The World of Fine Wine. 2010;28:122–9.

144. Morrot G, Brochet F, Dubourdieu D. The color of odors. Brain Lang. 2001;79:309–20.

145. Parr WV, White KG, Heatherbell D. The nose knows: influence of colour on perception of wine aroma. J Wine Res. 2003;14:79–101.

146. Broackes J. What do the color-blind see? In: Cohen J, Matthen M, editors. Color ontology and color science. Cambridge, MA: MIT Press; 2010. p. 291–389.

147. Bartoshuk LM. Comparing sensory experiences across individuals: recent psychophysical advances illuminate genetic variation in taste perception. Chem Senses. 2000;25:447–60.

148. Miller IJ, Reedy DP. Variations in human taste bud density and taste intensity perception. Physiol Behav. 1990;47:1213–9.

149. Ernst MO, Banks MS. Humans integrate visual and haptic information in a statistically optimal fashion. Nature. 2002;415:429–33.

150. Scanlon BA. Race differences in selection of cheese color. Percept Motor Skills. 1985;61:314.

151. Blackwell L. Visual clues and their effects on odour assessment. Nutrit Food Sci. 1995;5:24–8.

152. Davis RG. The role of nonolfactory context cues in odor identification. Percept Psychophys. 1981;30:83–9.

153. Michael GA, Galich H, Relland S, Prud'hon S. Hot colors: the nature and specificity of color-induced nasal thermal sensations. Behaviour Brain Res. 2010;207:418–28.

154. Petit CEF, Hollowood TA, Wulfert F, Hort J. Colour-coolant-aroma interactions and the impact of congruency and exposure on flavour perception. Food Qual Prefer. 2007;18:880–9.

155. Stevenson RJ, Oaten M. The effect of appropriate and inappropriate stimulus color on odor discrimination. Percept Psychophys. 2008;70:640–6.

156. Zellner DA, Bartoli AM, Eckard R. Influence of color on odor identification and liking ratings. Am J Psychol. 1991;104:547–61.

157. Zellner DA, Kautz MA. Color affects perceived odor intensity. J Exp Psychol: Hum Percept Perf. 1990;16:391–7.

158. Zellner DA, Whitten LA. The effect of color intensity and appropriateness on color-induced odor enhancement. Am J Psychol. 1999;112:585–604.

159. Zellner DA. Color-odor interactions: a review and model. Chemosens Percept. 2013;6:155–69.

160. Spence C, Piqueras-Fiszman B. The multisensory packaging of beverages. In: Kontominas MG, editor. Food packaging: Procedures, management and trends. Hauppauge NY: Nova Publishers; 2012. p. 187–233.

161. Oberfeld D, Hecht H, Allendorf U, Wickelmaier F. Ambient lighting modifies the flavor of wine. J Sens Stud. 2009;24:797–832.

162. Spence C, Velasco C, Knoeferle K. A large sample study on the influence of the multisensory environment on the wine drinking experience. Flavour. 2014;3:8.

163. Maga JA. Influence of freshness and color on potato chip sensory preferences. J Food Sci. 1973;38:1251–2.

164. Skrandies W, Reuther N. Match and mismatch of taste, odor, and color is reflected by electrical activity in the human brain. J Psychophysiol. 2008;22:175–84.

165. Österbauer RA, Matthews PM, Jenkinson M, Beckmann CF, Hansen PC, Calvert GA. Color of scents: chromatic stimuli modulate odor responses in the human brain. J Neurophysiol. 2005;93:3434–41.

166. Kappes SM, Schmidt SJ, Lee SY. Color halo/horns and halo-attribute dumping effects within descriptive analysis of carbonated beverages. J Food Sci. 2006;71:S590–5.

167. Shepherd GM. Neurogastronomy: how the brain creates flavor and why it matters. New York: Columbia University Press; 2012.

168. Francis FJ, Clydesdale FM. Food colorimetry: theory and applications. New York, NY: Van Nostrand Reinhold/AVI; 1975.

169. Boesveldt S, Frasnelli J, Gordon AR, Lündstrom JN. The fish is bad: negative food odors elicit faster and more accurate reactions than other odors. Biol Psychol. 2010;84:313–7.

170. Sawyer FM, Cardello AV, Prell PA. Consumer evaluation of the sensory properties of fish. J Food Sci. 1988;53:12–8. 24.

13

Subtle changes in the flavour and texture of a drink enhance expectations of satiety

Keri McCrickerd[1*], Lucy Chambers[1], Jeffrey M Brunstrom[2] and Martin R Yeomans[1]

Abstract

Background: The consumption of liquid calories has been implicated in the development of obesity and weight gain. Energy-containing drinks are often reported to have a weak satiety value: one explanation for this is that because of their fluid texture they are not expected to have much nutritional value. It is important to consider what features of these drinks can be manipulated to enhance their expected satiety value. Two studies investigated the perception of subtle changes in a drink's viscosity, and the extent to which thick texture and creamy flavour contribute to the generation of satiety expectations. Participants in the first study rated the sensory characteristics of 16 fruit yogurt drinks of increasing viscosity. In study two, a new set of participants evaluated eight versions of the fruit yogurt drink, which varied in thick texture, creamy flavour and energy content, for sensory and hedonic characteristics and satiety expectations.

Results: In study one, participants were able to perceive small changes in drink viscosity that were strongly related to the actual viscosity of the drinks. In study two, the thick versions of the drink were expected to be more filling and have a greater expected satiety value, independent of the drink's actual energy content. A creamy flavour enhanced the extent to which the drink was expected to be filling, but did not affect its expected satiety.

Conclusions: These results indicate that subtle manipulations of texture and creamy flavour can increase expectations that a fruit yogurt drink will be filling and suppress hunger, irrespective of the drink's energy content. A thicker texture enhanced expectations of satiety to a greater extent than a creamier flavour, and may be one way to improve the anticipated satiating value of energy-containing beverages.

Keywords: Beverage, Creamy flavour, Satiety expectations, Sensory characteristics, Viscosity

Background

In the UK, beverages account for approximately 18% of an adult's daily energy intake [1] and evidence that energy-yielding beverages have a weak satiety value suggests that the 'fluid calories' in our diet could be a quiet contributor to obesity and weight gain [2]. A variety of studies indicate that energy consumed in liquid form fails to suppress subjective appetite [3,4] or reduce subsequent food intake [5-7] to the same extent as equi-caloric solid food. However, other studies have reported no relationship between food form and its satiety value [8,9], although a general criticism of studies in this field is that they often compare dissimilar foods (for example, calorie-matched cola against cookies) across

a range of food contexts (for example, a beverage or a snack), and do not quantify differences in the cognitive and sensory evaluations of these foods [10,11]. Therefore, it is important to consider what it is about these features of energy-yielding liquids that limit their satiety value.

Because of their fluid nature, beverages require less oral processing time than semi-solid and solid caloric equivalents and as a result they are consumed fairly quickly, minimising oro-sensory exposure [12]. Although increasing oral processing time may not necessarily lead to a reduction in the amount of a food that is consumed [13], oro-sensory exposure is important for the development of satiety [14,15]: the thought, sight, smell and taste of food trigger a cascade of anticipatory salivary and gastrointestinal responses, which improve the efficiency of nutrient processing and enhance the experience of satiety [16-19].

* Correspondence: k.mccrickerd@sussex.ac.uk
[1]School of Psychology, Pevensey Building, University of Sussex, Brighton BN1 9QH, UK
Full list of author information is available at the end of the article

Oro-sensory exposure to food is thought to trigger anticipatory responses because animals, including humans, learn to associate the sensory characteristics of a food with its caloric value post-consumption [20-23], and these associations are likely to influence explicit expectations about the effect a food will have on appetite [24,25], including how filling a food is likely to be (expected satiation) and the extent to which it will stave off hunger until the next meal (expected satiety): such expectations have been shown to influence appetitive satisfaction and portion size selection [26-28] and seem to be more strongly influenced by certain sensory characteristics. For example, a food is expected to be more filling when it is perceived to be heavier [29] or thicker in texture [30]. One explanation for the reported weak satiety value of beverages is that because of their fluid texture they are not expected to have much nutritional value [2,11].

Studies indicate that 'thick' drinks suppress hunger to a greater extent than equi-caloric flavour matched 'thin' versions [31,32] and recent research suggests that the sensory characteristics of a beverage interact with its post-ingestive effects to influence satiety. Yeomans and Chambers [33] reported that a high-energy liquid pre-load suppressed intake at a later meal to a greater extent than a low-energy equivalent, but only when the beverage had a thick texture and a creamy flavour. Furthermore, when participants consumed the low-energy version with thick and creamy sensory characteristics they ate more at the test meal than after the low-energy version without the enhanced sensory context. The researchers argue that the thick and creamy sensory characteristics predicted the delivery of nutrients, generating expectations that these drinks would be filling, which acted to enhance the experience of satiety when energy had been consumed. Thus, when the sensory characteristics predicted nutrients that were not delivered (as with the low-energy version of the thick and creamy drink) the mismatch between the actual and expected nutrient delivery tended to result in rebound hunger.

According to the findings of Yeomans and Chambers [33], designing a high-energy drink to taste thick and creamy could be one way to increase its satiating capacity, but their results also suggest that designing a low-energy drink to taste thick and creamy might actually increase subsequent appetite. Presumably, this is because a drink that tastes thick and creamy will increase expectations of satiety, regardless of its actual energy content, which would only be determined post-consumption. However, the extent to which the sensory characteristics of a drink influence expectations of satiety is not clear, and it is important to consider this if these expectations interact with the energy content of a drink post-consumption.

To characterise the influence of sensory cues on such expectations, we investigated the role of satiety-relevant texture and flavour cues in the generation of satiety expectations in high- and low-energy drinks. In study one, we assessed the extent to which participants were sensitive to small changes in drink texture and how sensory perceptions relate to the actual viscosity of a drink: it is important to clarify the scale of textural manipulations and how they actually translate to physical differences within a liquid product, in order to make it easier to compare textural differences of drinks across studies. In study two, we examined whether small variations in the thick texture and creamy flavour influence expectations of satiety, irrespective of the drink's actual energy content. We assessed the role of texture and flavour as independent sensory cues and together in a combined sensory context (thick and creamy) to see how the two interact.

Results: study one

Participants who were not sensory panellists tasted and rated 16 fruit yogurt drinks of varying thickness, manipulated by the addition of small quantities of tara gum (0.0 to 0.47 g/100g of the drink, increasing in 0.03 g increments across the 16 drinks). Rheological measurements were taken and participants rated how thick, creamy, fruity, sticky, sweet and sour each sample was (0 = not at all, 100 = extremely) on two non-consecutive days. Perceived thickness was related to viscosity at a shear rate of ≈50 reciprocal seconds (1/s).

Viscosity
Viscosity significantly increased with the addition of tara gum across the 16 samples of fruit yogurt drink ($F_{(15, 176)} = 1552.17$, $P < 0.001$; linear contrast $P < 0.001$), see Figure 1.

Sensory evaluations of the test drinks
The mean sensory ratings are presented in Table 1. Perceived thickness ($F_{(6.5, 135.6)} = 65.38$, $P < 0.001$), creaminess ($F_{(4.8, 90.4)} = 20.53$, $P < 0.001$) and stickiness ($F_{(5.5, 104.1)} = 11.96$, $P < 0.001$) increased with the amount of tara gum in each sample (linear contrast $P < 0.001$ for all) but rated sweetness, sourness and fruitiness did not differ across samples ($P > 0.05$ for all). There was no effect of gender or test day on any of the ratings (all $P > 0.05$) except for sourness, where there was a small but significant gender × day × sensory interaction ($F_{(8.0, 168.7)} = 2.02$, $P = 0.047$): some of the 16 samples were rated as slightly more or less sour depending on the gender of the participant and the day the rating was made, although there was no clear pattern to this interaction, which is likely to be a spurious finding, given the large number of potential interactions.

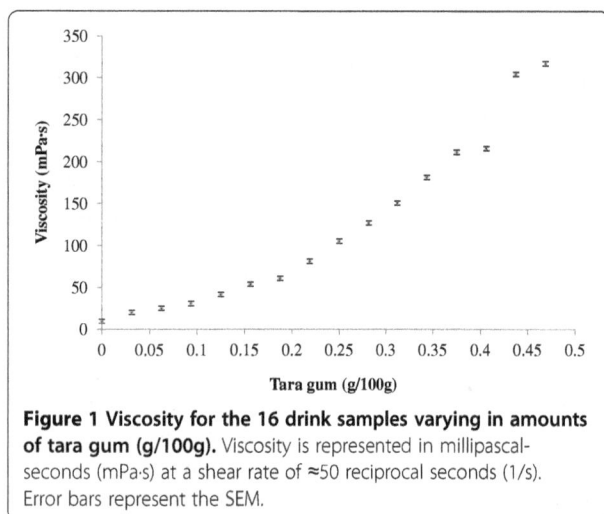

Figure 1 Viscosity for the 16 drink samples varying in amounts of tara gum (g/100g). Viscosity is represented in millipascal-seconds (mPa·s) at a shear rate of ≈50 reciprocal seconds (1/s). Error bars represent the SEM.

Relating sensory characteristics to viscosity

Table 1 details the correlations between the viscosity of each sample and their perceived sensory characteristics. Perceived thickness was strongly related to viscosity: as the viscosity of each sample increased, so did perceived thickness. Creaminess and stickiness ratings also increased with viscosity. There was a small but significant positive relationship between rated fruitiness and viscosity, indicating that there was a small increase in perceived fruitiness in the thicker samples, which was not picked up in the ANOVA (analysis of variance) on the fruitiness ratings. There was no relationship between the viscosity of the sample and perceived sweetness or sourness.

Summary

The results from study one indicate that participants, who are not trained sensory panellists, were able to perceive subtle differences in drink texture, and these differences were closely related to actual viscosity. This is in line with previous evidence that suggests that viscosity at a shear rate of 50 1/s relates to perceived thickness [34,35]. Small incremental increases in tara gum across the 16 drink samples produced measurable increases in viscosity (10 to 317 mPa·s, ranging from a fluid juice texture to a thicker yogurt drink texture, all consumed through a regular straw) and the participants perceived these subtle changes, although probably not at the level of every incremental increase. This sensitivity to subtle differences in viscosity is not surprising because texture is likely to be one sensory characteristic of food that reliably predicts the presence of nutrients, such as fat [36].

Results: study two

In study two, new participants, who were not trained sensory panellists, evaluated the sensory and hedonic characteristics of eight versions of a fruit yogurt drink, which varied in thickness (thin or thick), creamy flavour (low-creamy or creamy) and energy content (high- or low-energy). The participants also rated how filling they expected each drink to be (0 = not at all, 100 = extremely) and its expected satiety. In the expected satiety measure, participants indicated the extent to which they expected each drink to suppress hunger until the next meal by selecting a portion of pasta and sauce that they thought would have the same effect on their hunger. Selecting a larger portion of pasta and sauce (kcal) indicated that the drink was expected to be more satiating.

Filling ratings

The ANOVA revealed a significant effect of both thickness (F (1, 21) = 98.98, $P < 0.001$) and creamy flavour (F (1, 21) = 20.89, $P < 0.001$) on the extent to which the drinks were expected to be filling, independent of the drinks' energy content (interactions with energy all $P > 0.05$), see Figure 2. Averaged across energy versions,

Table 1 Sensory ratings for each fruit yogurt drink sample used in study one

	Fruit yogurt drink sample (tara gum g/100g)																
	0	0.03	0.06	0.09	0.13	0.16	0.19	0.22	0.25	0.28	0.31	0.34	0.38	0.41	0.44	0.47	Pearson's r
Thickness	10.5	15.9	23.7	26.5	30.1	30.9	41.0	40.6	56.7	54.8	64.3	64.0	73.4	81.5	83.1	84.9	0.92 [a]
	± 1.9	± 3.0	± 3.7	± 4.3	± 3.3	± 4.3	± 4.1	± 4.0	± 4.0	± 3.4	± 3.7	± 3.8	± 3.2	± 2.3	± 2.0	± 1.9	
Creaminess	24.8	31.5	41.1	41.9	34.7	47.7	51.9	50.6	55.6	60.3	62.9	59.8	70.8	70.1	72.9	76.9	0.92 [a]
	± 3.1	± 4.5	± 4.1	± 4.2	± 2.9	± 3.2	± 3.5	± 3.4	± 3.7	± 2.8	± 3.2	± 2.9	± 3.2	± 3.2	± 3.4	± 3.0	
Stickiness	21.1	25.2	24.8	26.7	31.3	29.7	34.8	33.6	35.7	40.6	47.9	44.6	51.9	53.7	51.1	53.8	0.95 [a]
	± 3.5	± 3.8	± 3.4	± 4.1	± 3.6	± 3.8	± 3.3	± 4.2	± 3.9	± 4.9	± 3.7	± 4.6	± 4.6	± 4.9	± 4.9	± 5.4	
Sweetness	58.7	57.0	58.0	61.7	59.7	59.1	61.2	59.1	65.2	62.4	59.8	61.0	61.3	61.1	58.6	62.2	0.29 *non*
	± 3.9	± 4.0	± 4.0	± 4.2	± 4.3	± 3.6	± 3.8	± 3.0	± 2.5	± 3.3	± 3.7	± 2.7	± 3.6	± 3.8	± 3.6	± 3.6	*significant*
Sourness	35.0	39.0	42.2	36.9	40.7	36.1	41.4	44.3	36.9	38.4	36.7	38.4	47.8	35.6	36.8	42.4	0.13 *non*
	± 4.7	± 5.1	± 4.3	± 4.8	± 4.4	± 4.3	± 4.6	± 4.7	± 4.3	± 4.2	± 4.8	± 4.3	± 5.1	± 5.3	± 4.3	± 5.0	*significant*
Fruitiness	55.4	60.4	57.8	59.3	63.7	63.0	60.0	60.4	63.6	66.8	66.2	61.8	64.2	64.0	64.6	61.2	0.50 [b]
	± 3.8	± 3.3	± 4.4	± 3.6	± 2.6	± 3.1	± 3.7	± 3.5	± 2.8	± 3.1	± 2.3	± 2.7	± 3.3	± 2.7	± 3.7	± 4.4	

Numbers represent the mean visual analogue scale (VAS) rating (0 = not at all, 100 = extremely) and associated SEM for each of the sensory evaluations across the 16 fruit yogurt drinks varying in the amount of tara gum /100g. Pearson's r shows the relationship between each sensory characteristic and the drink's measured viscosity. [a] Correlation coefficient is significant at $P < 0.001$. [b] Correlation coefficient is significant at $P < 0.05$.

the thick drinks (M = 64.6 ± 2.2) were expected to be more filling than the thin drinks (M = 41.8 ± 2.2) and the creamy versions of the drink (M = 57.0 ± 2.3) were expected to be more filling than the low-creamy versions (M = 49.4 ± 1.8). There was no thick × creamy interaction (F (1,21) = 0.62, P = 0.44): increasing drink thickness increased the filling rating, which was enhanced by the addition of creamy flavour similarly across the thick and thin versions (see Figure 2). There was no overall effect of the drink's energy content on ratings of how filling the drink was expected to be (F (1, 21) = 3.16, P = 0.09).

Expected satiety

There was also a significant effect of drink thickness on expected satiety judgements (F (1, 21) = 63.27, P < 0.001): the thick drinks had a greater expected satiety than the thin drinks, see Figure 3. However, the creamy versions of the drinks were not expected to suppress hunger any more than their low-creamy counterparts (F (1, 21) = 0.60, P = 0.45) and there was no thick × creamy interaction (F (1, 21) = 2.60, P = 0.12). There was no main effect of the drinks' energy content on expected satiety (F (1, 21) = 0.52, P = 0.48) but the analysis did reveal a significant thick × energy interaction (F (1, 21) = 12.73, P = 0.002). The interaction suggested that the high-energy thin drinks (M = 127.7 ± 16.4) had a lower expected satiety than the low-energy thin drinks (M = 148.4 ± 16.6), whereas the high-energy thick drinks (M = 269.2 ± 33.9) and low-energy thick drinks (M = 266.2 ± 33.9) were similarly expected to be the most satiating. However, Bonferroni adjusted comparisons revealed no significant difference in expected satiety between the high- and low-energy thin drinks (P = 0.42) or the high- and low-energy thick drinks

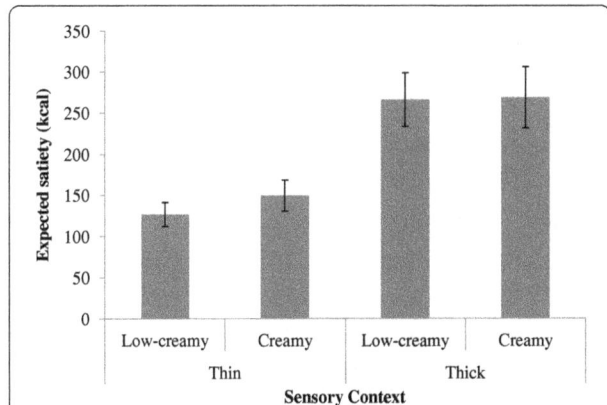

Figure 3 Mean portion of pasta and tomato sauce selected in the expected satiety task (kcal ± SEM) in study two, collapsed across drink energy content. The thick drinks had a significantly larger expected satiety than the thin drinks (P < 0.001) and the addition of creamy flavour did not increase this expectation, as the expected satiety was similar for the low-creamy and creamy versions (P > 0.05).

(P = 0.99), only a difference between the expected satiety value based on the drinks' thickness (all P < 0.001).

Relating the filling rating to expected satiety

We anticipated that the judgements measuring the extent to which the drinks were expected to be filling (VAS ratings) and the extent to which the drinks were expected to suppress hunger (expected satiety) would be related. Unexpectedly, Pearson's correlation indicated that for each of the eight drinks varying in thickness, creamy flavour and energy content, there was little relationship between the expectation that it would be filling and its expected satiety. Across the eight drinks, the two expectations were only significantly related for two of the drinks (for all others P > 0.05). For the high-energy thick and creamy drink, the more filling it was expected to be, the greater its expected satiety (r = 0.53, P = 0.011). However, the more filling the low-energy thick and low-creamy drink was expected to be, the lower its expected satiety (r = –0.57, P = 0.005). This suggests little relationship between the two expectations.

Sensory and hedonic evaluations of the drinks

ANOVAs revealed that the drinks differed on several sensory attributes (see Table 2). The thick drinks were rated as more thick (F (1, 21) = 170.79, P < 0.001), creamy (F (1, 21) = 52.48, P < 0.001) and sticky (F (1, 21) = 40.96, P < 0.001) than the thin drinks, and less fruity (F (1, 21) = 18.19, P < 0.001). Drink texture did not affect sweetness ratings. The creamy drinks were rated as creamier (F (1, 21) = 17.74, P > 0.001), thicker (F (1, 21) = 13.47, P = 0.001) and slightly sweeter (F (1, 21) = 6.40, P = 0.02) than the low-creamy drinks. The addition of creamy flavour

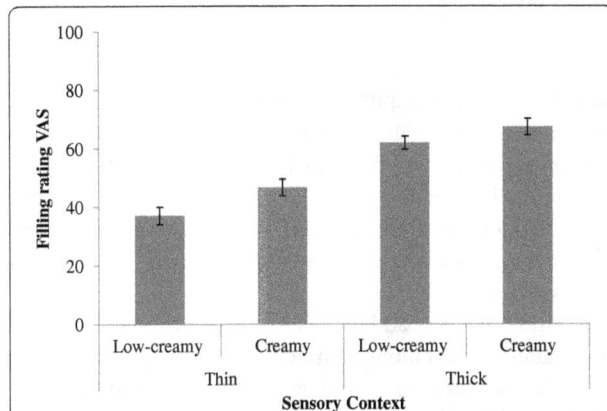

Figure 2 Filling VAS ratings (0 = not at all, 100 = extremely) ± SEM for the drinks used in study two in the four sensory contexts, collapsed across drink energy content. The thick drinks were expected to be significantly more filling than the thin drinks (P < 0.001). The addition of creamy flavour increased this expectation, as the creamy drinks were rated as significantly more filling than the low-creamy versions (P < 0.001).

Table 2 Sensory evaluations of drinks used in study two across each sensory context

	Thin		Thick	
	Low-creamy	Creamy	Low-creamy	Creamy
Creamy	35. ± 3.8	51.4 ± 3.5	66.9 ± 2.9	71.9 ± 2.3
Fruity	63.2 ± 3.5	66.4 ± 3.4	49.8 ± 4.3	53.8 ± 3.6
Pleasant	57.6 ± 3.6	63.0 ± 3.2	55.8 ± 4.0	57.4 ± 5.9
Sticky	33.4 ± 4.2	34.4 ± 3.4	52.2 ± 3.8	55.8 ± 4.0
Sweet	58.3 ± 3.1	65.3 ± 2.5	54.5 ± 3.2	61.4 ± 2.8
Thick	26.0 ± 3.0	37.3 ± 2.3	68.0 ± 3.0	73.2 ± 3.4

Evaluations are collapsed across high-energy and low-energy versions for the eight drinks and represent the mean VAS rating (0 = not at all, 100 = extremely) and associated ± SEM for the drinks in the four sensory contexts, varying in thickness and creamy flavour.

did not affect the perceived fruitiness or stickiness of the drinks. All the drinks were rated as similarly pleasant, regardless of thick texture, creamy flavour or energy content (all main effects and interactions $P > 0.05$). There was no thick × creamy interaction for any of the sensory characteristics (all $P > 0.05$). Overall, there was no main effect of drink energy on thick, creamy, sticky, fruity, sweet and pleasantness ratings for any of the drinks (all $P > 0.05$). However, there was a small but significant thick × energy interaction for the creamy ratings ($F (1, 21) = 4.77$, $P = 0.04$). Bonferroni adjusted comparisons revealed that the high-energy thick drinks ($M = 66.6 ± 2.8$) were rated as similarly creamy to the low-energy thick drinks ($M = 72.4: P = 0.35$), but the high-energy thin drinks ($M = 35.0 ± 3.7$) were rated as less creamy than the low-energy thin drinks ($M = 50.8 ± 3.1: P = 0.003$).

Hunger, fullness and thirst pre- and post-test

Rated hunger decreased ($F (1, 21) = 13.91$, $P = 0.001$) and rated fullness increased ($F (1, 21) = 110.70$, $P < 0.001$) from pre- to post-test. There was no difference in thirst from the beginning to the end of the session. Importantly, pre-test hunger ratings were not related to the filling ratings and expected satiety judgements across the eight drinks (all $P > 0.05$).

Summary

The results from study two indicate that sensory characteristics can influence satiety expectations of a drink, independent of its actual energy content. Both creamy flavour and thick texture enhanced the expectation that a drink would be filling (the anticipated satiation that is expected to be experienced straight after consumption), but thick texture influenced this expectation more so than creamy flavour. Thick texture alone influenced the expectation that the yogurt drink would suppress hunger over time, as the thick drinks had a greater expected satiety than the thin versions and there was no difference in expected satiety between the low-creamy and creamy

drinks. Interestingly, for each drink, the participants' expectations that it would be filling and its expected satiety value were generally not related, suggesting that participants used different strategies to make these two judgements.

Discussion

The results of these studies suggest that consumers are sensitive to subtle changes in the sensory characteristics of a drink and that thick texture and creamy flavour can be manipulated to enhance satiety expectations, but that their contributions are not equal. Our findings also indicate that beverages can differ in the extent to which they are expected to be satiating, regardless of the actual calories that they contain. This is important because, at least in the short term, manipulating the expected and not the actual calories of a product can influence subjective appetite [26], subsequent ghrelin response [37] and intake at a later meal [38,39]. Although this study did not measure the actual satiating value of the drinks used, Yeomans and Chambers [33] found that thick and creamy sensory characteristics enhance the satiety value of a drink, but only when those characteristics correctly predicted the delivery of nutrients. Taken together, this suggests that both high- and low-energy drinks that are made to taste thicker will be expected to be more satiating, but this expectation may have different effects on satiety, depending on the actual energy content that is delivered post-consumption.

So why then should thickness be a good predictor of satiety in a beverage? For one, human adults have already had a wealth of experience with foods across their lifetime and often liquids that are more viscous do have more calories (such as honey compared to water). For example, variation in the energy density of breast milk has been shown to correlate with viscosity [40] and this variability might lead to learnt associations between perceived thickness and satiety [41]. The natural flavour of milk would be expected to be part of this association but one possibility is that increased oral exposure experienced with more viscous liquids makes it easier to associate the sensory characteristics of a thicker beverage, such as flavour, with its postingestive consequences [42,43]; creamy flavour alone is not likely to increase oral exposure, which may make it a less effective cue for learning when it is independent of an increase in viscosity.

In study two, the addition of creamy flavour did not impact satiety expectations as much as a thick texture, so it is possible that creamy flavour is not a good predictor of a food's caloric value. Reduced-fat and 'diet' food products, such as low-fat yogurts, are often produced to have the same 'creamy' flavour as the full-calorie versions to increase satisfaction and palatability. An inconsistent relationship between the sensory characteristics of a food and its energetic value may weaken the associations formed between them [44-46]. We could have taken a measure of participants' reported previous experience with these types of diet food products to see if this affected the ability of the creamy flavour cue to generate satiety expectations. However, our results consistently indicated that, as the viscosity of a yogurt drink increased, it was perceived to be thicker but also creamier and stickier. It seems likely that rating the drinks as 'creamy' is simply not a sufficiently sensitive measure for the general consumer, and is confounded by the complex sensory profile of creamy dairy products, which is based on a combination of flavour and texture attributes [47]. Furthermore, the creamy drinks were not only rated as creamier than the low-creamy drinks, but also thicker, so we cannot rule out the possibility that the creamy drinks were instead expected to be more filling, based on their enhanced perceived thickness.

The complexity of the creamy sensory characteristic may have contributed to any discrepancies between the high- and low-energy versions of the drinks. Energy content was not predicted to influence satiety expectations, as the high- and low-energy versions of the drinks were designed to be matched in terms of perceived flavour and texture and the drink samples were only tasted and not consumed in full portions. However, there was evidence in the expected satiety measure that the low-energy thin drinks were expected to be more satiating than the high-energy thin drinks. This difference maps onto the finding that the low-energy thin drinks were also rated as creamier than the high-energy thin drinks, possibly because, overall, the low-energy drinks were slightly more viscous and contained slightly more fromage frais than the high-energy drinks (see study two 'test drinks' in the method section for viscosities and ingredients), and this difference may have been more noticeable in the thin versions. This highlights the importance of matching high- and low-energy versions of test food for characteristics, such as thickness and creaminess, in satiety studies.

Within a liquid context, thicker drinks have been shown to suppress hunger to a greater extent than a calorie-matched thin version [31,32] and this could be because the thicker drinks were expected to be more satiating. However, an alternative explanation for this could be that the thickener used to manipulate viscosity had a post-ingestive effect. If this is the case, the effect of increased satiety expectations generated by these texture cues may be redundant. Water-soluble polysaccharides used to increase liquid viscosity, such as tara gum and guar gum, also increase its dietary fibre content and the addition of a small quantity of fibre (0.82 to 1.5 g per 100g of a drink) has been shown to increase the short-term satiety value of a beverage, with delayed gastric emptying implicated as a possible mechanism [48,49]. However, what was not considered in these studies is that the addition of fibre also increases oral viscosity; moreover, the quantities of fibre used were larger than those used to manipulate thickness in the

current study. One possibility is that expectations of satiety generated by a thicker liquid actually contribute to the increased satiety value of these fibre-enhanced beverages. Expectations generated by the oral viscosity and anticipated gastric viscosity of a solid and liquid food have recently been shown to influence subjective appetite, intake and gastrointestinal function [10], highlighting the potential of the satiety-relevant expectations in influencing the post-ingestive development of satiety. It is unlikely that small differences in the viscosity of a beverage would persist post-ingestion owing to the influence of gastric dilution [49]; instead beliefs about the post-ingestive effects of the beverage may important.

An unexpected outcome of study two was the lack of relationship between the expectation that a drink will be filling and its expected satiety. There is evidence to suggest that people differ in the sensory information that they use to guide food intake [22] and one possibility is that our participants were using different strategies to make these two judgements. However, the way in which individuals differentially use flavour and texture cues to generate satiety expectations is not clear. In this study, it appears that both textural and flavour cues contributed to the extent to which the drinks were expected to be filling, whereas only drink thickness influenced expected satiety. In our measure of expected satiety, participants compared the anticipated satiating effect of a fruit yogurt drink to that of pasta and tomato sauce, whereas the expectation that the drink will be filling was measured on a rating scale. One possibility is that when the participants imagined the expected satiety of each drink sample in comparison to pasta and sauce, texture was a more relevant cue for satiety. Creamy flavour may have been overlooked because it is not a relevant sensory characteristic of pasta and sauce. Furthermore, participants may have found it harder to imagine a suppression of hunger in the expected satiety tasks than an increase in fullness in the rating measure. In future, it would be useful to measure the method of adjustment comparisons and VAS ratings for both types of expectations generated by sensory cues, to see how they compare.

Finally, it is important to note that this research had a repeated measures design and all the participants tasted each of the drinks during the session. It is possible that the influence of the drinks' sensory characteristics on satiety expectations was more pronounced, due to contrast effects, and from this study it is not clear how these subtle sensory differences would influence expectations in a single drink product day to day when not tasted alongside a similar product.

Conclusion

Overall, this research indicates that people are sensitive to subtle changes in the sensory quality of a drink and that these characteristics can increase the expectation that a drink will be filling (anticipated satiation) and suppress hunger over time (expected satiety). It appears that thick texture, rather than creamy flavour, had the biggest influence on satiety expectations and this was independent of the drink's actual energy content. Therefore, enhancing the texture of high-energy drinks to be more satiety relevant may be one way to increase their weak satiating capacity. These findings also highlight the importance of matching sensory characteristics, such as texture, in studies that manipulate the energy density of foods or the sensory context of energy-matched products.

Method: study one

Participants

Twenty-four (12 male) participants were recruited from a volunteer database of staff and students at the University of Sussex. Participants were aged between 19 and 26 years (mean = 21.0, SD = 2.0) and were non-obese (mean = 23.3 kg/m^2, SD = 2.8) with a mean dietary restraint score of 6.3 (SD = 3.6) for females and 4.2 (SD = 2.7) for males (measured using the three-factor eating questionnaire (TFEQ) [50]). Male and female participants did not differ in age, restraint or body mass index (BMI). They were selected to be healthy non-smokers, not currently dieting or taking prescription medication, with no eating disorders and without allergies or aversions to any of the test foods. The research was approved by the University of Sussex, Life Science Research Ethics Board. All participants gave consent to take part in a study entitled 'Investigating the interaction between mood and taste' and received £10 payment on completion.

Fruit yogurt drinks

All test drinks were designed and prepared in the Ingestive Behaviour Unit at the University of Sussex and consisted of two training drinks and 16 test drinks made from the same low-energy fruit yogurt base (see Table 3). Thickness was manipulated with the addition of tara gum (Kaly's Gastronomie, France), a naturally occurring non-ionic polysaccharide commonly used commercially as a thickening agent and stabiliser. The amount of tara gum ranged from 0.0 to 0.47 g/100g portion of the drink base, increasing in 0.03 g increments across the 16 drinks. The training drinks were an example of a 'thin' drink (water) and a 'thick' drink (the fruit yogurt drink with 0.63 g/100g tara gum added). All samples were kept at 1 to 5°C and used within 4 days of preparation.

Measures

Viscosity

Rheological measurements were taken at the University of Birmingham, Department of Chemical Engineering, at

5°C on a Bohlin Rotational Rheometer (Malvern Instruments Ltd) using parallel-plate geometry (60 mm diameter) and a gap size of 1.0 mm. Flow behaviour was measured at shear rates from 0.001 to 800 1/s and back down in reverse sequence for the same duration, with three repeats using a fresh sample each time. Tara gum solutions typically show non-Newtonian shear thinning behaviour [51], which means that their viscosity is not constant but is dependent on rate of flow (the shear rate) during measurement. For this reason, viscosity reported in the results section is an average of the data collected at a shear rate of 52.6 1/s (referred to as ≈50 1/s), which was the actual shear rate the rheometer achieved when aiming for 50 1/s, which is thought to best represent in-mouth viscosity [34,35]. Although shear rates of above 1000 1/s have been associated with in-mouth viscosity [52], the highest shear rate that could be obtained for the samples was 800 1/s, as all the samples were relatively thin and liable to run off the rheometer plate. Parallel-plate geometry was used to spread the force created under shear over a wider area, allowing a larger range of shear rates to be achieved accurately.

Sensory ratings

Sensory evaluations of the 16 samples were collected in the form of VAS ratings using the Sussex Ingestion Pattern Monitor (SIPM) [53] running on a Dell PC using the Windows XP professional operating system. Participants were asked 'How <target> is sample X?' with the targets 'thick', 'sweet', 'sour', 'sticky', 'fruity' and 'creamy'. Participants were instructed to indicate the extent that each sample was <target> by dragging a marker along a 100 mm line. The scale was always anchored with the words 'Not at all < target>'

Table 3 Ingredients and basic nutritional composition of the high- and low-energy fruit yogurt drink base

Ingredients per 100 g portion	Low-energy [a]		High-energy [a]	
	weight (g)	kcal	weight (g)	kcal
Peach and passion fruit juice [b]	31.3	14.4	31.3	14.4
Peach squash [c]	10.9	1.2	10.9	1.2
0.1% fat Fromage frais [b]	17.2	8.6	9.4	4.7
Water	40.6	0	31.2	0
Maltodextrin [d]	0	0	17.2	65.3
Aspartame [e]	0.03	0	0	0
Yellow colour [f]	3 drops	0	0	0
Red colour [f]	1 drop	0	0	0
Total	**100g**	**24.2**	**100g**	**85.6**

[a] Low-energy drinks were used in study one and both high- and low-energy drinks were used in study two.
[b] J Sainsbury's plc, London, UK.
[c] Robinsons, Britvic, UK.
[d] Cargill, UK.
[e] Aspartame Powder, Ajinomoto Sweeteners Europe.
[f] Silverspoon, British Sugar, UK.

(0) and 'Extremely <target>' (100). The presentation of each question was randomised.

Procedure

Test sessions were scheduled between 10.30 am and 12.00 noon or between 2.30 pm and 4.00 pm, Monday to Friday. To minimise differences in hunger, participants were instructed not to consume any food or drink (excluding water) for two hours before they were due in the laboratory. Participants then underwent a brief training task to introduce them to the idea of rating a drink's 'thickness' and provide a reference standard. In the training task, participants were presented with an example of the thickest and the thinnest sample they would taste throughout the session. Participants were instructed to take a small mouthful of a sample through a straw, to hold the sample in their mouth while they counted to three and then swallow. Some research suggests that samples should be swallowed immediately, to reduce dilution by saliva and temperature equilibration, which can affect rheological properties of the food [54]. However, this technique significantly reduces the sensory exposure and oro-sensory sensitivity of the participants [55]. By allowing participants three seconds of oral exposure, this allowed some degree of sensitivity whilst maintaining a level of standardisation across all samples and participants. After swallowing, participants rated the thickness of the sample and were then prompted to take a sip of water. All participants rated the thickest sample first.

Following the training, participants were presented with a tray of 16 samples of the yogurt fruit drink and were required to taste each sample, holding the drink in the mouth for three seconds before swallowing. The samples were presented in 25 g portions in a small clear glass with a straw and labelled A to P. After each taste, participants completed a series of VAS ratings, assessing the sensory characteristics of each sample. Participants were prompted to take a sip of water before moving on to the next sample. The order of presentation of the samples was randomised across all participants and sessions.

Due to the large number of samples to be tasted, participants completed the tasting session twice on two non-consecutive days, to check that their sensory evaluations were consistent. Each test session lasted 30 minutes and participants completed the two sessions at a similar time of day. After the final session, the participant's age, weight and height were recorded. Finally, participants completed questions pertaining to the purpose of the study, were debriefed, thanked and paid.

Data analysis

The main outcome measures were the actual viscosity of the samples thickened with tara gum measured using rheometry and the perceived sensory characteristics evaluated by volunteers. A one-factor independent sample ANOVA assessed the effect of tara gum on viscosity across the 16 test drinks.

A three-way mixed ANOVA was conducted for each sensory evaluation to assess the effect of added tara gum (16 levels) on the sensory judgements while controlling for test day (1 or 2) and gender (male or female participants). Where sphericity was not assumed Greenhouse-Geisser ($\varepsilon < 0.75$) or Huynd-Feldt ($\varepsilon > 0.75$) corrected degrees of freedom and P values are presented. Means and SEM are presented throughout. The relationship between viscosity at ≈ 50 1/s and each of the sensory evaluations were investigated using Pearson's correlations.

Method: study two
Participants

Twenty-five participants (9 male) were staff and students at the University of Sussex, recruited from the same volunteer database as study one and conformed to the same selection criteria but had not taken part in study one. Participants were aged 19 to 26 ($M = 21.0$, $SD = 2.7$), and were non-obese (mean BMI = 22.8 kg/m^2, SD = 3.3) with an average TFEQ restraint score of 5.7 (SD = 4.9) for males and 6.4 (SD = 3.9) for females; these characteristics were similar between males and females. The study was approved by the University of Sussex, Life Science Research Ethics Board. All participants gave written consent to take part in a study entitled 'Investigating the interaction between mood and taste' and received £6 payment on completion.

Test drinks

The fruit yogurt drinks were designed with four satiety-relevant sensory contexts varying in thickness (thin or thick) and creamy flavour (low-creamy or creamy) with high-energy (HE) and low-energy (LE) versions for each. Table 3 lists the ingredients and basic nutritional composition of the low-energy and high-energy fruit yogurt drink bases. Creamy flavour was enhanced by the addition of vanilla extract (Nielsen-Massey, NL: 19 drops/100g) and milk caramel flavouring (Synrise, DE: 0.16g/100g) and thickness was increased by manipulating the amount of tara gum (g/100g) in each drink (low-creamy/thin LE: 0.09g, low-creamy/thin HE: 0g, creamy/thin LE: 0.09g, creamy/thin HE: 0g, low-creamy/thick LE: 0.38g, low-creamy/thick HE: 0.31g, creamy/thick LE 0.38g, creamy/thick HE: 0.31g); more tara gum was added to the LE versions of the drinks, to account for the small increase in thickness caused by the addition of maltodextrin to the HE versions and rheological measurements were relatively well matched across high- and low-energy drinks in the thin (LE = 20.8 mPa·s, HE = 30.8 mPa·s) and thick (LE = 221.5 mPa·s, HE = 184.0 mPa·s) contexts. The thick drinks were similar in viscosity to the sample containing 0.34 to 0.40g/100g tara gum in study one, and the thin drinks were similar in viscosity to the sample containing 0.03to 0.09g/100g in study one. Colour was matched between all the drink samples by the addition of small quantities of natural food colouring (see Table 3).

Measures
Hunger, fullness and thirst
VAS ratings of appetite were collected using SIPM and had the same format as the sensory ratings in study one. Participants rated how 'hungry', 'full' and 'thirsty' they were from not at all (0) to extremely (100) and these ratings were embedded amongst other distracter 'mood' questions: calm, happy, clearheaded, anxious, nauseous, headachy, tired, energetic, and alert. Only the appetite questions were analysed and all questions were presented in a randomised order.

Sensory evaluations and filling rating
Participants also made VAS ratings of how 'sweet', 'thick', 'creamy', 'pleasant', 'sticky' and 'fruity' the drinks were, as well as rating the extent to which each sample was expected to be filling. All ratings were from 'not at all' (0) to 'extremely' (100) and were presented in a random order.

Expected satiety
The measurement of expected satiety was based on a computer-based methodology developed by Brunstrom et al. [25]. The program was written in Visual Basic software displayed on a Dell laptop computer running Windows 7, and all testing was in a windowless air-conditioned testing cubicle. Participants were presented with the set of eight drink samples and a 320 g portion of the drink base in a clear plastic bottle with a fastened lid representing a standard drink serving. Participants were prompted by on screen instructions to 'Take a sip of sample X' using the straw provided. Then, they were presented with an image of pasta and tomato sauce and participants were instructed, 'Imagine you are going to consume the whole bottle of sample X for lunch. How much pasta would you need to eat to match the effect of sample X on your hunger?' Participants used the left and right arrow keys on the keyboard to move through images and increase or decrease the amount of pasta and sauce displayed. There were 101 images of pasta and sauce in total ('Egg penne pasta': J Sainsbury plc, UK; 'Sundried stir-in tomato sauce': Mars Food, UK) ranging from 10 kcal in image 0 to 1000 kcal in image 100. Portion sizes increased across images in logarithmic steps, such that images 0, 20, 40, 60, 80, 100 showed 10 kcal, 25.1 kcal, 63.1 kcal, 158.5 kcal, 398.1 kcal, and 1000 kcal respectively. Participants selected enter when they had selected their required portion size. All images were taken by a high-resolution digital camera mounted

above a 255-mm diameter white plate and care was taken to maintain consistency of lighting and camera angle across each photograph. All participants confirmed that they had eaten pasta and tomato sauce before.

Procedure
Participants completed one test session that lasted approximately 45 minutes and was scheduled on a weekday between 10.30 am and 12.30 pm or 2.30 pm and 4.30pm. As in study one, participants were required to consume only water for 2 hours prior to attending the lab and they completed the session in an air-conditioned testing cubicle with a PC computer.

To begin, participants rated their subjective appetite disguised as a series of 'mood questions'. They were then presented with 25 g portions of the eight test drinks each in a small clear glass labelled A to H and were informed that they would taste each sample twice using the straws provided. Participants first tasted each sample to make the sensory VAS ratings and to rate how filling they expected it to be, and then tasted the samples for a second time to complete the expected satiety task. Half of the participants completed the two tasks in the reverse order and all were provided with water throughout. Once the tastings were finished, participants completed a final set of appetite ratings and were debriefed, thanked and received their compensatory payment.

Data analysis
Appetite ratings were taken before and after the tasks as a difference in subjective appetite prior to the test may have influenced task performance. A one-factor mixed ANOVA assessed the effect of time (pre- and post-test) on the three measures of appetite and a series of Pearson's correlations was used to assess the relationship between pre-test hunger, fullness and thirst to the anticipated fullness and expected satiety of the drinks.

A series of three-way mixed ANOVAs and Bonferroni adjusted comparisons contrasted the effect of drink thickness (thin or thick), creamy flavour (low-creamy or high-creamy) and energy context (high or low) on each of the expectations (anticipated fullness and expected satiety) and the sensory and hedonic ratings. The expected satiety scores represent the quantity (in kcal) of pasta and tomato sauce presented in the image selected by the participants. These data were log transformed to improve normality for the analysis. However, the descriptive data and mean values were presented in kcal to aid interpretation. It was predicted that the expectation that a drink would be filling would be strongly related to its expected satiety, and this was tested using a series of Pearson's correlations to assess the relationship between these two expectations across the eight test drinks.

Initially, these analyses also included task order (VAS ratings then expected satiety or expected satiety then VAS ratings) as a factor. However, as there was no significant effect or interactions with this factor it was removed from the final analysis. Twenty-five participants took part in the study but the data from three participants were removed as their expectation values (filling rating or expected satiety) were more than two standard deviations from the mean. Consequently, data from 22 participants were included, leaving 16 females and just 6 males in the final analysis. For this reason gender was not included as a factor, owing to an inadequate number of males. Means and SEM are presented throughout.

Abbreviations
ANOVA: analysis of variance; BMI: body mass index; HE: high energy; LE: low energy; SIPM: Sussex Ingestion Pattern Monitor; TFEQ: three-factor eating questionnaire; VAS: visual analogue scale.

Competing interests
The authors declare that they have no competing interests.

Authors' contributions
MY, LC and KMcC designed the study. KMcC prepared the study materials, collected and analysed the human data and viscometry, and drafted the manuscript. JB programmed the expected satiety task. All authors contributed to and approved the final manuscript.

Acknowledgements
This research was funded by the BBSRC DRINC initiative and conducted as part of a PhD studentship. The rheological measurements were conducted at the University Of Birmingham Department Of Chemical Engineering, with the help of Professor Ian Norton, Dr Tom Mills and Dr Jennifer Norton of The Microstructure Group. Dr Tom Mills helped to set up, run and interpret the viscosity measurements.

Author details
[1]School of Psychology, Pevensey Building, University of Sussex, Brighton BN1 9QH, UK. [2]Nutrition and Behaviour Unit, School of Experimental Psychology, University of Bristol, Priory Road, Bristol BS8 1TU, UK.

References
1. Ng SW, Ni Mhurchu C, Jebb SA, Popkin BM: **Pattern and trends of beverage consumption among children and adults in Great Britain, 1986-2009.** Br J Nutr 2012, 108(03):536-551.
2. Mattes RD: **Beverages and positive energy balance: the menace is the medium.** Int J Obes (Lond) 2006, 30:S60-S65.
3. Hulshof T, Degraaf C, Weststrate JA: **The effects of preloads varying in physical state and fat-content on satiety and energy-intake.** Appetite 1993, 21(3):273-286.
4. Leidy HJ, Apolzan JW, Mattes RD, Campbell WW: **Food form and portion size affect postprandial appetite sensations and hormonal responses in healthy, nonobese, older adults.** Obesity 2010, 18(2):293-299.
5. Mattes RD: **Dietary compensation by humans for supplemental energy provided as ethanol or carbohydrate in fluids.** Physiol Behav 1996, 59(1):179-187.
6. Mourao DM, Bressan J, Campbell WW, Mattes RD: **Effects of food form on appetite and energy intake in lean and obese young adults.** Int J Obes (Lond) 2007, 31(11):1688-1695.
7. Tournier A, Louis-Sylvestre J: **Effect of the physical state of a food on subsequent intake in human subjects.** Appetite 1991, 16(1):17-24.
8. Almiron-Roig E, Chen Y, Drewnowski A: **Liquid calories and the failure of satiety: how good is the evidence?** Obes Rev 2003, 4(4):201-212.
9. Almiron-Roig E, Flores SY, Drewnowski A: **No difference in satiety or in subsequent energy intakes between a beverage and a solid food.** Physiol Behav 2004, 82(4):671-677.
10. Cassady BA, Considine RV, Mattes RD: **Beverage consumption, appetite, and energy intake: what did you expect?** Am J Clin Nutr 2012, 95(3):587-593.
11. Mattes R: **Soup and satiety.** Physiol Behav 2005, 83(5):739-747.
12. Zijlstra N, Mars M, de Wijk RA, Westerterp-Plantenga MS, de Graaf C: **The effect of viscosity on ad libitum food intake.** Int J Obes (Lond) 2008, 32(4):76-683.
13. Martin CK, Anton SD, Walden H, Arnett C, Greenway FL, Williamson DA: **Slower eating rate reduces the food intake of men, but not women: implications for behavioral weight control.** Behav Res Ther 2007, 45(10):2349-2359.
14. Cecil JE, Francis J, Read NW: **Relative contributions of intestinal, gastric, oro-sensory influences and information to changes in appetite induced by the same liquid meal.** Appetite 1998, 31(3):377-390.
15. Cecil JE, Francis J, Read NW: **Comparison of the effects of a high-fat and high-carbohydrate soup delivered orally and intragastrically on gastric emptying, appetite, and eating behaviour.** Physiol Behav 1999, 67(2):299-306.
16. Giduck SA, Threatte RM, Kare MR: **Cephalic reflexes - their role in digestion and possible roles in absorption and metabolism.** J Nutr 1987, 117(7):1191-1196.
17. Mattes RD: **Physiologic responses to sensory stimulation by food: nutritional implications.** J Am Diet Assoc 1997, 97(4):406.
18. Mattes RD: **Orosensory considerations.** Obesity 2006, 14:164S-167S.
19. Woods SC: **The eating paradox - how we tolerate food.** Psychol Rev 1991, 98(4):488-505.
20. Birch LL, Deysher M: **Conditioned and unconditioned caloric compensation - evidence for self-regulation of food-intake in young-children.** Learn Motiv 1985, 16(3):341-355.
21. Booth DA, Mather P, Fuller J: **Starch content of ordinary foods associatively conditions human appetite and satiation, indexed by intake and eating pleasantness of starch-paired flavors.** Appetite 1982, 3(2):163-184.
22. Shaffer SE, Tepper BJ: **Effects of learned flavor cues on single meal and daily food-intake in humans.** Physiol Behav 1994, 55(6):979-986.
23. Yeomans MR, Weinberg L, James S: **Effects of palatability and learned satiety on energy density influences on breakfast intake in humans.** Physiol Behav 2005, 86(4):487-499.
24. Blundell J, de Graaf C, Hulshof T, Jebb S, Livingstone B, Lluch A, Mela D, Salah S, Schuring E, van der Knaap H, Westerterp M: **Appetite control: methodological aspects of the evaluation of foods.** Obes Rev 2010, 11(3):251-270.
25. Brunstrom JM, Shakeshaft NG, Scott-Samuel NE: **Measuring 'expected satiety' in a range of common foods using a method of constant stimuli.** Appetite 2008, 51(3):604-614.
26. Brunstrom JM, Brown S, Hinton EC, Rogers PJ, Fay SH: **'Expected satiety' changes hunger and fullness in the inter-meal interval.** Appetite 2011, 56(2):310-315.
27. Brunstrom JM, Rogers PJ: **How many calories are on our plate? Expected fullness, not liking, determines meal-size selection.** Obesity 2009, 17(10):1884-1890.
28. Brunstrom JM, Shakeshaft NSG: **Measuring affective (liking) and non-affective (expected satiety) determinants of portion size and food reward.** Appetite 2009, 52(1):108-114.
29. Piqueras-Fiszman B, Spence C: **The weight of the container influences expected satiety, perceived density, and subsequent expected fullness.** Appetite 2012, 58(2):559-562.
30. Hogenkamp PS, Stafleu A, Mars M, Brunstrom JM, de Graaf C: **Texture, not flavor, determines expected satiation of dairy products.** Appetite 2011, 57(3):635-641.
31. Mattes RD, Rothacker D: **Beverage viscosity is inversely related to postprandial hunger in humans.** Physiol Behav 2001, 74(4-5):551-557.
32. Zijlstra N, Mars M, de Wijk RA, Westerterp-Plantenga MS, Holst JJ, de Graaf C: **Effect of viscosity on appetite and gastro-intestinal hormones.** Physiol Behav 2009, 97(1):68-75.
33. Yeomans MR, Chambers L: **Satiety-relevant sensory qualities enhance the satiating effects of mixed carbohydrate-protein preloads.** Am J Clin Nutr 2011, 94(6):1410-1417.

34. Shama F, Sherman P: Identification of stimuli controlling the sensory evaluation of viscosity II oral methods. *J Texture Stud* 1973, **4**:111–118.

35. Sherman P: Hydrocolloid solutions and gels - sensory evaluation of some textural characteristics and their dependence on rheological properties. *Prog Food Nutr Sci* 1982, **6**(1–6):269–284.

36. Drewnowski A: The new fat replacements - a strategy for reducing fat consumption. *Postgrad Med* 1990, **87**(6):111.

37. Crum AJ, Corbin WR, Brownell KD, Salovey P: Mind over milkshakes: mindsets, not just nutrients, determine ghrelin response. *Health Psychol* 2011, **30**(4):424–429.

38. Shide DJ, Rolls BJ: Information about the fat-content of preloads influences energy-intake in healthy women. *J Am Diet Assoc* 1995, **95**(9):993–998.

39. Wooley OW, Wooley SC, Dunham RB: Can calories be perceived and do they affect hunger in obese and nonobese humans. *J Comp Physiol Psych* 1972, **80**(2):250.

40. Picciano MF: Human milk: nutritional aspects of a dynamic food. *Biol Neonate* 1998, **74**(2):84–93.

41. Davidson TL, Swithers SE: A Pavlovian approach to the problem of obesity. *Int J Obes (Lond)* 2004, **28**(7):933–935.

42. de Graaf C, Kok FJ: Slow food, fast food and the control of food intake. *Nat Rev Endocrinol* 2010, **6**(5):290–293.

43. Mars M, Hogenkamp PS, Gosses AM, Stafleu A, De Graaf C: Effect of viscosity on learned satiation. *Physiol Behav* 2009, **98**(1–2):60–66.

44. Rudenga KJ, Small DM: Amygdala response to sucrose consumption is inversely related to artificial sweetener use. *Appetite* 2012, **58**(2):504–507.

45. Swithers SE, Doerflinger A, Davidson TL: Consistent relationships between sensory properties of savory snack foods and calories influence food intake in rats. *Int J Obes (Lond)* 2006, **30**(11):1685–1692.

46. Swithers SE, Ogden SB, Davidson TL: Fat substitutes promote weight gain in rats consuming high-fat diets. *Behav Neurosci* 2011, **125**(4):512–518.

47. Kirkmeyer SV, Tepper BJ: Consumer reactions to creaminess and genetic sensitivity to 6-n-propylthiouracil: a multidimensional study. *Food Qual Prefer* 2005, **16**(6):545–556.

48. Ibrugger S, Kristensen M, Mikkelsen MS, Astrup A: Flaxseed dietary fiber supplements for suppression of appetite and food intake. *Appetite* 2012, **58**(2):490–495.

49. Marciani L, Gowland PA, Spiller RC, Manoj P, Moore RJ, Young P, Al-Sahab S, Bush D, Wright J, Fillery-Travis AJ: Gastric response to increased meal viscosity assessed by echo-planar magnetic resonance imaging in humans. *J Nutr* 2000, **130**(1):122–127.

50. Stunkard AJ, Messick S: The three-factor eating questionnaire to measure dietary restraint, disinhibition and hunger. *J Psychosom Res* 1985, **29**(1):71–83.

51. Wu Y, Cui W, Eskin NAM, Goff HD: An investigation of four commercial galactomannans on their emulsion and rheological properties. *Food Res Int* 2009, **42**(8):1141–1146.

52. Koliandris AL, Morris C, Hewson L, Hort J, Taylor AJ, Wolf B: Corresponding author contact information: Correlation between saltiness perception and shear flow behaviour for viscous solutions. *Food Hydrocolloid* 2010, **24**(8):792–799.

53. Yeomans MR: Rating changes over the course of meals: what do they tell us about motivation to eat? *Neurosci Biobehav R* 2000, **24**(2):249–259.

54. Bourne MC: Relationship between rheology and food texture. In *Engineering and Food for the 21st Century*. Boca Raton: CRC Press-Taylor & Francis Group; 2002:291–306.

55. de Wijk RA, Engelen L, Prinz JF: The role of intra-oral manipulation in the perception of sensory attributes. *Appetite* 2003, **40**(1):1–7.

14

Prevalence of cilantro (*Coriandrum sativum*) disliking among different ethnocultural groups

Lilli Mauer[1] and Ahmed El-Sohemy[2*]

Abstract

Background: Cilantro, the leaf of the *Coriandrum sativum* plant, is an herb that is widely consumed globally and has purported health benefits ranging from antibacterial to anticancer activities. Some individuals report an extreme dislike for cilantro, and this may explain the different cilantro consumption habits between populations. However, the prevalence of cilantro dislike has not previously been reported in any population. The objective of this study was to determine the prevalence of cilantro dislike among different ethnocultural groups from a population of young adults living in Canada. Subjects (n = 1,639) between the ages of 20 and 29 years were participants of the Toronto Nutrigenomics and Health Study. Individuals rated their preference for cilantro on a 9-point scale from 'dislike extremely' to 'like extremely'. Subjects also had the option to select 'have not tried' or 'would not try'. Subjects who selected 1 to 4 were classified as disliking cilantro.

Results: The prevalence of dislike ranged from 3 to 21%. The proportion of subjects classified as disliking cilantro was 21% for East Asians, 17% for Caucasians, 14% for those of African descent, 7% for South Asians, 4% for Hispanics, and 3% for Middle Eastern subjects.

Conclusions: These findings show that the prevalence of cilantro dislike differs widely between various ethnocultural groups.

Keywords: Cilantro, Coriander, Flavor perception, Food preference

Background

Cilantro is one of the most polarizing and divisive food ingredients known. It has been well documented that those who like or dislike the herb provide extremely different descriptions of its flavor [1-3]. Individuals who like cilantro may describe it as fresh, fragrant or citrusy, whereas those who dislike cilantro report that it tastes like soap, mold, dirt or bugs, among other descriptors [2,3]. Numerous websites and online communities have been created to voice pro- or anti-cilantro opinions. This segregation is not seen with many common foods, which is why cilantro is of great interest to sensory scientists [1-3]. Most flavors do exhibit some degree of polarity, though it is rarely as extreme as that observed with cilantro. It has been documented that hereditary factors, along with exposure, shape our food preferences [4-6]. With the rising concern of global obesity, it is prudent to

determine the factors that determine acceptance of healthy foods [7].

Numerous factors influence food preferences, such as socio-cultural factors and genetics. Familiarity with certain foods also influences preference, and can affect the likelihood of trying new foods. Flavor, however, is one of the most important factors influencing food selection [8-10]. Perception of bitter, sweet, salty, sour and umami taste is mediated by clusters of taste receptor cells on the tongue, palate, larynx, oropharynx, epiglottis and esophagus. These receptors are scattered across the epithelial surface, and are interspersed with one another. Contrary to what was previously believed, there does not exist a clear map of taste regions across the tongue [11]. The current understanding is that signals are transduced to numerous gustatory areas of the brain by the binding of tastants to specific taste receptors [12]. The cholinergic system is thought to primarily mediate gustatory signal transduction; however, glutamate signaling has also been shown to be involved in establishment of conditioned taste aversions [13]. The amygdala and insular cortex are two areas that have been

* Correspondence: a.el.sohemy@utoronto.ca
[2]Department of Nutritional Sciences, University of Toronto, Room 310, 150 College St, Toronto, ON M5S 3E2, Canada
Full list of author information is available at the end of the article

identified as being involved in consolidation and storage of gustatory memories, though research into the full extent of regions involved in this signaling is ongoing [14,15]. Formation of long-term memory involves new protein synthesis, whereas more rapid formation of short-term memory has been shown to be protein-independent. Formation of human taste memories is thus thought to involve novel protein synthesis; however, the time parameters of these processes are yet to be determined [16]. These memories are retrieved for comparison when a food is consumed, thus biological familiarity plays a role in taste preference. Similarly, olfaction is mediated by olfactory receptor neurons, which utilize a G-protein mechanism to transmit information about an odorant through the olfactory bulb to the olfactory cortex [17]. Odorant recognition helps the gustatory regions of the brain to identify the stimulus present in the oral cavity [12]. Perception of texture and consistency of foods is not mediated by one specific pathway, but could influence flavor intensity, likely by influencing the perception of tastants [18].

It is currently unknown whether strong reactions to the flavor of cilantro are a result of odorants or tastants. Whereas some research has investigated odorants [19], taste mechanisms have yet to be examined, although anecdotal evidence indicates that those who find cilantro offensive dislike the taste as well as the smell. Gas chromatography-olfactometry and CharmAnalysis have been used to identify the character-impact odorants in the oil of cilantro leaves [19]. Thirty-eight odor-active peaks were isolated from a sample of cilantro oil. Of those, the two trained panelists qualitatively described 33 eluted compounds in the same way, but two co-eluting odorant clusters were described differently by the two participants in the study [19]. Although only two individuals were involved in that study, the findings show that the odor profile is complex and perceived differently between individuals. Genetic factors are known to influence perception of certain odors and tastes [20], and twin studies have suggested strong heritability for cilantro preference [1,3]. However, no genetic factors associated with cilantro preference have yet been identified.

Anecdotally, the polarizing nature of cilantro has been well documented [1,3]; however, the prevalence of cilantro dislike remains unknown. This observational study aimed to determine the prevalence of cilantro dislike in different ethnocultural groups from a population of young adults.

Results

Table 1 shows the characteristics of the 1,381 subjects (419 male and 962 female) for which complete data were collected on all variables of interest. A total of 43% of females were Caucasian, which was significantly higher than the 40% of males who were Caucasian (P <0.0001). A total of 41% of females were East Asian, which was

Table 1 Subject characteristics[a]

Characteristic	Males (n = 419)	Females (n = 962)
Age, years	22.9 ± 2.5	22.6 ± 2.4
Ethnicity		
Caucasian	169 (40)	412 (43)
East Asian	148 (35)	392 (41)
South Asian	63 (15)	102 (11)
Middle Eastern	17 (4)	19 (2)
Hispanic	11 (3)	16 (2)
African descent	11 (3)	21 (2)

[a] Values are mean ± standard deviation (SD) for continuous variables, and number (%) for categorical variables.

significantly higher than the 35% of males who were East Asian (P <0.0001). Fifteen percent of males were South Asian compared to 11% of females who were South Asian (P = 0.002). No other ethnocultural groups had significantly different proportions of men and women.

Distribution of cilantro preference ratings in the population is shown in Table 2. The proportions of dislikers were not significantly different between men and women (P = 0.15), with 14% of females and 10% of males being dislikers. No significant differences in the proportions of dislikers were observed between men and women in any ethnocultural group. However, the overall response distributions differed significantly between men and women when examining either the Caucasian or East Asian groups individually (P = 0.02, P = 0.01). This was not the case with any other group, or in the population as a whole. The response distributions differed significantly between the ethnocultural groups (P <0.0001) with the Middle Eastern, Hispanic, and South Asian groups having the lowest proportions of dislikers (3%, 4% and 7%, respectively). The Hispanic and South Asian groups both also had significantly higher proportions of likers than any other groups (92% and 75%, respectively; P <0.001). A high proportion of East Asians, Caucasians and individuals of African descent had never tried cilantro (27%, 16% and 31%, respectively); these groups also had the highest prevalence of dislikers. The proportion of individuals who would not try cilantro was highest among East Asians at 1.1%.

Figure 1 shows the distribution of cilantro preference ratings on the 9-point scale for the three major ethnocultural groups: Caucasians, East Asians and South Asians. This histogram shows the specific breakdowns (the numeric responses selected) of *liker, neutral* and *disliker* categories.

Table 3 shows the distribution of leaf lettuce preferences among the ethnocultural groups. This demonstrates a typical preference distribution for a food that is considered non-polarizing. The most frequently selected preference response for this food was 7 (like moderately) within each

Table 2 Cilantro preference distributions between different ethnocultural groups

	Preference category[a]				
	Have tried[b]			Have not tried[c]	
	Like	Neutral	Dislike	Never tried	Would not try
Caucasian (n = 581)	311 (64)	88 (18)	85 (17)	96 (16)	1 (0.2)
East Asian (n = 540)	207 (53)	102 (26)	81 (21)	144 (27)	6 (1.1)
South Asian (n = 165)	119 (75)	27 (17)	11 (7)	8 (5)	0
Middle Eastern (n = 36)	8 (69)	20 (28)	1 (3)	7 (19)	0
Hispanic (n = 27)	24 (92)	1 (4)	1 (4)	1 (4)	0
African descent (n = 32)	13 (59)	6 (27)	3 (14)	10 (31)	0

[a] Subjects selecting 1 to 4 are classified as *dislikers*, 5 are *neutral*, 6 to 9 are *likers*;
[b] Values are n(%) of subjects who *have* tried cilantro.
[c] Values are n(%) of ethnocultural group.

ethnocultural group. The prevalence of dislikers ranged from 0 to 6%. A significantly lower proportion of individuals within each ethnocultural group reported disliking leaf lettuce as compared to cilantro (Caucasian: P <0.0001, East Asian: P <0.0001, South Asian: P = 0.02).

Discussion

Despite the well-recognized extreme differences in cilantro preference between individuals [1], no study has previously reported the prevalence of this trait in any population. In the present study, we examined the prevalence of cilantro dislike in different ethnocultural groups from a convenience sample of young Canadian adults recruited from the University of Toronto campus. We observed a difference in the distribution of preferences between the different ethnocultural groups as well as between men and women among certain ethnocultural groups, which may be attributed to both biological and social factors.

The Middle Eastern, Hispanic and South Asian groups had the lowest proportions of cilantro dislikers. This may be due to frequency of exposure, as cilantro is most popular in these styles of cuisine [21], and culture does modify

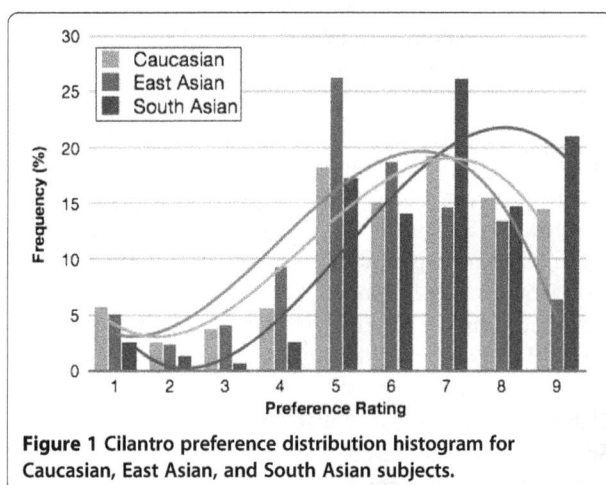

Figure 1 Cilantro preference distribution histogram for Caucasian, East Asian, and South Asian subjects.

food-related behaviors [22]. The lower prevalence of cilantro dislike among these groups could also be due to genetic differences influencing cilantro flavor perception. East Asians and Caucasians had the highest prevalence of cilantro dislikers. One limitation of our study was that the East Asian group included individuals of Thai, Korean, Japanese, Vietnamese and Chinese descent. Cilantro may be more widely used in certain East Asian cuisines, such as Thai and Vietnamese [23], and less so in others, which may have influenced our estimated proportions of East Asians who dislike or have never tried cilantro (21% and 27%, respectively). Furthermore, the Caucasian group also consisted of individuals from a wide variety of European countries. Dietary patterns vary greatly between the different regions of Europe and it was not possible to distinguish whether regional differences may have influenced cilantro preference responses in our large, heterogeneous Caucasian group. It should also be noted that the numbers of subjects within some of the ethnocultural groups was much smaller than for Caucasians and East Asians. Nonetheless, differences were observed between ethnicities. It has been suggested that genetic factors may be responsible for differential perception of the flavor of cilantro [1]. Genetic heterogeneity between ethnocultural groups may thus contribute to the different preference distributions.

Table 3 shows the preference distribution of leaf lettuce, an example of a common food that is considered to be non-polarizing. Among each ethnocultural group, the response distribution curves were normal, with peaks at 7 (like moderately). Similar findings would be expected when examining most common foods. While leaf lettuce likers and dislikers seem to exist, reactions are not extreme. This underscores the unusual, divisive nature of cilantro.

Because qualitative descriptions of the flavor of cilantro differ considerably between those who like and dislike it, differences in perception of the flavor are likely driving the observed differences in preference. Whether this is due to differential perception of an odorant or tastant, or both, is currently unknown [1,3]. It may be that individuals who

Table 3 Leaf lettuce preference distributions between different ethnocultural groups

| | Preference category[a] | | | | |
| | Have tried[b] | | | Have not tried[c] | |
	Like	Neutral	Dislike	Never tried	Would not try
Caucasian (n = 581)	518 (89)	50 (9)	11 (2)	2 (0.3)	0
East Asian (n = 540)	441 (82)	78 (15)	17 (3)	4 (0.7)	0
South Asian (n = 165)	133 (84)	22 (14)	3 (2)	7 (4.2)	0
Middle Eastern (n = 36)	36 (100)	0	0	0	0
Hispanic (n = 27)	26 (96)	0	1 (4)	0	0
African Descent (n = 32)	23 (74)	6 (19)	2 (6)	1 (3)	0

[a] Subjects selecting 1 to 4 are classified as *dislikers*, 5 are *neutral*, 6 to 9 are *likers*;
[b] Values are n(%) of subjects who *have* tried leaf lettuce.
[c] Values are n(%) of ethnocultural group.

dislike cilantro are anosmic to one or more of the pleasant smelling compounds found in cilantro. Alternatively, those who like cilantro may be anosmic to an unpleasant smelling compound - perhaps an aldehyde that, alone, smells of soap [3]. E-(2)-Decenal has been proposed as a candidate compound, as it is emitted by stink bugs and other insects in defensive secretions [24,25]. It has been suggested that this may be one of the compounds in cilantro that individuals find unpleasant, although this has not yet been tested. Because of the complex chemical composition of the oil of cilantro leaves, there are many potential candidates. Considering there are approximately 350 olfactory receptor genes and another 300 or more olfactory receptor pseudogenes of unknown function [17,26], there are many potential candidates that could explain inter-individual differences in cilantro preference. The interaction between taste and olfaction is well-established [27], but it remains unclear whether one is more influential than the other with respect to cilantro preference.

Although differences in flavor perception, possibly attributable to genetic differences between ethnocultural groups, is likely responsible for the different distributions of cilantro preference seen, we cannot rule out the possibility of differences in exposure and use of cilantro in the traditional cuisines of different ethnocultural groups driving differences in preference. The relationship between flavor perception, familiarity and exposure, and preference is complex and cannot be further explored in the present study. Future studies will need to be conducted to determine the cause of the different preference distributions observed in this study.

It should be noted that the current study consists of a convenience sample of young adults recruited from the University of Toronto campus and results may not reflect older adults or the Canadian population in general. Further studies will be required to assess the prevalence of cilantro liking and disliking in the broader population and among other ethnocultural groups.

Conclusions

In summary, we report that cilantro dislike varies from 3% to 21% in this population of young adults depending on the ethnocultural group. The contribution of individual genetic differences to this trait remains to be determined.

Methods
Subjects

Participants (n = 1,639; 1,117 women and 522 men) were enrolled in the Toronto Nutrigenomics and Health Study, which is a cross-sectional study investigating gene-diet interactions and biomarkers of chronic disease, as well as genetic determinants of eating behaviors. Subjects between 20 and 29 years of age were recruited from the University of Toronto campus. Subjects were excluded if they were pregnant or breastfeeding, due to metabolic and dietary changes that take place during this period. Subjects who could not communicate in English, or who did not provide a 12-hour fasting venous blood sample were also excluded. Smokers (n = 105) were excluded from the present analysis because of the known effects of smoking on taste and odor perception [17]. Subjects with any missing data were also excluded (n = 10). At the time of screening, subjects identified the ethnocultural group(s) they belong to. Subjects who listed more than one ethnicity (n = 143) or any group with fewer than 20 subjects were excluded from the current analyses, and the remaining individuals were classified into one of six groups (Caucasian, n = 581; East Asian, n = 540; South Asian, n = 165; Middle Eastern, n = 36; African descent, n = 32; and Hispanic, n = 27). After exclusions, the final sample population consisted of 1,381 subjects (962 women and 419 men). All subjects provided written informed consent, and the University of Toronto Research Ethics Board approved the study protocol.

Cilantro preference data collection

Subjects completed a 63-item food preference checklist, which included a range of common foods and beverages, as well as food garnishes and condiments. Participants gave each item a rating from 1 (dislike extremely) to 9 (like extremely). Alternatively, subjects had the option of selecting 'never tried' or 'would not try'.

Statistical analysis

All statistical analyses were conducted using Statistical Analysis Systems software (SAS version 9.2; SAS Institute, Cary, NC, USA). The frequency procedure was used to compare preference responses between ethnocultural groups, and χ^2 tests were used to examine differences between preference distributions. *Dislikers* were defined as those reporting 1, 2, 3 or 4 (dislike extremely, dislike very much, dislike moderately, dislike slightly) on the 9-point scale. Those selecting 5 (neither like nor dislike) were classified as *neutral*, and those selecting 6, 7, 8 or 9 (like slightly, like moderately, like very much, like extremely) were classified as *likers*. The mean and median ratings fell to the right of the arithmetic center of the scale (6.08 and 6, respectively), suggesting a slightly skewed distribution, which was confirmed using a Shapiro-Wilk test for normality. Those selecting 'never tried' were included in the analyses in order to examine the ethnocultural breakdown of this group. Those selecting 'would not try' were also included in the analyses since some of these individuals may dislike the odor so strongly that they would never consume cilantro. For comparison, leaf lettuce preference distributions were examined using the same methods. Leaf lettuce is a food commonly used as a garnish, but is not known to elicit the same polarizing responses as cilantro.

Competing interests
The authors declare they have no competing interests.

Authors' contributions
LM completed statistical analyses and prepared the first draft of the manuscript. AE obtained funding and provided supervision. Both authors contributed to data interpretation and critical review of the manuscript for important intellectual content. Both authors read and approved the final manuscript.

Authors' information
AE holds a Canada Research Chair in Nutrigenomics.

Acknowledgements
Grant funding was supplied by the Advanced Foods and Materials Network Centre of Excellence (AFMNet). The funding body had no role in the design, collection, analysis, interpretation, writing or publication of this manuscript.

Author details
[1]Department of Nutritional Sciences, University of Toronto, Toronto, Canada. [2]Department of Nutritional Sciences, University of Toronto, Room 310, 150 College St, Toronto, ON M5S 3E2, Canada.

References
1. Herz RS: *I know what i like: understanding odor preferences, The Smell Culture Reader.* Providence, RI: Oxford: Berg; 2004:190–203.
2. McGee H: *Cilantro haters, it's not your fault.* New York City: The New York Times; 2010:1.
3. Rubenstein S: *Across the land, people are fuming over an herb (no, not that one).* New York City: The Wall Street Journal; 2009:1.
4. Capaldi EP: *(Ed): Why We Eat What We Eat: the Psychology Of Eating.* Washington: American Psychological Association; 1996.
5. Cowart BJ: **Development of taste perception in humans: Sensitivity and preference throughout the life span.** *Psychol Bull* 1981, **90**:43–73.
6. Rozin P: **Development in the food domain.** *Dev Psychol* 1990, **26**:555–562.
7. Eertmans A, Baeyens F, Van den Bergh O: **Food likes and their relative importance in human eating behavior: review and preliminary suggestions for health promotion.** *Health Educ Res* 2001, **16**:443–456.
8. Fallon AE, Rozin P: **The psychological bases of food rejections by humans.** *Ecol Food Nutr* 1983, **13**:15–26.
9. Birch LL: **Development of food preferences.** *Ann Rev Nutr* 1999, **19**:41–62.
10. Drewnowski A: **Taste preferences and food intake.** *Ann Rev Nutr* 1997, **17**:237–253.
11. Chandrashekar J, Hoon M, Ryba N, Zuker C: **The receptors and cells for mammalian taste.** *Nature* 2006, **444**:288–294.
12. Kobayashi M: **Functional organization of the human gustatory cortex.** *J Oral Biosci* 2006, **48**:244–260.
13. Bermúdez-Rattoni F, Ramírez-Lugo L, Gutiérrez R, Miranda MI: **Molecular signals into the insular cortex and amygdala during aversive gustatory memory formation.** *Cell Mol Neurobiol* 2004, **24**:25–36.
14. Bermudez-Rattoni F, Nunez-Jaramillo L, Balderas I: **Neurobiology of Taste-recognition Memory Formation.** *Chem Senses* 2005, **30**:i156–i157.
15. Behrens M, Meyerhof W: **Gustatory and extragustatory functions of mammalian taste receptors.** *Physiol Behav* 2011, **105**:4–13.
16. Houpt TA, Berlin R: **Rapid, labile, and protein synthesis - independent short-term memory in conditioned taste aversion.** *Learn Mem* 1999, **6**:37–46.
17. Buck LB: **Olfactory receptors and odor coding in mammals.** *Nutr Rev* 2004, **62**:184–188. discussion S224-141.
18. Hollowood TA, Linforth RST, Taylor AJ: **The effect of viscosity on the perception of flavour.** *Chem Senses* 2002, **27**:583–591.
19. Eyres G, Dufour J-P, Hallifax G, Sotheeswaran S, Marriott PJ: **Identification of character-impact odorants in coriander and wild coriander leaves using gas chromatography-olfactometry (GCO) and comprehensive two-dimensional gas chromatography–time-of-flight mass spectrometry (GC×GC-TOFMS).** *J Sep Sci* 2005, **28**:1061–1074.
20. Reed DR, Knaapila A: **Genetics of taste and smell poisons and pleasures.** *Prog Mol Biol Transl Sci* 2010, **94**:213–240.
21. Wong PYY, Kitts DD: **Studies on the dual antioxidant and antibacterial properties of parsley (*Petroselinum crispum*) and cilantro (*Coriandrum sativum*) extracts.** *Food Chem* 2006, **97**:505–515.
22. Axelson ML: **The impact of culture on food-related behavior.** *Ann Rev Nutr* 1986, **6**:345–363.
23. Cadwallader Keith R, Benitez D, Pojjanapimol S, Suriyaphan O, Singh T: *Characteristic aroma components of the cilantro mimics, Natural Flavors and Fragrances.* Washington, DC: American Chemical Society: Frey C, Rouseff R; 2005:117–128.
24. Potter TL: **Essential oil composition of cilantro.** *J Agric Food Chem* 1996, **44**:1824–1826.
25. Borges M, Aldrich JR: **Instar-specific defensive secretions of stink bugs (Heteroptera: Pentatomidae).** *Cell Mol Life Sci* 1992, **48**:893–896.
26. Malnic B, Godfrey PA, Buck LB: **The human olfactory receptor gene family.** *Proc Natl Acad Sci U S A* 2004, **101**:2584–2589.
27. Murphy C, Cain WS: **Taste and olfaction: independence vs interaction.** *Physiol Behav* 1980, **24**:601–605.

15

A taste of Kandinsky: assessing the influence of the artistic visual presentation of food on the dining experience

Charles Michel*, Carlos Velasco, Elia Gatti and Charles Spence

Abstract

Background: Researchers have demonstrated that a variety of visual factors, such as the colour and balance of the elements on a plate, can influence a diner's perception of, and response to, food. Here, we report on a study designed to assess whether placing the culinary elements of a dish in an art-inspired manner would modify the diner's expectations and hence their experience of food. The dish, a salad, was arranged in one of three different presentations: One simply plated (with all of the elements of the salad tossed together), another with the elements arranged to look like one of Kandinsky's paintings, and a third arrangement in which the elements were organized in a neat (but non-artistic) manner. The participants answered two questionnaires, one presented prior to and the other after eating the dish, to evaluate their expectations and actual sensory experience.

Results: Prior to consumption, the art-inspired presentation resulted in the food being considered as more artistic, more complex, and more liked than either of the other presentations. The participants were also willing to pay more for the Kandinsky-inspired plating. Interestingly, after consumption, the results revealed higher tastiness ratings for the art-inspired presentation.

Conclusions: These results support the idea that presenting food in an aesthetically pleasing manner can enhance the experience of a dish. In particular, the use of artistic (visual) influences can enhance a diner's rating of the flavour of a dish. These results are consistent with previous findings, suggesting that visual display of a food can influence both a person's expectations and their subsequent experience of a dish, and with the common assumption that we eat with our eyes first.

Keywords: Food, Art, Perception, Multisensory, Experience, Plating

Background

'I try to interpret the artist's message and to make it mine, to translate it in my life and in the dishes.' ([1], Massimo Bottura, Chef at Osteria Francescana).

People perceive and appreciate food in a manner that is multisensory [2-4]; that is, information from the different senses is integrated at both the perceptual and semantic levels in order to give rise to specific multisensory experiences. Just imagine, for instance, a typical meal and the

* Correspondence: charles.michel@psy.ox.ac.uk
Crossmodal Research Laboratory, Department of Experimental Psychology, University of Oxford, South Parks Road, OX1 3UD, Oxford, UK

variety of factors that play a role in modulating the diner's overall experience [5-7]. These include, amongst other things, the presence of other people [8], the atmosphere or the environment in which the food is consumed [9,10], the cutlery with which we happen to be eating [11,12], and the plateware from which we are eating [13-15].

What people see also exerts a substantial influence over their perception of food and drink [16]. Visual cues such as colour [17] and texture [18] have been shown to exert a significant influence on the perceived flavour and acceptance of foods [19], and techniques typically belonging to the realm of painting and visual communication design have been theorized to be useful and resourceful tools when it comes to designing food experiences [20]. A food's visual features not only affect the perception of the food itself but also play a crucial role in driving our food-related expectations [21] and guiding our food choices [5].

Delwiche [22] recently reiterated an oft-made claim that people eat first with their eyes (see Apicius [23] for one of the earliest documented claims of this type). Although the complex visual arrangements of the various elements in a dish play an irrefutably important role in determining a diner's overall perception, there are still not many insights from the scientific literature on this matter available to culinary practitioners that would help them enhance the experience of their guests. In one of the few studies to have been published in this area, Debra Zellner and her colleagues assessed the influence of the balance and complexity of the elements in a dish on the perceived attractiveness, willingness to try, and liking [24]. Their results revealed that manipulating the interaction between complexity (increased by the addition of colour) and balance exerted a significant effect on the perceived attractiveness of the presentation and their participants' willingness to try the food. That said, they did not find any effect of these variables on their participants' liking for the food's flavour.

In a follow-up study, Zellner and her colleagues went on to demonstrate that people prefer food when it is presented in a neat, as compared to a messy, arrangement [25]. The neat visual presentation also exerted a positive influence on their participants' willingness to pay and their judgments of perceived quality. While the results of this previous research represent an interesting contribution to the study of how the visual arrangement of food can influence people's perception, there is still a need for researchers to further assess the influence that aesthetic dishes (the plating typically found in fine dining restaurants[a]) exert on dinners. When taken together with Zellner et al.'s studies, the present study helps to highlight different aspects of how the visual presentation of a dish can change the way the diner/consumer will perceive the food.

Specifically, in the present study, we assessed any influence of an abstract-art based dish design on people's food expectations and on their subsequent experience. We compared people's experience of a dish presented in a simple manner, with a dish whose presentation had been inspired by one of Kandinsky's paintings, and a dish in which the elements were arranged neatly, but without any artistic pretensions.

Methods
Participants
Sixty participants (mean age of 27.7 years, SD = 7.2; ranging from 18 to 58 years), 30 males and 30 females took part in the study. Upon arrival at the laboratory, the participants had to fill in a consent form and a questionnaire in order to assess the existence of any sensory dysfunctions, allergies, or food intolerances. A small number of the participants reported being allergic to, or disliking, certain ingredients, none of which were used in the present

study. The experiment was approved by the Ethics Committee of the Department of Experimental Psychology at the University of Oxford. The participants were compensated with five British pounds for their time.

Apparatus and materials
The stimuli consisted of the same set of ingredients presented in one of three different visual arrangements. Importantly, the visual arrangements characterizing the three conditions contained the exact same quantity of exactly the same ingredients. The 'regular' presentation condition consisted of a mix of the ingredients, which were simply placed in the middle of the plate. In the 'neat' presentation condition, the ingredients together with the sauces were placed side by side without touching each another. Lastly, for the 'art-inspired' condition, the ingredients were placed on the plate in a very specific manner, inspired by one of Wassily Kandinsky's abstract paintings [26]. The painting that served as the inspiration for this dish was 'Painting number 201' (see Figure 1), and was arbitrarily chosen by the authors[b]. It was described as 'nonobjective painting' by the artist himself, a landscape of colour free of descriptive devices [27]. Kandinsky's theories on colour and harmony could supposedly be applicable to any matter, or medium [28].

Before being placed on the plate, the vegetables and condiments were prepared in exactly the same manner for all three presentations. While the sauces were specifically laid out on the plate for the neat and art-inspired presentations, they were mixed with all the elements of the salad for the regular presentation. The plate on which the food was served consisted of a white rectangle of cardboard (dimensions of 270 × 180 mm).

The food consisted of a relatively complex salad with 17 distinct components made up of a total of 30 ingredients. They included three types of elements: vegetables, sauces (purees and a reduction), and condiments. The 17 components of the dish were as follows:

- Vegetables: seared Portobello slice, shimeji mushrooms (briefly boiled with a sweet vinegar marinade), cooked and raw broccoli sprouts, a variety of endive salad, raw red and yellow pepper cut into fine brunoises, one slice of raw red pepper, three slices of red pepper skin fine julienne, half a slice of raw yellow pepper, raw cauliflower sprouts, five slices of mange-tout fine julienne, and half a mange-tout.
- Sauces: beet purée, carrot purée, cauliflower and lemongrass crème, mushroom essence with squid ink, and, finally, pepperoncino oil.
- Condiments: Spanish olive oil, and Maldon sea salt.

A more detailed description of how to prepare each of the elements can be found in the culinary worksheet presented in Additional file 1.

A. "Painting #201" by Kandinsky. B. Kandinsky-inspired artistic presentation of food. C. Regular presentation D. Neat presentation

Figure 1 The Kandinsky painting used as the inspiration for the dish (A), and the three different visual arrangements presented (B, C, D). Note that the three arrangements consisted of the same quantity of the same ingredients.

Procedure

A between-participants experimental design was used. The experimental setting, which was the same for all participants, was designed to replicate a typical restaurant table (see Figure 2) in a dark room, isolated by means of a curtain. On the table and over a white tablecloth were placed a fork, a knife, a paper napkin and a glass of water. The only lighting in the room, a small lamp, was directed at the dish.

The three conditions were randomized across the various testing times (between 10:00 and 17:00 hrs) and gender was balanced for each condition. The experiment lasted for approximately twenty minutes. Upon completing the consent form, the participants were seated at the table and told the procedure by the experimenter. The participants were also instructed that they would be presented with a plate of food, a salad, and asked to eat it. Before they could start eating, they were asked to complete a questionnaire con-

cerning the visual aspects of the salad. Moreover, the participants were informed that after completion of the first questionnaire, they would be allowed to eat as much of the salad as they liked and that after they had finished they would be given another questionnaire to complete. While the experimenter explained this procedure, the dish was plated in an adjacent room. None of the participants were aware of the existence of different visual presentations and no further information was given concerning the aims of the study or the food they were about to eat and its preparation. When the dish was ready, it was placed on the table in front of the participant as shown on Figure 1, together with the first questionnaire. The participants were left alone while eating the food and completing the questionnaires.

All of the questions were presented using 10-point Likert scales. The first questionnaire was designed to assess the visual appeal of the dish and the participant's

Figure 2 Setting in which the experiment was conducted.

expectations. The second questionnaire assessed the perception of intensity of different taste attributes (saltiness, bitterness, sourness, and sweetness) and again the same questions as asked in the first questionnaire (liking, tastiness and willingness to pay), this time testing the actual experience of the food rather than merely the participants' expectations about it. For a complete list of the questions before and after consumption, see Table 1.

While similar questions were asked before and after consumption, we assumed that the preliminary judgments were based solely on the visual attributes of the food, while the latter judgments would provide information about the eating experience and the impression the food left in the mind of the participant.

Results

The effect of the three visual arrangements on participants' responses to each of the questions in the two questionnaires (pre- and post-consumption) was analyzed using a mixed model (to fit the data), including participants as a random factor (in order to account for any between-participants variability). Furthermore, we included the following as fixed factors in the model: gender, age, whether the participant considered his/herself to be a 'foodie' or not, how much they enjoyed eating vegetables, and how interested they were in the visual arts, in order to control for the effect of such variables on the results of each question.

Three of the variables tested (willingness to pay for the dish, liking, and tastiness) were assessed before and after consumption. The participants' ratings concerning these three variables were analyzed by pooling together the data, including consumption of the food (that is: whether the data belonged to the questionnaire presented before or after consumption) as a further factor in the model. Note that the effect of food consumption could vary depending on the dish presented. The significant factors resulting from the analysis performed on the three variables are shown in Table 2.

Post-hoc t-tests (Bonferroni corrected, alpha = .05/3, df = 19) were used in order to assess the difference between the ratings given by the participants for each presentation. In particular, statistically significant effects between the type of presentation were found for five of the items examined (see Figure 3): The art-inspired dish was considered as being presented in a more 'complex' and 'artistic' manner, and was liked more than either of the other two presentations. The expected tastiness of the food was also affected by the presentation, with the art-inspired dish associated with significantly higher ratings as compared to both the neat and the regular presentations. The participants were also keen to pay twice as much for the artistically presented dish than for the dishes in the other presentations. It is worth mentioning that the ratings concerning the regular and the neat presentations did not reveal any significant differences for any of the questions.

Only one variable was significantly affected by the consumption of the dish. The results revealed an increase of 18% in the tastiness ratings for the art-inspired presentation (6.8 ± 1.8 before consumption, 8.3 ± 1.5 after consumption, this difference was statistically significant: $t = 2.7$, $P < .01$), while the ratings decreased slightly in the regular presentation (6.0 ± 1.8 before consumption, 5.6 ± 2.2 after consumption). The latter difference was, however, not statistically significant ($t = -.5$, $P = .5$) as shown by the interaction plot (see Figure 4; coefficient value of the interaction: $-.74$, $P < .05$). For the neat presentation, the consumption did not have any effect on the tastiness ratings.

Eating the food led to an increase in ratings of the tastiness of the food in the case of the art-inspired dish, likely showing that the aesthetic value of this visual arrangement

Table 1 Questionnaires used in this study

Questionnaire 1. It was presented simultaneously with the dish, and aimed to measure the visual appeal of the food and the participant's expectations	Questionnaire 2. It was presented after the participant had finished eating, aiming to measure different cues of the experience the participant had eating the dish and other general impressions
How complex does this dish appear to be?	How salty was the food?
How much do you like the presentation of the food?	How bitter was the food?
How much would you be willing to pay for this dish (in British Pounds)?	How sour was the food?
Please rate how artistically arranged you think this plate is?	How sweet was the food?
How tasty does this dish look?	How much did you like this dish?
How healthy do you think this dish is?	How much would you be willing to pay for the dish (in British pounds)?
	How tasty did you find the dish?
	How full are you after eating this plate?
	Do you generally enjoy eating vegetables?
	How many ingredients do you think the dish contained?
	Do you consider yourself to be a 'foodie'?
	How interested are you in the visual arts?

Table 2 The left part of the table shows the significant factors in the mixed models highlighting the significant effects of the fixed variables, the right part shows the post-hoc test (Bonferroni corrected) significance values, the relevant means and the standard deviations, values in bold indicate significant results in the comparisons

Question	Significant factors	Coefficient value	t -value	P values	Means ± standard deviations for each presentation	Art versus Regular	Art versus Neat	Neat versus Regular
Complexity	Presentation	-.97	-2.83	.0064	Art-inspired (7.5 ± 1.95)	t = 2.28	t = 2.51	t = .65
					Regular (5.7 ± 1.7)	**P** value < .05	**P** value < .01	**P** value = .51
					Neat (5.2 ± 2.5)			
Artistic presentation	Presentation	-.99	-2.61	.01	Art-inspired (7.9 ± 2.3)	t = 3.04	t = 2.4	t = -.71
					Regular (5.7 ± 2.2)	**P** value < .01	**P** value < .05	**P** value = .47
					Neat (6.2 ± 2.12)			
Liking	Presentation	-1.18	-3.36	.0014	Art-inspired (8.0 ± 1.8)	t = 5.5	t = 5.3	t = .01
					Regular (5.6 ± 2.0)	**P** value < .001	**P** value < .01	**P** value = .90
					Neat (5.5 ± 2.3)			
Willingness to pay	Presentation	-116.9	-2.30	.02	Art-inspired (425 ± 511)	t = 3	t = 3.2	t = 0.3
					Regular (208 ± 283)	**P** value < .01	**P** value < .01	**P** value = .74
					Neat (214 ± 268)			
Tastiness	Presentation	-.70	-2.12	.03	Art-inspired (7.5 ± 1.8)	t = 3.9	t = 4.5	t = .73
					Regular (5.8 ± 2)	**P** value < .01	**P** value < .01	**P** value = .46
					Neat (5.4 ± 2.2)			
	Consumption	1.8	2.40	.01				
	Consumption: presentation	-.74	-2.07	.04				

made the food more enjoyable to eat. The food might also have been tastier than expected; even though the different perceived tastes (in terms of the rated saltiness, sweetness, bitterness, and sourness of the food) did not differ significantly between the three conditions.

The participants' appetite level was assessed before displaying the food, using a 10-point Likert scale. No significant differences were found between the three groups ('Art-inspired' group: M = 5.05, SD = 2.03; 'Regular' group: M = 5.05, SD = 1.90; 'Neat' group: M = 5.85, SD = 1.87). A Kolmogorov-Smirnov test revealed that the data from the three groups was normally distributed (P = .2 for the 'Art-inspired' group, P = .058 for the 'Regular' group, and P = .082 for the 'Neat' group).

Discussion

We compared an art-inspired food presentation to a condition where the same ingredients were arranged in a more regular manner, or else in a neat (and hence effortful) but non-artistic manner. Before the participants had tasted the food, the artistic plating was liked more than both the regular and neat food presentations, as well as being recognized as more artistic and complex than either of the other two presentations, even though the participants were not informed that the dish was supposed to mimic a work of art.

After eating, participants rated the food presented in the art-inspired as being more flavourful.

Art-infused food design

The fact that the participants in the present study liked the art-inspired dish more presumably reflects that they were actually able to recognize an artistic pattern in the food intuitively. The debate concerning what can be considered as 'art' has, for a long time, involved philosophers, aesthetes, psychologists, and, more recently, cognitive neuroscientists [29,30]. It is reasonable to assume that since art involves, at least in part, the ability to communicate feelings and sensations [30,31], the art-inspired presentation of the food could have been an edible rendition of the message originally intended by Kandinsky on the canvas 'Painting number 201'. Indeed, the differences between ratings of 'liking', 'artistic value' and 'complexity' could be attributed to an absolute aesthetic value that would have been transferred from the painting to the food design. According to another point of view, however, one could simply argue that art is that which viewers categorize as such [32,33].

The concept of an identifiable pattern is not an unusual idea in the field of art [34], and this could have led the participants to define the dish as being more artistic because patterns (of colour and shape) were easy to

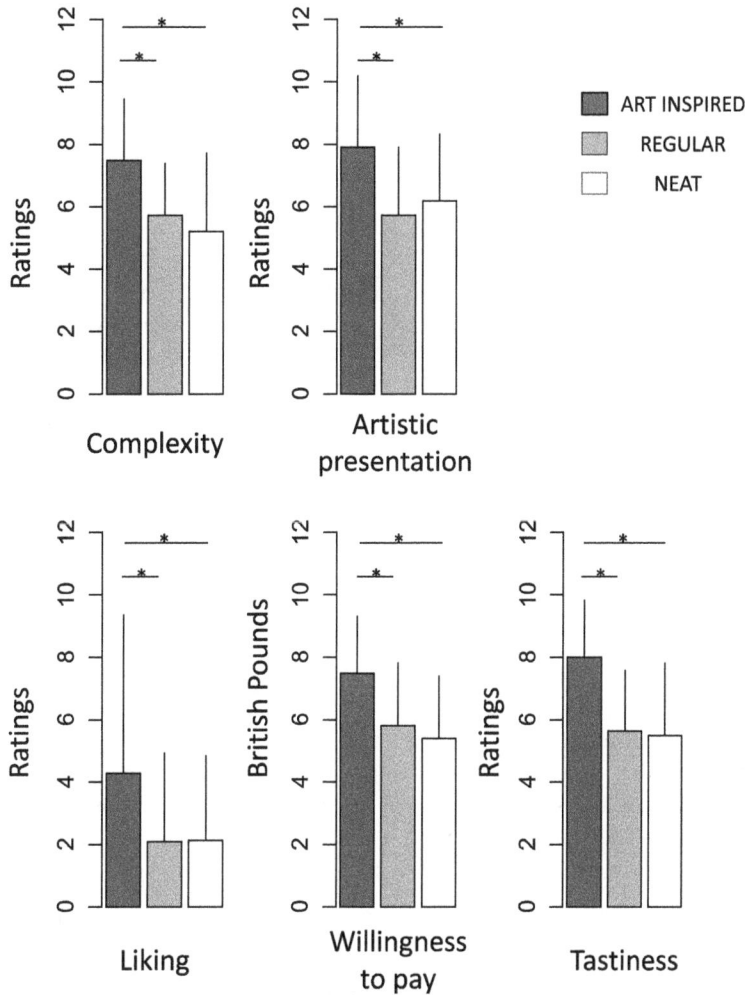

Figure 3 Bar graphs show the mean values and the standard deviations of the ratings for each of the variables that showed a statistically significant effect of the presentation (complexity, liking, artistically arrangement, tastiness, and willingness to pay). Statistically significant differences are highlighted (*P < .05) between the art-inspired presentation, the regular plating, and the neat presentation of the food.

identify. If this were to have been the case, the identified patterns could also have influenced the participants' liking judgments. Although participants' personal preferences (or taste) in art could have influenced our results, it should be noted that the display of art has been shown to activate reward systems in the human brain [35]. Results reported by Ramachandran [36] show that people may experience some sort of reward when processing visually complex stimuli (as, indeed, the art-inspired presentation was perceived by our participants). Furthermore, in one experiment, both art experts and novices were found to rate more complex artistic stimuli as being more interesting [31]. Indeed, the way the diners' interest is cultivated in high-end restaurants through highly complex food preparations and presentations, amongst other factors, is probably a key aspect of designing pleasurable food experiences.

A different perspective on the effect of plating on participants' responses to the food would be to consider the 'Art-Infusion' phenomenon as advanced by Hagtvedt and Patrick [37]. According to this theory, consumers evaluate products more favourably when they are associated with art. In this case, the art-inspired dish might have implicitly suggested a connotation of higher value (or effort) through the visual display, value that might have helped to deliver a more pleasurable eating experience.

Our participants were willing to pay more for the artistic presentation of the dish, both before and after tasting it (note that the consumption of the food did not modify people's price estimation). These results are consistent with previous research suggesting that the aesthetic presentation of food can result in people being willing to pay more for it [25]. It is important to assess any potential explanation as to why people may

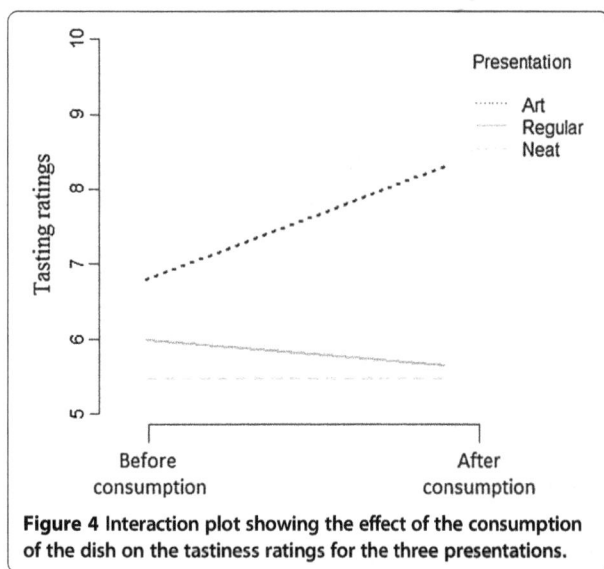

Figure 4 Interaction plot showing the effect of the consumption of the dish on the tastiness ratings for the three presentations.

pay more for an art-inspired dish (for example, a plate of salad inspired by Kandinsky); for instance, we might consider how the 'effort' involved in preparing a dish can be appreciated by a diner and, thus, change the perceived value of the dish. Neatness and complexity might be some of the elements that people are ready to pay more for as well; as the philosopher Denis Dutton puts it, the value of an artwork is rooted in the assumptions about the human performance underlying its creation [38].

A taste of Kandinsky

When the participants in the present study were asked to rate the expected tastiness of the dishes before they had sampled the food, no difference was reported between the three conditions. Even if the visual properties of the art-inspired condition received higher ratings than the other two conditions, participants were not expecting it to taste better. Interestingly, after consumption, the art-inspired presentation was rated as significantly tastier (up to 18% more) than the other two, even though they were composed of the same quantity of the same ingredients (see Figure 4). A higher rating for the experienced tastiness[c] [39,40] of the Kandinsky-inspired dish clearly shows that plating can have an important effect on flavour perception. This observation is consistent with previous findings [22], confirming that what we see can indeed influence what we taste. In addition to the arguments discussed in the previous sections, we would argue that such percept could be the result of more enjoyment elicited by the act of consuming an aesthetically pleasing product, whose creation requires a more skilful and effortful act. This result supports the theory that cultivating uniqueness in plating and presentation could be central to delivering pleasurable food experiences.

Art or novelty?

The higher ratings given to the art-inspired presentation of food could also be an effect of novelty; one might wonder if any salad that is not just mixed and placed on a plate would seem to be more 'artistic'. However, the elements of a salad placed in any form would not necessarily seem more 'artistic', given the risk that the plating might end up being considered as messy, and therefore less appealing. Indeed, Zellner and her colleagues have shown that people are more willing to try and tend to like a neat presentation more rather than a messy one [25].

Different food designs could be used in future research on plating, to understand how artistry and novelty are processed and evaluated by diners, and how this can impact on the eating experience. For instance, a novel and artistic plating could be compared to a novel but non-artistic one.

Limitations of present research and directions for future research

There are a number of limitations with the present study that should be borne in mind: First, it is important to note that mixed culinary elements on a plate (for example, two different sauces, or one sauce and one garnish) may have merged to create a new flavour. In this sense, the way in which the three dishes are arranged may, in fact, have led to their having different flavours (physically - as opposed to any effect that they have due to the psychological impact of the dish). The participants were not asked to eat all of the food, but rather to eat as much as they wanted. Interestingly, those participants in the art-inspired and regular presentations tended to eat all of the food, while they left more when presented the neat presentation; some of the participants would try the various elements in the dish without necessarily eating them all. It would be interesting to know the extent to which seeing the various components visually 'integrated', rather than separated, affected consumption behaviour and probably how flavourful the food ended up being perceived.

It is also important to note that the present study was conducted in a laboratory setting, a most unusual place in which to eat, granted, and with most of the participants being students [41]. This might explain the large standard deviation found for how much participants were ready to pay for the dish (with a few of the participants being willing to pay much more than the average for the dish). Although the experimental set-up attempted to replicate a restaurant table, the contextual variables of the laboratory setting may, for instance, have influenced how much the participants were willing to pay (presumably being lower than would have been expected had the same dish been served in a restaurant context)[d]. Moreover, another possible bias in the results regards the time of day at which the participants consumed the food. Indeed, it would be likely that higher ratings would have been obtained from hungry participants and lower ratings for participants performing the experiment in the early afternoon (for example, right after lunch). However, since conditions were randomly distributed across the time of the day, it is reasonable to assume that the groups had an equal number of 'hungry' and 'full' participants. As a consequence, this might have affected the variance of the data within groups, leaving untouched the differences between groups.

The fact that our participants received monetary compensation for taking part in this study might also have influenced their experience, as it is unusual experience for most of us to be paid to eat. It would also be intriguing to see whether preparing/presenting the dish in different contexts (for example, science lab/gastronomy event/restaurant) would change the way in which people respond to it as well[e].

Future research may, in turn, take into account the differences that arise in the perceived attributes of an art-inspired food when exactly the same dish is served in one context versus another [9,42,43]. One question that still needs to be addressed is whether it is possible to spot different visual patterns in the dish, and if so, which ones in particular may lead people to consider a dish as being artistic or not. It would also be interesting to test how more specific cues borrowed from the visual arts interact (here we are thinking of balance, symmetry, or colour associations). Future studies may also attempt to understand how knowing (before tasting) the story about the dish and its inspiration can impact the perception of the food. In addition, there is a need to develop objective measures of the resulting complexity of food presentations, since the various aspects of a visual arrangement can affect the resulting complexity of an image, in terms of its various components and their interaction.

Conclusions

The results of the study reported here provide evidence for the idea that there are differences in the expectations and consumption experience of a dish as a result of the various elements having been artistically arranged on the plate. Diners intuitively attribute an artistic value to the food, find it more complex and like it more when the culinary elements are arranged to look like an abstract-art painting. More importantly, people are ready to pay more for the food when it is presented in an aesthetically pleasing manner, both before and after trying it. Interestingly, consuming the artistically arranged dish enhanced people's assessment of the palatability of the food.

Taking these results into account, it could be assumed that the diner's hedonic and sensory perception of a dish is influenced by the expectations that have been established by visual cues. Complexity and neatness could be key aspects to produce an aesthetic display of ingredients. Furthermore, the positive values set by visual cues seem to be transferred to the perceived flavour of foods.

Here, we argue that using artistic inspiration in the design of the culinary experience, even when used implicitly, can indeed enhance the enjoyment of food.

The visual appeal of food has been, and will always be, an important matter to entice the appetite, ultimately enhancing the flavours of culinary creations. While chefs rely mostly on their intuition and expertise to plate their dishes [7], we suggest that studying food presentations under the lens of psychology and sensory science could give precious insights to the so far empirical, art of plating.

'Color is forever a part of our food, a visual element to which human eyes, minds, emotions and palates are sensitive. Perhaps through eons of time, man has come to build up strong and intuitive associations between what he sees and what he eats. A good meal, to say the least, is always a beautiful sight to behold.'
(Birren, 1963 [44]).

Endnotes

[a]Chefs in high-end restaurants tend to put a lot of thought in the design of the visual appearance of their dishes to enhance the experience of flavour [7], making this an opportunity for scientist to observe the contextual effects of plating on flavour perception.

[b]We wanted to prove the effect of an aesthetic food display on flavour perception. We thought that rather than designing a food display that would look 'nice' to us, we would choose a visual display widely established as being aesthetic, as the work of Kandinsky is recognised to be, and transform it into a dish of food.

[c]We assume that the term 'tastiness' could be interpreted in the sense of 'flavourful' or 'delicious'.

[d]In fact, we used the same questionnaire to assess the perception of the art-inspired dish in the *'Food and Wine Matching: A South American Perspective'* event held at the University of London, UK, 7 February 2013, as part of the London Enology Series. The results of this event were consistent with those obtained in the main experiment reported here. Compared to the main study, statistically significant differences ($P < .05$) were found regarding people's willingness to pay for the dish (£3.53 ± 3.1 in the main experiment, £7.60 ± 3.3 in the event in London), as well as the estimated number of ingredients needed to create the dish (11.75 ± 5.9 in the main experiment versus 16.89 ± 5.9 in the event in London). Such differences may be attributable to the different public that tasted the food (mostly students in the laboratory, people interested in wine and food in this event), but also to the different experimental settings in which the two studies were conducted. Indeed, in the London event, the participants were aware of the facts concerning the preparation of the dish, such as the painting that had inspired it and the cook preparing the food. This knowledge could have biased their judgment

on the actual difficulty of preparing the dish, resulting in their giving a higher estimate of the price and number of ingredients in comparison with the main experimental setting. (Of course, it could also be that food simply costs more in London than in Oxford where the main study was conducted.) This evidence suggests the importance of the context in which people eat, as well as the importance of the information regarding the food that they are eating, in their evaluation of a dish.

[e]In the original culinary version of this art-inspired dish, the chef offers truffle-oil scented paintbrushes to eat the salad. While people are comfortable eating with the use of a paintbrush in a large social setting, preliminary experiments have revealed that many of our participants were somewhat reluctant to use such a 'modernist' cutlery when we tested them individually in the laboratory.

Additional file

Additional file 1: Culinary worksheet for the 'Taste of Kandinsky' dish.

Competing interests
The authors confirm that there are no conflicts of interest.

Authors' contributions
CM created the design of the 'Salad with a taste of Kandinsky' dish used in this study. CV, EG, and CM worked on the experimental design of the study, under the supervision of CS. CM, CV, and EG conducted the experiment at the Crossmodal Research Laboratory, Oxford University. CM effectuated all the culinary preparations. EG and CV analysed the data. CM, CV, EG, and CS were all involved in the writing of the manuscript. All authors read and approved the final version of the manuscript.

Authors' information
CV DPhil candidate at the department of Experimental Psychology, University of Oxford.
CM is a classically trained professional cook and researcher. He is the first Chef in Residence at the Crossmodal Research Laboratory.
EG is a researcher working at Politecnico di Milano, interested in understanding sensory experience.
CS is a Professor of Experimental Psychology, and Head of the Crossmodal Research Laboratory at the Department of Experimental Psychology, University of Oxford.

Acknowledgements
CS is funded by the AHRC grant 'Rethinking the Senses' within the Science in Culture theme. CM would like to thank 'Comes Cake' for the picture used as cover image. The authors confirm that they received no external funding for this research.

References
1. Barba EB: My cuisine is tradition in evolution. [http://www.swide.com/food-travel/chef-interview/michelin-starred-chef-an-interview-with-massimo-bottura/2013/4/23]
2. Spence C: Multisensory flavour perception. Curr Biol 2013, 23:365–369.
3. Stevenson RJ: The Psychology Of Flavour. Oxford: Oxford University Press; 2009.
4. Verhagen JV, Engelen L: The neurocognitive bases of human multimodal food perception: Sensory integration. Neurosci Biobehav Rev 2006, 30:613–650.
5. Köster EP: Diversity in the determinants of food choice: A psychological perspective. Food Qual Prefer 2009, 20:70–82.
6. Meiselman HL: Dimensions of the meal: The science, culture, business, and art of eating. Gaithersburg, MA: Aspen Publishers; 2000.
7. Spence C, Piqueras-Fiszman B: The Perfect Meal: The Multisensory Science of Food and Dining. Oxford: Wiley-Blackwell; in press.
8. Herman CP, Roth DA, Polivy J: Effects of the presence of others on food intake: A normative interpretation. Psychol Bull 2003, 129:873–886.
9. Bell R, Meiselman HL, Pierson BJ, Reeve WG: Effects of adding an Italian theme to a restaurant on perceived ethnicity, acceptability, and selection of foods. Appetite 1994, 22:11–24.
10. Wansink B, Van Ittersum K: Fast food restaurant lighting and music can reduce calorie intake and increase satisfaction. Psychol Rep 2012, 111:1–5.
11. Harrar V, Spence C: The taste of cutlery. Flavour 2013, 2:21.
12. Spence C, Harrar V, Piqueras-Fiszman B: Assessing the impact of the tableware and other contextual variables on multisensory flavour perception. Flavour 2012, 1:1–12.
13. Piqueras-Fiszman B, Alcaide J, Roura E, Spence C: Is it the plate or is it the food? Assessing the influence of the color (black or white) and shape of the plate on the perception of the food placed on it. Food Qual Prefer 2012, 24:205–208.
14. Piqueras-Fiszman B, Spence C: The influence of the color of the cup on consumers' perception of a hot beverage. J Sens Stud 2012, 27:324–331.
15. Stewart P, Goss E: Plate shape and colour interact to influence taste and quality judgments. Flavour 2013, 2:27.
16. Delwiche J: The impact of perceptual interactions on perceived flavour. Food Qual Prefer 2004, 15:137–146.
17. Spence C, Levitan CA, Shankar MU, Zampini M: Does colour influence taste perception in humans? Chemosens Percept 2010, 3:68–84.
18. Okajima K, Spence C: Effects of visual texture on taste perception. i-Perception 2011, 2:966.
19. Imram N: The role of visual cues in consumer perception and acceptance of a food product. Nutr Food Sci 1999, 99:224–230.
20. Watz B: The entirety of the meal: A designer's perspective. J Foodservice 2008, 19:96–104.
21. Yeomans MR, Chambers L, Blumenthal H, Blake A: The role of expectancy in sensory and hedonic evaluation: The case of smoked salmon ice-cream. Food Qual Prefer 2008, 19:565–573.
22. Delwiche JF: You eat with your eyes first. Physiol Behav 2012, 107:502–504.
23. Apicius: Cooking and Dining in Imperial Rome. (c. 1st Century). Translated by Vehling JD. Chicago: University of Chicago Press; 1936.
24. Zellner DA, Lankford M, Ambrose L, Locher P: Art on the plate: Effect of balance and color on attractiveness of, willingness to try and liking for food. Food Qual Prefer 2010, 21:575–578.
25. Zellner DA, Siemers E, Teran V, Conroy R, Lankford M, Agrafiotis A, Ambrose L, Locher P: Neatness counts. How plating affects liking for the taste of food. Appetite 2011, 57:642–648.
26. Schapiro M: The nature of abstract art. In Modern art, 19th and 20th Centuries: Selected papers. New York: George Braziller; 1937:77–98.
27. Gallery label text, MoMA museum. [http://www.moma.org/collection/browse_results.php?object_id=79452]
28. Kandinsky W: Concerning the Spiritual in Art, Especially in Painting (1914). Translated by Sadler MTH. New York: Dover Publications; 1977.
29. Carey J: What Good Are the Arts? London: Faber & Faber; 2005.
30. Zeki S: Inner Vision: An Exploration of Art and the Brain. Oxford: Oxford University Press; 2000.
31. Silvia PJ: Emotional responses to art: From collation and arousal to cognition and emotion. Rev Gen Psychol 2005, 9:342–357.
32. Bourdieu P, Darbel A: The Love of Art: European Art Museums and Their Public. Oxford: Blackwell; 1997.
33. Dewey J: Having an experience. In The Later Works, 1925–1953: Art as Experience. Edited by Boydston JA, Dewey J. Carbondale: Southern Illinois University Press; 1989:42–63.
34. Behrens RR: Art, design and Gestalt theory. Leonardo 1998, 31:299–303.
35. Lacey S, Hagtvedt H, Patrick VM, Anderson A, Stilla R, Deshpande G, Hu X, Sato JR, Reddy S, Sathian K: Art for reward's sake: Visual art recruits the ventral striatum. NeuroImage 2011, 55:420–433.
36. Ramachandran VS: The Emerging Mind. London: Profile Books; 2003.
37. Hagtvedt H, Patrick VM: Art infusion: The influence of visual art on the perception and evaluation of consumer products. J Mark Res 2008, 45:379–389.
38. Dutton D: The Art Instinct, Beauty, Pleasure, and Human Evolution. New York: Bloomsbury Press; 2009.

39. Spence C, Smith B, Auvray M: **Confusing tastes and flavours**. In *The Senses*. Edited by Matthen M, Stokes D. Oxford: Oxford University Press; in press.

40. Rozin P: **"Taste–smell confusions" and the duality of the olfactory sense**. *Percept Psychophys* 1982, **31:**397–401.

41. Henrich J, Heine SJ, Norenzayan A: **The weirdest people in the world?** *Behav Brain Sci* 2010, **33:**61–135.

42. Green DM, Butts JS: **Factors affecting acceptability of meals served in the air.** *J Am Diet Assoc* 1945, **21:**415–419.

43. Meiselman HL, Johnson JL, Reeve W, Crouch JE: **Demonstrations of the influence of the eating environment on food acceptance.** *Appetite* 2000, **35:**231–237.

44. Birren F: **Color and human appetite.** *Food Technol* 1963, **17:**45–47.

16

Enhancing saltiness in emulsion based foods

Mita Lad[1,2], Louise Hewson[1] and Bettina Wolf[1*]

Abstract

Background: The concept of enhancing saltiness perception in emulsions and a liquid food formulated with the emulsions (ambient vegetable soup) through increasing salt concentration in the continuous phase while retaining the fat content of the (aqueous continuous) product was evaluated. This was accomplished by increasing the droplet phase volume using duplex emulsion technology. Viscosity and droplet size distribution was measured. Saltiness evaluation was based on simple paired comparison testing (2-Alternate Forced Choice tests, BS ISO 5495:2007).

Results: Single and duplex emulsions and emulsion-based products had comparable mean oil droplet diameters (25 to 30 μm); however, viscosity of the duplex emulsion systems was considerably higher. Sensory assessment of saltiness of emulsion pairs (2AFC) indicated duplex technology enhanced saltiness perception compared to a single emulsion product at the same salt content (6.3 g/100 g) in both simple emulsions and the formulated food product ($P = 0.0596$ and 0.0004 respectively) although assessors noted the increased viscosity of the duplex systems. The formulated food product also contained pea starch particles which may have aided product mixing with saliva and thus accelerated tastant transport to the taste buds. Lowering salt content in the duplex systems (to levels of aqueous phase salt concentration similar to the level in the single systems) resulted in duplex systems being perceived as less salty than the single system. It appears that the higher viscosity of the duplex systems could not be "overruled" by enhanced mixing through increased droplet phase volume at lowered salt content.

Conclusions: The results showed that salt reduction may be possible despite the added technology of duplex systems increasing the overall measured viscosity of the product. The changes in viscosity behavior impact mouthfeel, which may be exploitable in addition to the contribution towards salt reduction. With a view to applying this technology to real processed foods, it needs to be tested for the product in question but it should be considered as part of a salt reduction tool box.

Keywords: Salt, Health, Duplex emulsions, Microstructure, Fat

Background

Salt (NaCl) has been traditionally used as a food preservative to reduce microbial growth, thereby preventing food spoilage. Salt also plays a crucial role in imparting saltiness to foods, and enhancing or masking flavors, as well as being essential for textural attributes of some products [1]. Reported intake of salt by adults in most Western countries has been estimated to be 9 to 12 g per day [2], which is much greater than the level currently recommended by the World Health Organization (WHO) of a maximum of 6 g per day [3]. The negative health consequences of high salt intake have been widely reported. Of most concern is the direct link to the development of hypertension [4,5] which subsequently increases the chance of developing cardiovascular disease [6,7] and renal disease [8,9].

Excessive intake of salt through the Western diet has been attributed to the relatively large amounts of salt contained in prepared/pre-cooked foods. It has been estimated that over 75% of daily salt intake comes from processed foods [10]. Reduction of the overall amount of salt added to processed food is one possible strategy to reduce salt intake [11]. However, simply reducing the amount of salt added to foods not only represents technological difficulties and, in case of some foods, a safety risk, it inevitably means a compromise in product taste and flavor with negative impact on consumer acceptability. Despite these restraints, overall salt reduction

* Correspondence: Bettina.Wolf@nottingham.ac.uk
[1]Food Sciences, Sutton Bonington Campus, The University of Nottingham, Loughborough LE12 5RD, UK
Full list of author information is available at the end of the article

has been successfully applied to staple foodstuffs, such as bread, using gradual step-wise reduction [12]. This approach is time consuming, with levels being reduced slowly over several years. An alternative strategy is to use salts with an alternative cation, as it is the sodium causing the negative health impact. Replacement of NaCl with potassium chloride (KCl) has had mixed results as KCl is often reported to be bitter and can produce a metallic aftertaste [13-15]. The presence of low levels of NaCl [16,17] and substances imparting umami taste [18], for example in savory soups [18-20], can mask this bitter taste, rendering low-level substitution of NaCl with KCl feasible. In [1], it has been reported that naturally brewed soy sauce can replace NaCl without lowering overall taste intensity or decreasing consumer acceptance. Generally, adding components such as arginine [21], lysine [11] or glutamate [22], which enhance the umami character of food, may raise saltiness perception and may be a suitable approach in certain food types. Another method to enhance saltiness perception includes the use of specific odor-taste interactions in the presence of expected flavors [23,24].

The approach adopted in this study on salt reduction was based on enhancing salt concentration in the continuous product phase and is applicable to aqueous-based foods. Food emulsions, with the droplets acting as inert filler particles excluding salt from the volume corresponding to the dispersed phase volume of the emulsion, were evaluated. This approach has been taken in the past; increase in saltiness perception from oil-in-water emulsions with increasing oil phase volume and constant total amount of salt has been reported [25,26]. It was also found that with increasing oil phase volume for emulsions of constant *continuous phase* salt concentration, salt perception declined as the presence of oil obstructed the salt from reaching the taste buds [25,26]. Recently, a corresponding approach using air bubbles as "tastant excluded fillers" in a water-based gel has been reported [27]. It was described as offering scope for the reduction of sodium chloride and sucrose. In this case, no effect of filler phase volume on taste perception in samples with constant continuous phase tastant concentration was found as air would simply escape leaving the tastant to reach the taste buds. There was, however, an impact on sample texture and appearance due to the presence of 40% volume fraction of air bubbles.

In the literature cited thus far, the salt was dissolved in the aqueous product phase corresponding to the continuous product phase, in which it was equally distributed. Strategies to enhance saltiness perception at equi-salt stimuli by creating localized concentration differences have also been explored. Formulation of products with salt gradients is less of a challenge for low moisture content foods, such as bread, as diffusion is restricted, and it has been successfully applied [28]. The magnitude of saltiness enhancement was shown to depend on the total salt content (highest for lowest level) and sensory contrast (increased with increasing heterogeneity). For liquid foods this approach can only be realized by encapsulating the stimulant and facilitating release during oral processing. Whereas design of suitable food microstructures remains a challenge, the hypothesis of increased taste perception by temporal changes of tastant concentration using pulsed delivery systems has been tested for sweetness [29] and saltiness perception [20,30,31]. With respect to saltiness perception, success has been shown to be mixed depending on the time scale of the experiment. Spiking salt concentration within a delivery period of 15 s had no effect on saltiness perception and overall saltiness perception depended on the overall delivered amount of salt [31]. A 30 s long delivery period on the other hand was shown to increase salt perception based on short and intense stimulus delivery periods [30].

Here we report the results of a study undertaken to assess salt reduction in a simple ambient vegetable soup. The product was based on emulsion technology and the fat content of the food product was retained constant through application of duplex emulsion technology. Thereby, we specifically refer to oil-in-water emulsions with an included water phase in the internal oil phase. Such duplex emulsions are known as water-in-oil-in-water emulsions, or in short, w/o/w. Based on [26], samples with higher dispersed phase volume were expected to be perceived as saltier when the overall salt concentration was kept constant, although the emulsion system evaluated in the current study was more complex. Then the hypothesis was tested that reducing salt concentration in the product with the higher dispersed phase volume would lead to no significant difference in perceived saltiness up to a certain level of salt reduction. Salt concentrations used in these studies were chosen to be within the range presently found in commercial soups.

Results and discussion
Microstructure and viscosity
Figure 1 shows optical micrographs of emulsions and vegetable soup samples containing 0.63 g salt/100 g; gelatinized pea starch granules are highlighted. Figure 2 shows the corresponding particle size distributions The $d_{4,3}$ of the single emulsion was 23.7 ± 0.5 µm, 30.2 ± 0.2 µm for the duplex emulsion, 26.1 ± 0.1 µm for the single emulsion based vegetable soup and the duplex emulsion based vegetable soup had a $d_{4,3}$ of 31.2 ± 0.1 µm.

The micrographs of the emulsion samples (Figure 1a, 1c) show the typical microstructure of an unflocculated emulsion with a broad droplet size spectrum. The droplets of the duplex emulsion appear dark due to the

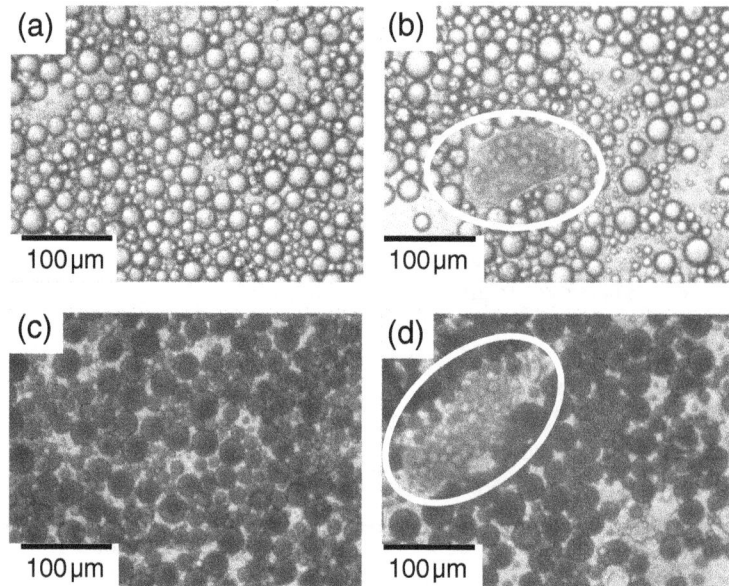

Figure 1 a) o/w, b) o/w based soup, c) w/o/w and d) w/o/w based soup.

enclosed water droplets corresponding to the droplet phase of the initially prepared w/o emulsion. The dark appearance of duplex emulsion droplets has previously been reported in the literature [32-34]. The droplet size spectra as well as the $d_{4,3}$ values of both types of emulsions compare well and, hence, droplet size can be removed as a variable which may influence saltiness perception.

Addition of pea purée to the emulsions to form a simple vegetable soup introduced an additional structure component as can clearly be seen in Figure 1b, 1d in which gelatinized pea starch granules have been highlighted. Their presence is also reflected as an additional peak in the particle size data (Figure 2), which was verified by sizing of the pea purée alone (results not shown).

Figure 3 shows the viscosity behavior of the four samples. Immediately obvious is the difference in viscosity between the single and the duplex systems for both emulsion and soup. The viscosity difference is much smaller comparing an emulsion and the corresponding

soup. These observations are not surprising. Shear viscosity of an emulsion can be estimated with the Krieger-Dougherty equation [35]:

$$\eta_r = \left[1 - \frac{\phi}{\phi_m}\right]^{-2.5\phi_m}$$

with η_r = relative viscosity [1] defined as η_e/η_c whereby η_e = emulsion viscosity [Pa s] and η_c = viscosity of continuous emulsion phase [Pa s], ϕ = dispersed volume fraction [1], ϕ_m = maximum packing fraction [1]. The maximum packing fraction for polydisperse emulsions is between 0.65 and 0.8 (depending on packing structure). The dispersed volume fraction for the single emulsion containing 32.64 g oil/100 g is 0.35 and it is 0.51 for the duplex emulsion (calculated with a density of 917 kg.m^{-3} for sunflower oil). Thus the viscosity of the duplex emulsion could be expected to be a factor of 2.4 to 3.5 higher than the viscosity of the single emulsion. Experimentally, a factor of 4 was found (viscosity at 10 s^{-1}: 6 ± 1 mPa s

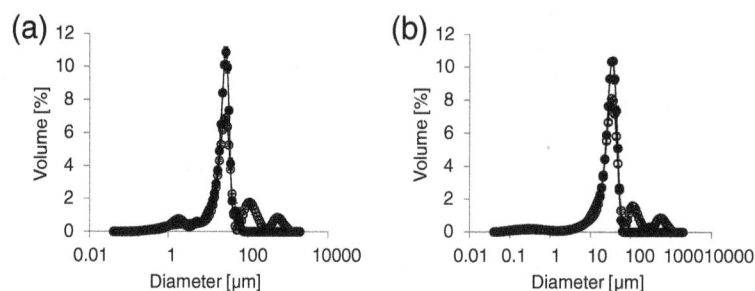

Figure 2 a) o/w and b) w/o/w systems. Filled symbols are for emulsions and open symbols for soups.

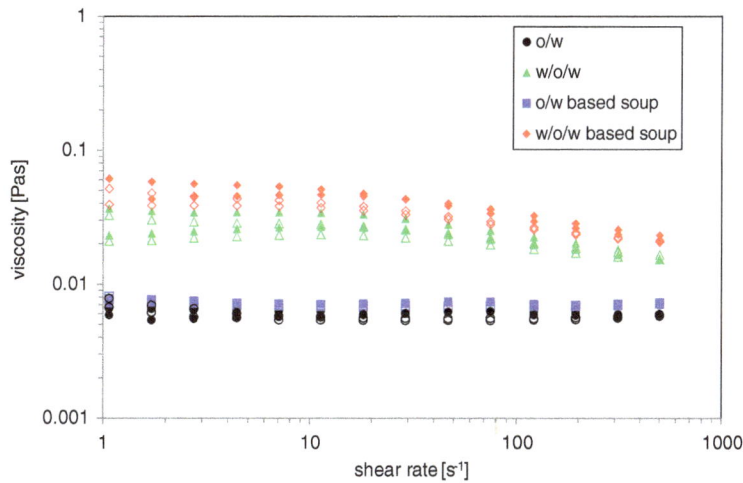

Figure 3 Viscosity profiles.

and 24 ± 3 mPa s for single and duplex emulsion respectively). Conducting the same analysis for the soup samples requires an estimate of the particulate dispersed phase volume added to the sample with the pea purée. Assuming a dispersed volume fraction increase by 0.1 returns a larger increase in viscosity for both types of samples than found experimentally. 7 ± 1 mPa s and 35 ± 3 mPa s was the viscosity of the single emulsion based soup and the duplex emulsion based soup, respectively (at 10 s^{-1}). However, the experimental values are not too different from the predicted values indicating that the Krieger-Dougherty equation is a useful tool to predict viscosity also for complex emulsions.

The increased viscosity of the duplex samples could be expected to affect saltiness perception. Previous studies have shown that viscosity increase leads to a decrease in saltiness perception [36-41]. On the other hand, it has also been demonstrated that including particles in a liquid food may enhance saltiness perception [42]. This is proposed to be due to the solid particles increasing mixing efficiency with saliva and thus salt delivery to the taste buds. Here the two higher viscous samples contain a larger volume fraction of dispersed particles which may promote in-mouth mixing.

Saltiness perception

Results of paired comparison tests, indicating which sample of each pair was perceived as the saltiest and calculated significance values, are included in Table 1. In PC1, the two samples of water contained the same amount of salt as the samples of the other two pairs evaluated in this session. Results indicated assessors clearly identified the saltier sample in the water-based pair evaluated in PC1 (P-value < 0.0001). Thus, it can be assumed that a perceivable difference existed in the salt concentration between the two emulsion samples

provided in PC2 if, indeed, the salt concentration of the continuous aqueous phase was the key driver of saltiness perception. However, as Table 1 shows, in PC2 results only indicated a trend for the duplex emulsion to be saltier than the single emulsion ($P = 0.0595$). Interestingly, sensory assessment of the same emulsions with the addition of 10 g pea purée showed that the duplex emulsion based soup was perceived to be significantly saltier than the single emulsion sample (PC3). This result seems to reinforce the trend found in PC2. Instrumental measurements indicated significant viscosity differences between the single and duplex emulsion and soup samples; indeed, several assessors commented that both duplex emulsion-based samples were thicker and creamier than the single. Thus, despite working with viscosity disparate samples, the result for the more complex sample pair (complex emulsion/soup-base) was as expected based on [26], indicating the aqueous phase tastant concentration is important in perception. It is possible that the higher dispersed phase volume enhanced mixing with saliva and thus accelerated transport of salt to the taste buds, as previously argued for liquid foods containing solid particles [42], or even that the higher salt concentration in the aqueous phase of the duplex emulsion created a sensory contrast effect greater than in the single emulsion. It is not likely that the present finding is a consequence of peas containing sodium and potassium known to impart saltiness [13-15]. Typical levels in fresh green peas are 6 mg Na/100 g and 350 mg K/100 g [43,44], which is very low compared to the amount of added salt in the studied samples.

PC4 and PC5 evaluated whether the overall salt content could be reduced in the duplex emulsion-based product without reducing perception of saltiness when compared to the single emulsion-based product. In PC4, the overall salt concentration of the duplex emulsion-

Table 1 Sample pairs presented to assessors

Test	Sample	g salt/100 g product	Continuous product phase salt concentration/mol/L	Number of assessors judging this sample to be the saltier of the pair	P-value
PC1	Water	0.94	0.160	4	<0.0001
	Water	1.21	0.207	37	
PC2	o/w	0.63	0.160	14	0.0596
	w/o/w	0.63	0.207	27	
PC3	o/w soup	0.63	0.160	9	0.0004
	w/o/w soup	0.63	0.207	32	
PC4	o/w soup	0.63	0.160	25	0.1081
	w/o/w soup	0.57	0.187	14	
PC5	o/w soup	0.63	0.160	28	0.0095
	w/o/w soup	0.50	0.166	11	

Samples for which $P < 0.05$ are considered significantly different.

based soup was reduced to 0.57 g salt/100 g compared to 0.63 g salt/100 g for the single emulsion. Paired comparison testing returned insufficient evidence to suggest a significant difference in saltiness between the two samples as shown in Table 1. However, there was a trend in the data pointing towards higher perceived saltiness of the sample in the pair that contained the overall higher salt concentration, which was the single emulsion/soup based product. Indeed, further analysis of the resultant data from PC4 for similarity testing (BS 5495:2007) indicated the sample pair cannot be concluded to be perceptually identical in terms of saltiness. Further salt reduction in the duplex emulsion/soup based product to 0.5 g salt/100 g led to the single emulsion/soup product being perceived as significantly saltier (Table 1). Assessors commented again that the duplex samples were thicker and creamier than the single emulsion samples.

The results of the actual salt reduction tests conducted in this study are complex but to some degree confirm previous findings [26] that perception of saltiness is a complex function of salt concentration in the aqueous phase and aqueous phase volume. In the present study, the use of single and duplex emulsion systems led to comparison of samples which differed significantly in viscosity. As mentioned previously, studies have shown increasing viscosity commonly decreases tastant perception [36-41], and this may result in a reduced perception of saltiness in duplex emulsions compared to single emulsions irrespective of salt content. It appears that the disparity in viscosity is not "overcome" by the increased dispersed phase volume in the duplex emulsion promoting in-mouth mixing and acceleration of tastant transport to the taste buds. However, this observation depends on the difference in salt content. Thus, to further clarify use of duplex microstructure in salt reduction technology, control of product viscosity and product microstructure would be

key to understanding the primary cause of altered taste perception. On the other hand, reducing the fat content of the type of emulsion-based product tested by incorporating water into the droplets of the dispersed phase appeared to have no effect on saltiness perception other than increasing the dispersed phase volume of the emulsion. The study of the viscosity behavior of duplex emulsions has received little attention compared to single emulsions [45], and it cannot be excluded that the included water droplets affect the viscosity behavior of the duplex system.

Conclusions

This study aimed to provide initial evaluation to assess whether duplex emulsion technology presents an opportunity to reduce salt in emulsion-based liquid or semi-liquid foods while keeping the fat content constant. Emulsions and a more complex emulsion-based ambient vegetable soup were tested using paired comparison tests to identify the sample perceived as overall saltier of each pair. The results showed that salt reduction may be possible despite the added technology increasing the overall measured viscosity of the product. The changes in viscosity behavior impact mouthfeel, which may be exploitable in addition to the contribution towards salt reduction.

To conclude, in view to applying this technology to real processed foods, it needs to be tested for the product in question, but given the interesting novel data presented it should be considered as part of a salt reduction tool box.

Methods

All materials used to prepare the emulsions were of food grade quality and used as obtained. Sunflower oil was purchased from a local supermarket, pea protein isolate (PPI) from Myprotein (Manchester, UK), sodium chloride

(+99%) from Sigma-Aldrich (Gillingham, UK), potassium chloride (+99%) from Fisher Scientific (Loughborough, UK) and polyglycerol polyricinoleate (Crestamul PR) from AAK Bakery Services (Oldham, UK). The vegetable base was prepared from frozen peas purchased in a local supermarket and bottled water (Evian, Danone, France) was used to prepare the foods.

Emulsions were formed so that the fat content remained constant and comminution conditions were selected to obtain comparable droplet size distributions in single and double emulsions with a mean diameter of 25 to 35 μm. Emulsion interfaces with oil on the inside were stabilized with protein. The internal interface of the w/o/w duplex emulsions was stabilized with polyglycerol polyricinoleate (PGPR). To minimize microstructure instability through diffusion of the internal water phase into the external water phase due to osmotic pressure gradients, potassium chloride was added to the internal water phase. However, "emptying out" was still observed when the samples were kept for longer than one week, probably due to differences in chemical potential between the two aqueous phases as recently demonstrated [32].

Emulsions and vegetable soup were prepared and analyzed as detailed in the following sections. Preparation and analysis were within the space of 48 h.

Single emulsions

A total of 100 g of single emulsion contained 1 g of PPI, 66.36 g of aqueous NaCl solution and 32.64 g of sunflower oil. The molar concentration of the NaCl solution was chosen to obtain the desired salt concentration in the final sample. Batches of emulsion (100 g) were prepared by immersing a high shear overhead mixer (L5M, emulsor screen, Silverson, Chesham, UK) into a glass beaker containing the aqueous emulsion phase, initially mixing for one minute at 3,000 rpm during which the oil phase was added followed by shearing for four minutes at 4,000 rpm. Temperature was not controlled.

Duplex emulsions

Duplex emulsions of the w/o/w type were formed in two stages by initially preparing a w/o emulsion with 30 g 0.2 M aqueous KCl solution/100 g. The oil phase contained 2.86 g PGPR/100 g. One hundred-gram batches of this w/o emulsion were formed by adding both phases to a glass beaker followed by mixing at 7,000 rpm for two minutes. The duplex emulsion was then obtained by emulsifying 48 g w/o emulsion into 52 g aqueous phase containing 1 g/100 g PPI and NaCl at the level required to obtain the desired salt concentration in the final sample. The comminution conditions were the same as in the single emulsion.

Vegetable soup

Vegetable soups were formulated based on emulsion ingredient and pea puree prepared as described earlier. One hundred grams of soup contained 90 g emulsion (single or duplex) and 10 g pea purée. The two ingredients were mixed manually and it was verified that emulsion microstructure remained intact by assessing the samples microscopically. Pea purée was prepared by initially washing thawed peas with water followed by microwave cooking (800 W, 8 minutes) of 200 g peas in excess water. Once cooked, the water content was adjusted to 25 g water/100 g peas by draining the cooking water and adding fresh water. This composition was then blended using a hand blender until a smooth mixture was obtained, which was then passed through a sieve with fine holes.

Light microscopy

The microstructure of the samples was evaluated using a light microscope (Leitz Diaplan, Stuttgart, Germany) applying bright field illumination. Micrographs were taken with a digital camera (Pixalink PLA662, Ottawa, ON, Canada).

Particle size

Particle size analysis was carried out with a laser diffraction particle size analyzer (Beckman Coulter LS 13230, Miami, FL, USA). All samples were measured in aqueous dispersion and measurement data were analyzed based on the Fraunhofer model using the instrument's software. Measurements were performed in triplicate and results are shown as volume based density distribution as well as average volume weighted mean diameter $d_{4,3}$.

Emulsion viscosity

Emulsion viscosity was measured in steady shear with a rotational rheometer (MCR301, Anton Paar, Graz, A) fitted with concentric cylinder geometry (C27) at 20°C. Shear rate was stepwise increased from 1 to 500 s^{-1} and then decreased to 1 s^{-1}. Samples were analyzed in duplicate and both data traces are shown in the Results section. Viscosity values in the Results section are reported as averaged data acquired at 10 s^{-1} on the increasing and decreasing shear rate ramp.

Sensory evaluation

Sensory assessment of the products was performed using simple paired comparison tests (2-Alternate Forced Choice tests, BS ISO 5495:2007) requiring assessors to determine which of two samples was perceived as saltier. In a first series of paired comparison tests (Table 1), samples (water-based, emulsion-based and complex emulsion/soup-based) with constant salt concentration in the continuous product phase were evaluated. In a second series of paired comparison tests, the overall concentration of salt in the duplex emulsion product was reduced and compared to the single emulsion product (complex emulsion/soup-base). Here, the samples of each pair had a different overall salt content and the concentration in the continuous product phase was increasingly similar. Assessors were recruited by advertisement (soup usage was not a prerequisite), from the students and staff of the University of Nottingham. A total of 41 assessors attended the first session which consisted of tests PC1, PC2 and PC3, and 39 assessors attended the second session of tests PC4 and PC5 (Table 1). Each sample pair was assessed once by each assessor. Sensory evaluation of samples was conducted approximately 24 h after sample preparation. For each pair, samples of 15 mL were presented in a randomized, balanced order across the panel, in plastic containers labeled with a random three-digit code. The order of the three tests was also randomized across the panel.

All testing was carried out in a quiet, air-conditioned room (18°C) and in individual booths with northern hemisphere lighting. A full explanation of the test was provided prior to testing. Assessors were told they would be given a series of pairs of samples. For each pair, assessors were instructed to taste the samples in the order presented and to determine which sample was the highest in "overall saltiness". "No difference" answers were not allowed. Assessors were informed that other attributes, such as texture, may differ between the samples and they were asked to focus only on the overall saltiness perceived. In addition, space was provided on the test form to give any comments about the samples they wished. Assessors were instructed to cleanse their palate before and between samples with unsalted crackers (99% Fat Free, Rakusen's, Leeds, UK) and mineral water (Evian).

Data were acquired and results were analyzed as individual two-tailed paired tests ($P < 0.05$) using Fizz software (Biosystèmes, Couternon, France).

Ethics approval
This study was approved by the Ethics Committee in the Medical School of The University of Nottingham.

Abbreviations
2AFC: Sensory assessment of saltiness of emulsion pairs; PGPR: Polyglycerol polyricinoleate; WHO: World Health Organization.

Competing interests
The authors declare that they have no competing interests.

Acknowledgments
The authors wish to thank the UK Technology Strategy Board for funding this project (TP/6/DAM/6/S/K3004C), the panelists for attending the sessions and the TSB project team, Joanne Hort and Cécile Morris for discussion.

Author details
[1]Food Sciences, Sutton Bonington Campus, The University of Nottingham, Loughborough LE12 5RD, UK. [2]Riddet Institute, Massey University, Private Bag 11222, Palmerston North, New Zealand.

Authors' contributions
ML conceived the idea for the study, formulated the products and prepared the first draft of the manuscript. LH designed, supervised and analyzed the sensory evaluation of the products. ML conducted the droplet size and viscosity measurements, which were analyzed by BW. LH and BW developed the first draft manuscript into the final version suitable for publication. All authors read and approved the final manuscript.

References
1. Kremer S, Mojet J, Shimojo R: **Salt reduction in foods using naturally brewed soy sauce.** *J Food Sci* 2009, **74**:S255–S262.
2. Brown IJ, Tzoulaki I, Candeias V, Elliott P: **Salt intakes around the world: implications for public health.** *Int J Epidemiol* 2009, **38**:791–813.
3. WHO: *Reducing salt intake in populations. Report of a WHO forum and technical meeting.* Geneva: WHO Document Production Services.: World Health Organization; 2007.
4. He FJ, MacGregor GA: **A comprehensive review on salt and health and current experience of worldwide salt reduction programmes.** *J Hum Hypertens* 2009, **23**:363–384.
5. Mohan S, Campbell NRC: **Salt and high blood pressure.** *Clin Sci* 2009, **117**:1–11.
6. Weinsier RL: **Overview - salt and development of essential hypertension.** *Preventive Med* 1976, **5**:7–14.
7. He J, Ogden LG, Bazzano LA, Vupputuri S, Loria C, Whelton PK: **Dietary sodium intake and incidence of congestive heart failure in overweight US men and women - First National Health and Nutrition Examination Survey Epidemiologic Follow-up Study.** *Arch Intern Med* 2002, **162**:1619–1624.
8. Cianciaruso B, Bellizzi V, Minutolo R, Tavera A, Capuano A, Conte G, De Nicola L: **Salt intake and renal outcome in patients with progressive renal disease.** *MinerElectrolyte Metab* 1998, **24**:296–301.
9. Cappuccio FP, Kalaitzidis R, Duneclift S, Eastwood JB: **Unravelling the links between calcium excretion, salt intake, hypertension, kidney stones and bone metabolism.** *J Nephrol* 2000, **13**:169–177.
10. James WP, Ralph A, Sanchez-Castillo CP: **The dominance of salt in manufactured foods in the sodium-intake of affluent societies.** *Lancet* 1987, **1**:426–429.
11. Dotsch M, Busch J, Batenburg M, Liem G, Tareilus E, Mueller R, Meijer G: **strategies to reduce sodium consumption: a food industry perspective.** *Crit RevFood Sci Nutr* 2009, **49**:841–851.
12. Girgis S, Neal B, Prescott J, Prendergast J, Dumbrell S, Turner C, Woodward M: **A one-quarter reduction in the salt content of bread can be made without detection.** *Eur J Clin Nutr* 2003, **57**:616–620.
13. Frank RL, Mickelsen O: **Sodium-potassium chloride mixtures as table salt.** *Am J Clin Nutr* 1969, **22**:464–470.
14. Vanderklaauw NJ, Smith DV: **Taste quality profiles for 15 organic and inorganic salts.** *Physiol Behav* 1995, **58**:295–306.
15. McGregor R: **Taste modification in the biotech era.** *Food Technol* 2004, **58**:24–30.
16. Breslin PAS, Beauchamp GK: **Salt enhances flavour by suppressing bitterness.** *Nature* 1997, **387**:563–563.
17. Beauchamp GK: **Salt preference in humans.** In *Encylopedia of Human Biology.* Edited by Dulbecco R. San Diego: Academic; 1997:669–675.
18. Prescott J: **Taste hedonics and the role of umami.** *Food Australia* 2001, **53**:550–554.
19. Roininen K, Lahteenmaki L, Tuorila H: **Effect of umami taste on pleasantness of low-salt soups during repeated testing.** *Physiol Behav* 1996, **60**:953–958.
20. Morris C, Labarre C, Koliandris AL, Hewson L, Wolf B, Taylor AJ, Hort J: **Effect of pulsed delivery and bouillon base on saltiness and bitterness perceptions of salt delivery profiles partially substituted with KCl.** *Food QualPrefer* 2010, **21**:489–494.

21. Angus F, Phelps T, Clegg S, Narein C, Ridder Den C, Kilcast D: *Salt in processed foods.* Leatherhead, UK: Leatherhead Food International; 2005. N0 00193.

22. Guerrero A, Kwon SS-Y, Vadehra DV: **Salt enhanced foods.** *Societe des Products Nestle* 1995, (Nestle SdP EP 0677249.

23. Lawrence G, Salles C, Palicki O, Septier C, Busch J, Thomas-Danguin T: **Using cross-modal interactions to counterbalance salt reduction in solid foods.** *Int Dairy J* 2011, 21:103–110.

24. Lawrence G, Salles C, Septier C, Busch J, Thomas-Danguin T: **Odour-taste interactions: a way to enhance saltiness in low-salt content solutions.** *Food Qual Pref* 2009, 20:241–248.

25. Yamamoto Y, Nakabayashi M: **Enhancing effect of an oil phase on the sensory intensity of salt taste of NaCl in oil/water emulsions.** *J Texture Stud* 1999, 30:581–590.

26. Malone ME, Appelqvist IAM, Norton IT: **Oral behaviour of food hydrocolloids and emulsions. Part 2. Taste and aroma release.** *Food Hydrocolloid* 2003, 17:775–784.

27. Goh SM, Leroux B, Groeneschild CAG, Busch J: **On the effect of tastant excluded fillers on sweetness and saltiness of a model food.** *J Food Sci* 2010, 75:S245–S249.

28. Noort MWJ, Bult JHF, Stieger M, Hamer RJ: **Saltiness enhancement in bread by inhomogeneous spatial distribution of sodium chloride.** *J Cereal Sci* 2010, 52:378–386.

29. Burseg KMM, Brattinga C, de Kok PMT, Bult JHF: **Sweet taste enhancement through pulsatile stimulation depends on pulsation period not on conscious pulse perception.** *Physiol Behav* 2010, 100:327–331.

30. Busch J, Tournier C, Knoop JE, Kooyman G, Smit G: **Temporal contrast of salt delivery in mouth increases salt perception.** *Chem Senses* 2009, 34:341–348.

31. Morris C, Koliandris AL, Wolf B, Hort J, Taylor A: **Effect of pulsed or continuous delivery of salt on sensory perception over short time intervals.** *Chemosens Percept* 2009, 2:1–8.

32. Pawlik A, Cox PW, Norton IT: **Food grade duplex emulsions designed and stabilised with different osmotic pressures.** *J Colloid Interface Sci* 2010, 352:59–67.

33. Garti N, Aserin A, Cohen Y: **Mechanistic considerations on the release of electrolytes from multiple emulsions stablized by BSA and nonionic surfactants.** *J Controlled Release* 1994, 29:41–51.

34. Rojas EC, Staton JA, John VT, Papadopoulos KD: **Temperature-induced protein release from water-in-oil-in-water double emulsions.** *Langmuir* 2008, 24:7154–7160.

35. Krieger IM, Dougherty TJ: **A mechamism for non-Newtonian flow in suspensions of rigid spheres.** *Trans Soc Rheol* 1959, 3:137–152.

36. Baines ZV, Morris ER: **Flavour/taste perception in thickened systems: the effect of guar gum above and below c.** *Food Hydrocolloid* 1987, 1:197–205.

37. Christensen CM: **Effects of solution viscosity on perceived saltiness and sweetness.** *PerceptPsychophys* 1980, 28:347–353.

38. Cook DJ, Hollowood TA, Linforth RST, Taylor AJ: **Perception of taste intensity in solutions of random-coil polysaccharides above and below c.** *Food Qual Pref* 2002, 13:473–480.

39. Moskowitz HR, Arabie P: **Taste intensity as a function of stimulus concentration and solvent viscosity.** *J Texture Stud* 1970, 1:502–510.

40. Pangborn RM, Traube IM: **Effect of hydrocolloids on oral viscosity and basic taste intensities.** *J Texture Stud* 1973, 4:224–241.

41. Koliandris AL, Morris C, Hewson L, Hort J, Taylor AJ, Wolf B: **Correlation between saltiness perception and shear flow behaviour for viscous solutions.** *Food Hydrocolloid* 2010, 24:792–799.

42. Ferry AL, Hort J, Mitchell JR, Cook DJ, Lagarrigue S, Pamies BV: **Viscosity and flavour perception: Why is starch different from hydrocolloids?** *Food Hydrocolloid* 2006, 20:855–862.

43. Duke JA: *Handbook of Legumes of World Economic Importance.* New York: Plenum Press; 1981.

44. Hulse JH: **Nature, composition and utilization of food legumes.** In *Expanding the Production and Use of Cool Season Food Legumes.* Edited by Muehlbauer FJ, Kaiser WJ. Dordrecht: Kluwer Academic Publisher; 1994:77–97.

45. Pal R: **Rheology of simple and multiple emulsions.** *Curr OpinColloid Interface Sci* 2011, 16:41–60.

17

The edible cocktail: the effect of sugar and alcohol impregnation on the crunchiness of fruit

Elke Scholten[1*] and Miriam Peters[1,2]

Abstract

Background: Vacuum impregnation is seen as a valuable technique for flavor pairing in the catering industry. One of the applications of this technique is the creation of edible cocktails by impregnating fruits with liquors, leading to an interplay of different flavors. However, the effect of the impregnation of sugar and alcohol into the fruit will affect the texture of the fruit and therefore its crunchiness. Thus, the positive effect of flavor pairing might be inhibited by a negative effect in texture changes.

Results: This investigation focused on the change in crunchiness as a result of the impregnation of different sugar and alcohol containing solutions. When hypotonic solutions were used, the impregnation resulted in the rupture of the cells, thereby leading to a decrease in crunchiness. When hypertonic solutions were used, the cells shrunk, which also resulted in a decrease in crunchiness. Isotonic solutions resulted in crunchiness comparable to its fresh version. When alcohol was used, the crunchiness decreased at all concentrations investigated.

Conclusions: Crunchiness of fruit can only be maintained when impregnated with isotonic sugar solutions. When the sugar or alcohol content deviates from that in the fruit, impregnation of these liquids will lead to a decrease in crunchiness. This has consequences for the creation of edible cocktails: for an optimal crunchiness, the sugar content of the impregnation liquid has to be equal to the sugar content of the fruit or vegetable.

Background

The impregnation of different solutes into porous materials is an important process in the food industry. As the demand for healthy and natural products is increasing, processes are being used to adjust food system formulations by processes like osmotic dehydration, impregnation and ultrasound treatments [1-7]. Examples can be found where fruits and vegetables are enriched with cryoprotectants, calcium or zinc [8-10]. The amount of solutes and liquid that can be absorbed by the porous products depends on two diffusion processes: molecular and capillary diffusion [2,11,12], which are influenced by parameters such as surface tension, pore size, porosity, and others. The process is often slow and absorption of liquids is limited. To overcome these limitations, vacuum impregnation is more often used [2-4,10]. This technique applies a reduced pressure to porous materials, which expands the air in the open spaces and forces it to leave the porous

material [13]. As the air flows out of the pores, any surrounding liquid is allowed to flow in through the capillary pores, a phenomenon called hydrodynamic mechanism (HDM). The pressure change can promote deformations in the product due to the viscoelastic properties of the matrix, which leads to the coupling of the HDM with deformation-relaxation phenomena (DRP) [14]. Due to the simultaneous expansion of air and flow of liquid, air in porous products can be substituted with surrounding liquids. This technique therefore allows the uptake of more liquid in a shorter period of time [13]. It offers the possibility to impregnate any porous matter with any liquid, which makes pairing of a large variety of food materials with different liquids possible.

Vacuum impregnation is a very simple technique to fill porous materials with liquids and therefore offers opportunities in restaurants, bars and the retail sector. Recently, restaurants and bars have been exploring the options of using these industrial techniques to create different textures and combinations of flavors in their dishes [15,16]. Almost all food products that are used in these sectors are porous materials, such as meat, fish, fruit and vegetables,

---
* Correspondence: elke.scholten@wur.nl
[1]Food Physics Group, Department of Agrotechnology and Food Sciences, Wageningen University, Bomenweg 26703HDWageningen, the Netherlands
Full list of author information is available at the end of the article

and can be impregnated or marinated with oils and flavored liquids. Combinations of chocolate milk in strawberries, orange juice in apple and vodka in cucumber are
just a few examples of the many combinations possible.
Liquor-filled pieces of fruit are described as edible cocktails, and combine the flavor of the liquor with the texture
of fruit. Besides affecting the flavor of the fruit, the impreg-
nations of these sweet and alcoholic liquids also have an
effect on the texture of the fruits and vegetables. The type
of incorporated liquid can lead to dramatic structural
changes of the fruit, which lead to changes in sensory perception. Although the change in flavor might have beneficial effects, the dramatic changes in the texture of the fruit
might lead to detrimental effects in these edible cocktails.
This work focuses on the structural changes of the edible
cocktails when impregnated with sugar and alcohol solutions. The change in texture is examined by investigation
of the crunchiness (toughness) of the fruit.

Results and discussion
Weight gain of impregnated fruit
The impregnation of different solutes into fruits and vegetables has an effect on texture. During the impregnation,
air pockets or interstitial spaces between cells are filled by
the impregnation liquid. This will give rise to a difference
in osmotic pressure within the porous material [13]. As a
result, osmosis will occur that will lead to the flow of
liquids and solutes. The osmotic treatment can have different effects on the food material, depending on the nature
of the osmotic solutions. Three types of solutions can be
distinguished: isotonic, a solution containing a similar concentration of solutes as the food material; hypotonic, a solution containing less solute molecules than the food
material; and hypertonic, a solution containing more solute molecules than the food material [13]. Osmosis will
lead to a change in pressure of the cells (turgor pressure)
and will have an effect on the cell size and the weight gain
of the fruit samples. Figure 1 shows the weight gain for an
apple, melon and cucumber impregnated in sugar solutions with different concentrations.

When isotonic solutions are used, no net liquid flow
will occur and the plant cell sizes will not change. Total
weight gain of the samples at these concentrations can
be regarded as the weight gain due to the filling of the
interstitial spaces only, and can be used to calculate the
porosity of the fruit. Isotonic solutions are sugar solutions with an equal concentration of sugar as the fruit
itself. In fruits, the sugar concentration is given as a brix
value (°Bx), which contains the contributions of the different sugars present in fruits and vegetables (for example, sucrose and fructose). Refractive index
measurement results in 12°Bx for apples (equivalent to
12% wt/wt sucrose), 6°Bx for melon (equivalent to 6%
wt/wt sucrose) and negligible values for cucumber. This

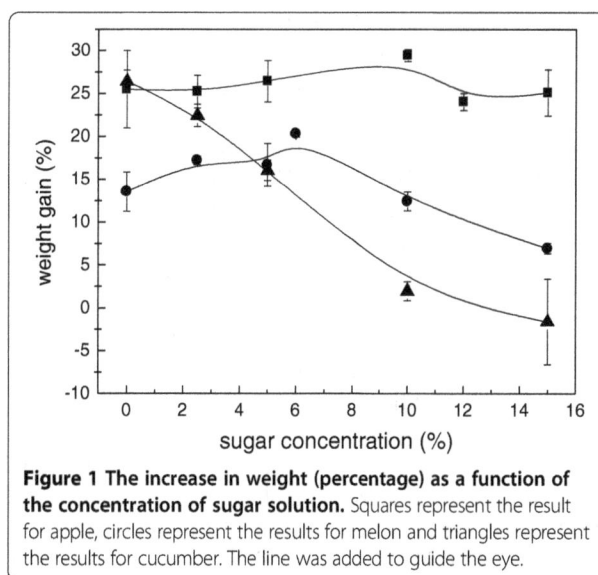

**Figure 1 The increase in weight (percentage) as a function of
the concentration of sugar solution.** Squares represent the result
for apple, circles represent the results for melon and triangles represent
the results for cucumber. The line was added to guide the eye.

is in agreement with values found previously [17]. This
brix value is used in the calculation of porosity, which
works out as 30% for apple, comparable to results found
previously [4], 20% for melon and 25% for cucumber.

At sucrose concentrations different from the brix
values of the fruit itself, osmotic differences will lead to
changes in the cell structure as a flow of liquid and
solutes is generated. In hypotonic solutions (low sugar
concentration), the osmotic pressure within the cells is
higher than in its surroundings. To decrease the difference in pressure, water will flow into the cells and
solutes diffuse to the outside. This will lead to the swelling of the cells, and may lead to rupture, thereby
destroying the cell wall. For the melon, we see that the
weight gain at low sugar concentrations is decreased in
comparison to in an isotonic solution. As melon has a
soft structure, it is reasonable to assume that the cell
swelling leads to rupture of the cells. As the liquid is
now allowed to flow from the cells, the total weight gain
decreases as the amount of intact cells is reduced. This
effect is less visible in the case of apple, which has a
firmer structure. As the cucumber does not contain any
sugar, no hypotonic osmosis is observed.

On the other hand, when placed in hypertonic solutions, cells will shrink as water will be pulled out of the
cells to decrease the pressure in the surrounding solutions. At the same time, the sugar diffuses into the plant
cells. However, as the sugar molecules are much larger
than water molecules, diffusion will be limited as the
molecules have to diffuse through the network of the cell
wall. The composition of the cell wall determines the
total amount of solute that can penetrate through the
cell. For both melon and cucumber, we see a large decrease in weight gain in hypertonic impregnation solutions compared to in isotonic solutions. This is a result

of the shrinkage of the cells that leads to a collapse of the structure. Although we see large differences for melon and cucumber, the weight gain for apple does not show dramatic changes. However, as apple contains a large amount of sugar, this effect will only occur at sugar concentrations higher than 12% wt/wt, and the effect is therefore very limited in this concentration range. As the results for the melon and cucumber show, hypertonic solutions lead to water removal from the cells and are therefore often used to dehydrate fruit samples [3,6,18].

These results show that the weight gain for the melon and the cucumber is largest when the samples are placed in an isotonic solution; no change in plant cell size has occurred. When the samples are placed in either a hyper- or hypotonic solution, we see a decrease in the weight gain, and therefore a dramatic change in the structure of the fruit. However, impregnation of the apple does not result in large differences in weight gain regardless of the type of solution. Apparently, the structure is not that much affected by the solutions compared to the much softer melon and cucumber. Overall, the degree of shrinkage for concentrated sugar solutions (15% wt/wt) is related to the sugar content of the fruit; apple, containing the highest concentration of sugar, shows the least effect; cucumber, containing the lowest concentration, shows the largest effect.

Figure 2 shows the weight gain of apple, melon and cucumber impregnated in ethanol solutions. In the case of impregnation with alcohol solutions, both water and alcohol will flow into the samples and sugar diffuses out. The rate of diffusion will depend on the size of the different molecules. As water is the smallest molecule, it will have the largest diffusion capacity. Sugar is the largest and will therefore have the lowest diffusion capacity through the

cell wall. The ratio between the different diffusion coefficients and the immersion time will determine the relative changes of the solution and solutes. The interplay between the three diffusion processes will determine the total flow, and depends on the osmolarities of the solutes and the permeability of the cell walls. As can be seen, the weight of the sugar-containing apple and melon does not change dramatically. The diffusion rates for water, ethanol and sugar are in balance to some extent. When placed in high concentrations of alcohol, a slightly larger decrease in overall weight gain is observed: water is being pulled out of the cells. Apparently, the water flux from inside to outside of the cells is larger than the ethanol flux into the cells. Even though the apple and the melon do not show large differences in weight gain, the cucumber does. Similar to the impregnation with sugar solutions, the weight gain for the cucumber samples decreases with an increase in ethanol concentration. However, the total decrease in weight gain for higher ethanol concentrations is less than in the case of concentrated sugar solutions. As the osmolarity of an ethanol solution is approximately three times larger than the osmolarity of a sugar solution, the driving force for water extraction would be much higher using ethanol solutions, and should therefore result in larger cell shrinkage. As this is not the case, we can conclude that osmolarity by itself is not the only driving force for liquid flow. Apparently, the diffusion of ethanol determines to a great extent the exchange of liquids and solutes in the cells. As can be seen, the diffusion of ethanol into the cells is much faster than the diffusion of sugar. The diffusion coefficient (and thus the size of the solute) determines the total water flow. The larger the solute, the slower the diffusion through the cells, the more water is pushed out before equilibrium is reached, which leads to a larger weight loss (compared to isotonic conditions).

Fracture toughness measurements

The mechanical properties of the fruit are mainly responsible for the fracture behavior of these materials and will define their sensory attributes. Vincent and coworkers showed that the stress needed to crack a certain fruit sample is closely related to the sensory attributes of hardness and crunchiness [19,20]. The general fracture behavior of fruit can be ascribed to two different failure modes within the sample: cell separation and cell rupture [8,21]. The type of failure mode depends largely on the turgor pressure of the cells; cell debonding is more likely to occur at low turgor pressure and cell rupture at higher turgor pressure. The fracture (mechanical) properties of the impregnated fruit samples were measured with a three point bending test. The samples were placed on the support and loaded with a cell. Force displacements curves were recorded for all samples, for which the maximum load, P_c, before fracture was noted. Examples for

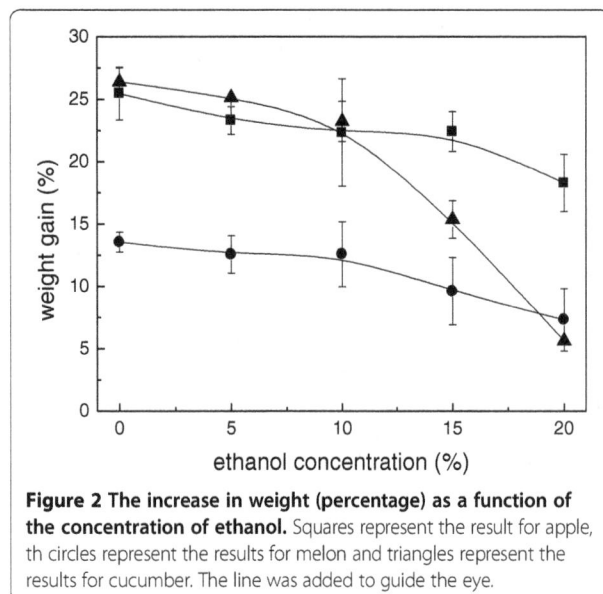

Figure 2 The increase in weight (percentage) as a function of the concentration of ethanol. Squares represent the result for apple, th circles represent the results for melon and triangles represent the results for cucumber. The line was added to guide the eye.

force displacement curves for untreated and water-impregnated fruit samples are shown in Figure 3.

These force displacement curves are approximately linear up to a certain critical load, at which one of the cracks propagates. As the crack gets arrested, the load is increased again, and decreased the moment another crack appears. The fracture toughness of the samples was calculated according to Equation 2. The maximal load used, needed to calculate the fracture toughness, is very dependent on differences in the samples, and therefore the test was done three times to obtain an averaged value. The solid lines in Figure 3 show the fracture behavior for the untreated samples. As can be seen, the fracture toughness of a fresh apple is highest, but impregnation with water leads to a large decrease in the maximal load. The maximal load for impregnating cucumber, on the other hand, does not differ significantly between untreated and impregnated samples.

Overall, the impregnation of the fruits had a negative effect on the fracture toughness compared to fresh fruit. The cucumber impregnated with 15% wt/wt of sugar does not even fracture, becoming too soft. A similar result is obtained for the melon with 5% wt/wt of ethanol.

Figure 4 shows the result of the fracture toughness of the samples impregnated with sugar solutions. As can be seen, the error bars of the measurements are quite large, which reflects the difficulty of working with real fruit systems. Fruit tissues are often anisotropic and therefore large fluctuations in fracture can be observed [22]. Fracture will occur along the weakest pathway of resistance, and will therefore largely depend on the tissue structure and the orientation of the network. Even though the samples were all loaded in the same direction, differences in fracture behavior are still visible.

The fracture toughness reflects the fracture behavior of the fruits through cell rupture and cell separation, which is mainly influenced by the cell turgor pressure and the composition of the cell wall and the lamellae. Despite these large fluctuations in the fracture toughness, a clear trend can be observed for the three fruit types. The largest fracture toughness is found in isotonic solutions; around 12% wt/wt for the apple, 6% wt/wt for the melon and 0% for the cucumber. These values are the most comparable with the values for the untreated fresh sample. At these concentrations of sugar, the cell turgor pressure does not change, and no large changes in fracture toughness due to cell rupture will be present. However, when hypo- and hypertonic solutions are used, the turgor pressure on the cell wall does change. In hypotonic solutions, the cells swell, and according to the weight gain experiments, cell rupture has

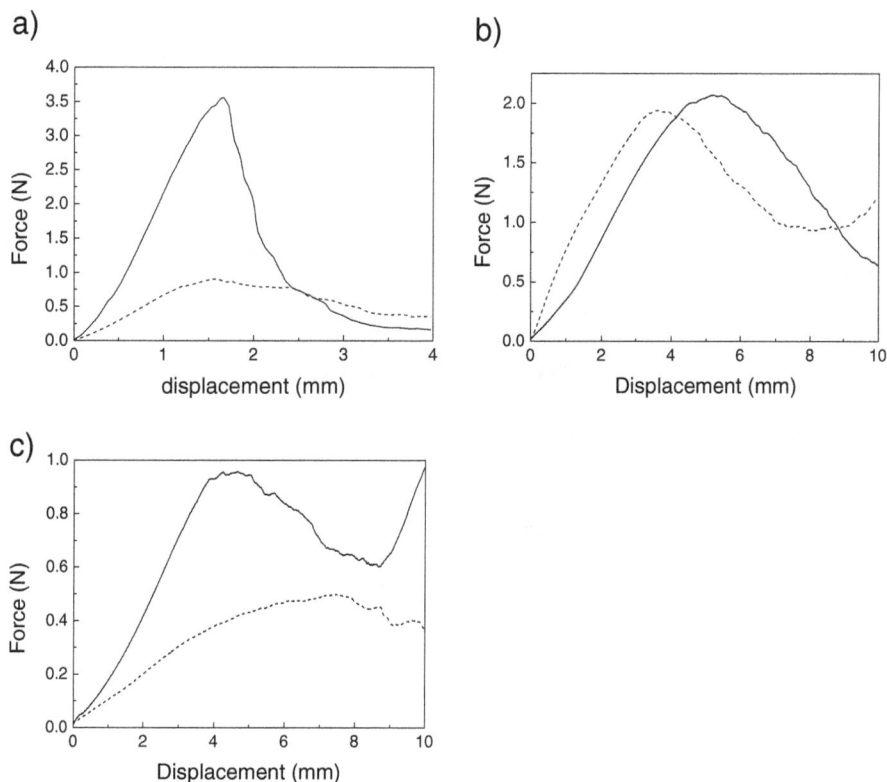

Figure 3 Force displacement curves for loaded fruit samples. (a) apple; **(b)** cucumber; **(c)** melon. Solid lines refer to the untreated fresh fruit. Dashed lines refer to the treated samples with water (no solutes).

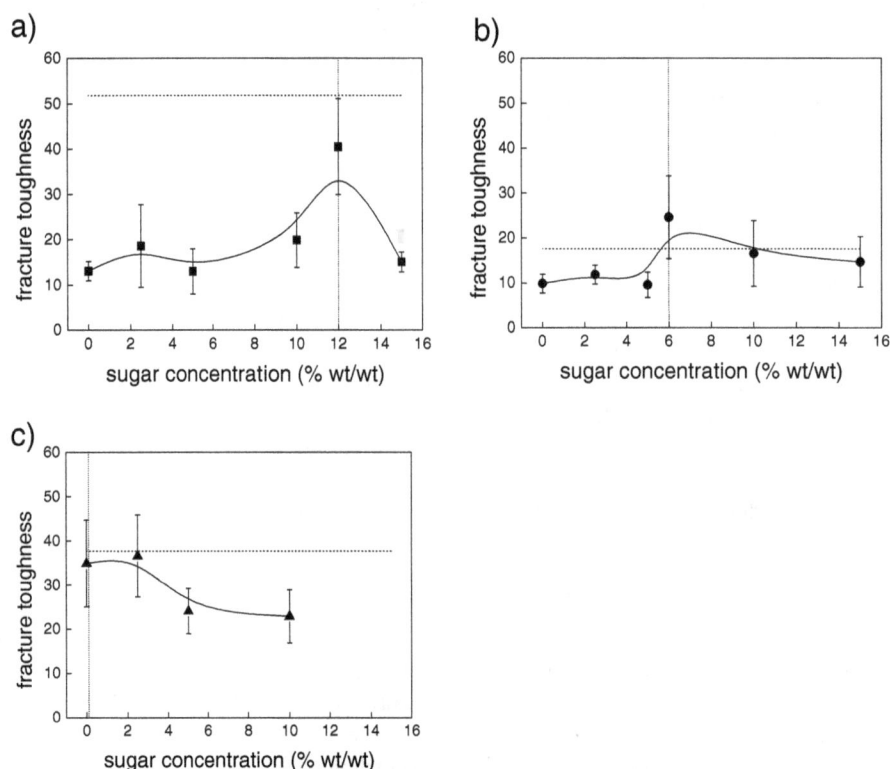

Figure 4 The fracture toughness as a function of the concentration of sugar. (a) Squares represent the result for apple; **(b)** circles represent the results for melon; **(c)** triangles represent the results for cucumber. The line was added to guide the eye. Dashed horizontal line refers to the toughness of an untreated sample. Dotted vertical line refers to the isotonic solution.

probably occurred. As the cell ruptures, fracture toughness decreases, which is clearly visible in Figure 5. Had cell rupture not occurred, cell fracture would have been more difficult, which would have led to higher fracture toughness. In the case of hypertonic solutions, water is extracted from the cells, which causes a loss in turgor pressure (shrinkage of the cells) and a decrease in cell wall elasticity. This leads to a decrease in the fracture toughness.

Although the pressure seems to be the most important parameter responsible for the mechanical properties of fruit samples, other attributes have also been shown to contribute: cell wall resistance, cell bonding, cell density and porosity [8]. Cell separation, especially, is an important factor [21]. The impregnation solutions have an effect on the cell walls and the lamellae between the cells. The solutes in the solutions have the ability to change the composition of the cell wall. Plant cell walls consist of polysaccharides, such as pectins, celluloses, hemicelluloses, starch and galactomannans, which can be found in the primary cell wall and the middle lamellae [23]. These polysaccharides can be degraded (by age or solutes) and determine the composition and, therefore, the strength of the cell wall. Pectin, one of the polysaccharide, has been found to have the largest effect on fruit softening, as the dissolution of pectin leads to weaker cell walls and the dissolution of

middle lamellae, leading to separation of the plant cells [7,10,18,23-26]. The incorporation of liquid sugar solutions leads to solubilization of the pectin and decreases the cell adhesion. This leads to a decrease in fracture toughness, which is visible for all fruits as the fracture toughness of all the impregnated samples was lower than the fracture toughness of fresh fruit.

Figure 5 shows the fracture toughness for the samples impregnated with the alcohol solutions. During the impregnation, water diffuses out of the cells while ethanol will diffuse into the cells, until equilibrium in osmotic pressure is reached. The exchange of water for ethanol has already been shown to cause shrinkage of cells in the case of apples [27].

There was a large decrease in fracture toughness for apple and melon in ethanol solutions compared to sugar solutions. For cucumber, we see a similar negative effect as in the case of the sugar impregnation. Even though the ethanol solution will have an osmotic pressure comparable to the fruit's cells at a certain concentration, it appears that the presence of alcohol has a larger effect than just the effect on turgor pressure. This is probably due to cell separation instead of cell rupture. When the ethanol is drawn into the cells, it changes the composition of the cell walls and the strength of the lamellae. The cell wall

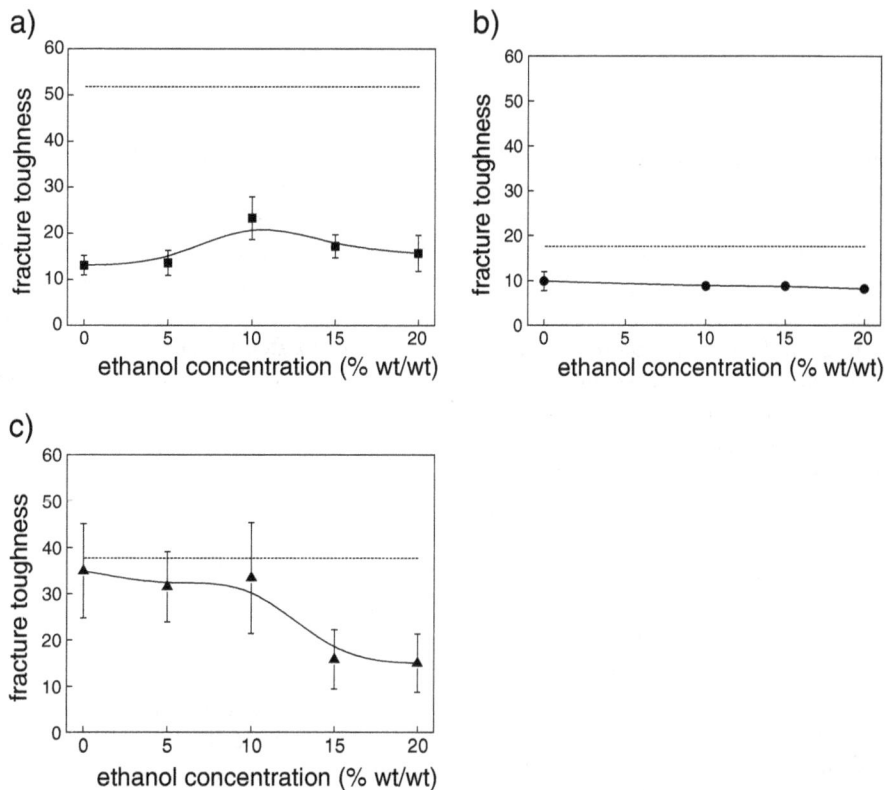

Figure 5 The fracture toughness as a function of the concentration of alcohol. (a) Squares represent the result for apple; **(b)** circles represent the results for melon; **(c)** triangles represent the results for cucumber. The line was added to guide the eye. Dashed horizontal line refers to the toughness of an untreated sample.

polysaccharides are soluble in water, but mostly insoluble in ethanol so they precipitate out. This precipitation distorts the network and its strength. For apple and melon, we see that the fracture toughness for ethanol-impregnated samples is much lower than for sugar-impregnated samples. The presence of ethanol therefore seems to have a negative effect on the cell wall strength. Apparently, the precipitation of the polysaccharides leads to a loss in cell wall strength and, therefore, increased ease of cell separation or debonding. However, the exact contributions of cell rupture and cell separation are not clear. Although research can be found on the cell tissue structure for osmotic hydrated fruit, the effect of alcohol on the cell structure has not been studied extensively.

The effect of impregnation of liquors into fruit: The edible cocktails, practical implications

Vacuum impregnation has proven to be an effective method to incorporate liquids into porous fruits and vegetables. This provides the opportunity to combine different flavors of fruit with additional flavored liquids. In the catering and beverage industry, it can be used to combine fruits, such as apples and cucumber, with alcoholic beverages, such as vodka and martini, to create impregnated fruit samples, also known as edible cocktails. These edible cocktails can be made by placing a small piece of fruit in a liquid-filled container (a combination of liquor, sugar solution and flavored water), and then placing the container in a vacuum sealer. The vacuum should be applied for roughly 30 minutes to allow the air to escape, after which the vacuum showed be released slowly to allow the liquid to enter the fruit gradually and fill all available pores. When this exercise is performed, the liquids and its flavors are infused into the fruit. However, not only the flavors of these products will change, but also the texture of the fruit and therefore the sensory perception. Using this method for flavor pairing, one therefore has to keep in mind that it comes with a change in texture, which might be detrimental to consumers' perception. If one would like to keep the crunchiness and hardness of the fruit, the fruit should be infused with isotonic solutions. The sugar concentration of the fruit can be determined using a Brix meter. One degree of Brix is equivalent to a 1% sugar solution. For example, a fruit with a Brix value of 10 will best maintain its crunchiness and hardness with liquid that contains 10% sugar (wt/wt). Large deviations from the sugar content of the fruit will likely lead to a loss of texture. The impregnation of alcohol always leads to a decrease in the texture and will make the fruit much softer.

Conclusion

Vacuum impregnation has been evaluated as a new technique for flavor pairing in restaurants and bars. The technique can be used to infuse different liquids into porous materials, such as meat, fish, fruit and vegetables. This research focused on the impregnation of different sugar- and ethanol-containing liquids into fruits. Although the combination of flavors can be a positive contribution to a dish, the change in texture could be detrimental to consumers' perception. This research therefore focused on the effect on texture of impregnation with different solutes into apple, melon and cucumber. The impregnation of sugar and alcohol solutions into the three different types of fruit resulted in a large change in weight gain and fracture toughness. When isotonic solutions were used, maximum weight gain and maximum fracture toughness were found. This indicated an unchanged cell structure; fracture behavior was maintained. When impregnated with hypotonic solutions, a lower weight gain and decreased fracture toughness was found. This could be attributed to cell swelling and rupture. Hypertonic solutions induced cell shrinkage and cell debonding, which also led to a lower weight gain and decreased fracture toughness. The impregnation of alcohol led to a decrease in fracture toughness in all cases. These results show that the impregnation of fruit samples with different liquids, either sugar- or alcohol-containing, has a large influence on the crunchiness of the fruit.

Methods

Materials

The fruits used in this study were apple (Elstar), melon (Galia) and cucumber and were purchased from a local store. They were selected on the basis of size and visual absence of damages. They were stored at room temperature and used within one day of purchase. The solutions for impregnation were prepared with sucrose and ethanol. The sucrose was obtained from Suikerunie (Oud-Gastel, the Netherlands). Ethanol was obtained from Sigma-Aldrich (Zwijndrecht, the Netherlands). The solutions were prepared with demineralized water.

Sample preparation for impregnation treatments

The apple, melon and cucumber were cut in small pieces of roughly $4.0 \times 1.0 \times 0.8$ cm. Pieces were cut in identical directions within the fruit tissue to avoid differences in cell build-up. Skin and seed areas were avoided to create homogeneous textures and the pieces were weighed and measured before use. Three pieces of each kind were placed in a container that contained 50 mL of the impregnation solution. These solutions contained either

sugar (0%, 2.5%, 5%, 6%, 10%, 12% and 15% wt/wt) or alcohol (0%, 5%, 10%, 15% or 20% wt/wt). The containers were placed in a vacuum oven at 25 °C and the pressure was reduced to near vacuum. The samples were kept at this pressure for 30 minutes, after which atmospheric pressure was restored slowly. The impregnated pieces of fruit were removed from the container and the excess solution removed from the surface. The samples were weighed again and the uptake of the solution was noted. Assuming that isotonic solutions do not have an effect on the structure of the cells, the amount of liquid that was incorporated is equal to the effective porosity, which can be interpreted as the fraction of the pores that are available for HDM. The effective porosity, ϵ, is related to the weight gain as below:

$$\epsilon = \frac{M_f - M_i}{\rho_s V_0} \tag{1}$$

where M_f is the final mass of the material, M_i is the initial mass of the material, ρ is the density of the solution and V_0 is the initial volume of the material. This only applies to fruit types that do not show a large DRP due to viscoelastic effects, as this would influence the structure of the cells. Previous studies have shown that firm fruits, such as apples and melons, suffer minor deformations, indicating that DRP can be neglected [4].

Mechanical test

Mechanical tests were carried out using a Texture Analyzer, equipped with a load cell of 50 N and a homemade razor blade. To measure the fracture toughness of the impregnated samples, a three point bending test was used, which has been shown to give good results for the determination of fracture toughness [28,29]. The samples were placed horizontally on the supporting carriers of the set-up. Prior to testing, the samples were notched with a razor blade to halfway through the sample, and the notch in the sample was placed facing down underneath the loading point. The ratio between the depth of this notch

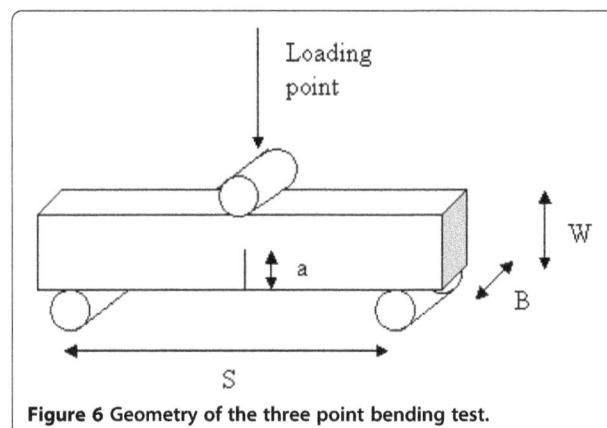

Figure 6 Geometry of the three point bending test.

and the total width of the sample (a/W) was between 0.45 and 0.55. The razor blade was then lowered into the samples with a speed of 1 mm/s. The maximal force that corresponded to the onset of crack propagation was recorded for all samples. The measurements were performed three times, and the fracture toughness, K_{IC} was calculated where P_c is the maximal load before propagation starts, S is the distance between the supports, according to:

$$K_{IC} = \frac{P_c S}{B W^{3/2}} \cdot f\left(\frac{a}{W}\right) \qquad (2)$$

B is the breadth of the specimen, W is the width of the specimen and a is the crack length. $F(a/W)$ is a correction factor, which is equal to 1.5 [30]. The geometry for the experimental set-up can be found in Figure 6. The subscript IC refers to the mode I loading conditions.

Competing interests

The authors declare that they have no competing interests.

Author details

[1]Food Physics Group, Department of Agrotechnology and Food Sciences, Wageningen University, Bomenweg 26703HDWageningen, the Netherlands. [2]FrieslandCampina Cheese & Butter, Nieuwe Kanaal 7R6709PAWageningen, the Netherlands.

Authors' contributions

ES has made the major contribution to the design and drafted the article. MP performed the experimental work. Both authors read and approved the final manuscript.

Authors' information

ES works as an assistant professor at Wageningen University. One of the interests of the Food Physics group is Molecular Gastronomy, which is part of the educational program. This field of science is focused on the link between science and real applications and/or innovations in restaurants and bars.

References

1. Torreggiani D, Bertolo G: Osmotic pre-treatments in fruit processing: chemical, physical and structural effects. *J Food Eng* 2001, **49**:247–253.
2. Guillemin A, Degraeve P, Noel C, Saurel R: Influence of impregnation solution viscosity and osmolarity on solute uptake during vacuum impregnation of apple cubes (var. Granny Smith). *J Food Eng* 2008, **86**:475–483.
3. Mujica-Paz H, Valdez-Fragoso A, Lopez-Malo A, Palou E, Welti-Chanes J: Impregnation and osmotic dehydration of some fruits: effect of the vacuum pressure and syrup concentration. *J Food Eng* 2003, **57**:305–314.
4. Mujica-Paz H, Valdez-Fragoso A, Lopez-Malo A, Palou E, Welti-Chanes J: Impregnation properties of some fruits at vacuum pressure. *J Food Eng* 2003, **56**:307–314.
5. Valdez-Fragoso A, Soto-Caballero MC, Blanda G, Welti-Chanes J, Mujica-Paz H: Firmness changes of impregnated whole peeled prickly pear. *J Text Studies* 2009, **40**:571–583.
6. Lewicki PP, Porzecka-Pawlak R: Effect of osmotic dewatering on apple tissue structure. *J Food Eng* 2005, **66**:43–50.
7. Fernandes FAN, Gallao MI, Rodrigues S: Effect of osmotic dehydration and ultrasound pre-treatment on cell structure: melon dehydration. *LWT-Food Science and Technology* 2008, **41**:604–610.
8. Chiralt A, Martinez-Navarrete M, Martinez-Monzo J, Talens P, Moraga G, Ayala A, Fito P: Changes in mechanical properties throughout osmotic processes - Cryoprotectant effect. *J Food Eng* 2001, **49**:129–135.
9. Sousa MB, Canet W, Alvarez MD, Fernandez C: Effect of processing on the texture and sensory attributes of raspberry (cv. Heritage) and blackberry (cv. Thornfree). *JJ Food Eng* 2007, **78**:9–21.
10. Xie J, Zhao Y: Nutritional enrichment of fresh apple (Royal Gala) by vacuum impregnation. *Int J Food Sci Nutr* 2003, **54**:387–398.
11. Chiralt A, Fito P: Transport mechanisms in osmotic dehydration: the role of the structure. *Food Sci Tech Int* 2003, **9**:179–186.
12. Fito P, Chiralt A: Food matrix engineering: The use of the water-structure-functionality ensemble in dried food product development. *Food Sci Tech Int* 2003, **9**:151–156.
13. Zhao YY, Xie J: Practical applications of vacuum impregnation in fruit and vegetable processing. *Trends Food Sci Tech* 2004, **15**:434–451.
14. Salvatori D, Andres A, Chiralt A, Fito P: The response of some properties of fruits to vacuum impregnation. *J Food Process Eng* 1998, **21**:59–73.
15. Vega C, Ubbink J: Molecular gastronomy: a food fad or science supporting innovative cuisine? *Trends Food Sci Tech* 2008, **19**:372–382.
16. Cousins J, O'Gorman K, Stierand M: Molecular gastronomy: cuisine innovation or modern day alchemy? *Int J Cont Hosp Manage* 2010, **22**:399–415.
17. McFeeters RF, Lovdal LA: Sugar composition of cucumber cell-walls during fruit development. *J Food Sci* 1987, **52**:996–1001.
18. Fernandes FAN, Gallao MI, Rodrigues S: Effect of osmosis and ultrasound on pineapple cell tissue structure during dehydration. *J Food Eng* 2009, **90**:186–190.
19. Vincent JFV: Application of fracture mechanics to the texture of food. *Eng Fail Anal* 2004, **11**:695–704.
20. Vincent JFV, Saunders DEJ, Beyts P: The use of critical stress intensity factor to quantify "hardness" and "crunchiness" objectively. *J Text Studies* 2002, **33**:149–159.
21. Harker FR, White A, Gunson FA, Hallett IC, De Silva HN: Instrumental measurement of apple texture: a comparison of the single-edge notched bend test and the penetrometer. *Postharvest Biol Tech* 2006, **39**:185–192.
22. Khan AA, Vincent JFV: Anisotropy in the fracture properties of apple flesh as investigated by crack-opening tests. *J Mat Sci* 1993, **28**:45–51.
23. Prasanna V, Prabha TN, Tharanathan RN: Fruit ripening phenomena - an overview. *Crit Rev Food Sci Nutr* 2007, **47**:1–19.
24. Martinez VY, Nieto AB, Castro MA, Salvatori D, Alzamora SM: Viscoelastic characteristics of Granny Smith apple during glucose osmotic dehydration. *J Food Eng* 2007, **83**:394–403.
25. Chassagne-Berces S, Poirier C, Devaux MF, Fonseca F, Lahaye M, Pigorini G, Girault C, Marin M, Guillon F: Changes in texture, cellular structure and cell wall composition in apple tissue as a result of freezing. *Food Res Int* 2009, **42**:788–797.
26. Redgwell RJ, Curti D, Gehin-Delval C: Role of pectic polysaccharides in structural integrity of apple cell wall material. *Eur Food Res Techn* 2008, **227**:1025–1033.
27. Kunzek H, Kabbert R, Gloyna D: Aspects of material science in food processing: changes in plant cell walls of fruits and vegetables. *Eur Food Res Tech Z Lebensm Unters Forsch* 1999, **208**:233–250.
28. Pitts MJ, Davis DC, Cavalieri RP: Three-point bending: an alternative method to measure tensile properties in fruit and vegetables. *Postharvest Bio Tech* 2008, **48**:63–69.
29. Alvarez MD, Saunders DEJ, Vincent JFV: Fracture properties of stored fresh and osmotically manipulated apple tissue. *Eur Food Res Tech* 2000, **211**:284–290.
30. Wang CH: *Introduction to fracture mechanics*. Melbourne, Australia: DSTO Aeronautical and Maritime Research Laboratory; 1996.

18

A genetic variant near olfactory receptor genes influences cilantro preference

Nicholas Eriksson*, Shirley Wu, Chuong B Do, Amy K Kiefer, Joyce Y Tung, Joanna L Mountain, David A Hinds and Uta Francke

Abstract

Background: The leaves of the *Coriandrum sativum* plant, known as cilantro or coriander, are widely used in many cuisines around the world. However, far from being a benign culinary herb, cilantro can be polarizing—many people love it while others claim that it tastes or smells foul, often like soap or dirt. This soapy or pungent aroma is largely attributed to several aldehydes present in cilantro. Cilantro preference is suspected to have a genetic component, yet to date nothing is known about specific mechanisms.

Results: Here, we present the results of a genome-wide association study among 14,604 participants of European ancestry who reported whether cilantro tasted soapy, with replication in a distinct set of 11,851 participants who declared whether they liked cilantro. We find a single-nucleotide polymorphism (SNP) significantly associated with soapy-taste detection that is confirmed in the cilantro preference group. This SNP, rs72921001 ($p = 6.4 \times 10^{-9}$, odds ratio 0.81 per A allele), lies within a cluster of olfactory receptor genes on chromosome 11. Among these olfactory receptor genes is *OR6A2*, which has a high binding specificity for several of the aldehydes that give cilantro its characteristic odor. We also estimate the heritability of cilantro soapy-taste detection in our cohort, showing that the heritability tagged by common SNPs is low, about 0.087.

Conclusions: These results confirm that there is a genetic component to cilantro taste perception and suggest that cilantro dislike may stem from genetic variants in olfactory receptors. We propose that one of a cluster of olfactory receptor genes, perhaps *OR6A2*, may be the olfactory receptor that contributes to the detection of a soapy smell from cilantro in European populations.

Keywords: Cilantro, Coriander, Olfactory receptor, Genetics of taste and smell

Background

The *Coriandrum sativum* plant has been cultivated since at least the second millennium BCE [1]. Its fruits (commonly called coriander seeds) and leaves (called cilantro or coriander) are important components of many cuisines. In particular, South Asian cuisines use both the leaves and the seeds prominently, and Latin American food often incorporates the leaves.

The desirability of cilantro has been debated for centuries. Pliny claimed that coriander had important medicinal properties: '*vis magna ad refrigerandos ardores viridi*' ('while green, it is possessed of very cooling and refreshing properties') [2]. The Romans used the leaves and seeds in many dishes, including moretum (a herb, cheese, and garlic spread similar to today's pesto) [3]; the Mandarin word for cilantro, 香菜 (*xiāngcài*), literally means 'fragrant greens.' However, the leaves in particular have long inspired passionate hatred as well, e.g., John Gerard called it a 'very stinking herbe' with leaves of 'venemous quality' [4,5].

It is not known why cilantro is so differentially perceived. The proportion of people who dislike cilantro varies widely by ancestry [6]; however, it is not clear to what extent this may be explained by differences in environmental factors, such as frequency of exposure. In a twin study, the heritability of cilantro dislike has been estimated as 0.38 (confidence interval (CI) 0.22–0.52) for odor and 0.52 (CI 0.38–0.63) for flavor [7].

The smell of cilantro is often described as pungent or soapy. It is suspected, although not proven, that cilantro dislike is largely driven by the odor rather than the taste.

*Correspondence: nick@23andme.com
23andMe, Inc., Mountain View, CA, 94043, USA

Table 1 Summary of the cohorts used in the analysis

	N	Proportion female	Age (SD)
Tastes soapy	1,994	0.566	49.0 (15.0)
Does not taste soapy	12,610	0.489	48.3 (15.2)
Total	14,604	0.500	48.4 (15.2)
Dislikes cilantro	3,181	0.487	47.1 (16.6)
Likes cilantro	8,906	0.420	43.8 (14.5)
Total	12,087	0.438	44.7 (15.1)

The key aroma components in cilantro consist of various aldehydes, in particular (E)-2-alkenals and n-aldehydes [8,9]. The unsaturated aldehydes (mostly decanal and dodecanal) in cilantro are described as fruity, green, and pungent; the (E)-2-alkenals (mostly (E)-2-decenal and (E)-2-dodecenal) as soapy, fatty, 'like cilantro,' or pungent [8,9].

Several families of genes are important for taste and smell. The TAS1R and TAS2R families form sweet, umami, and bitter taste receptors [10,11]. The olfactory receptor family contains about 400 functional genes in the human genome. Each receptor binds to a set of chemicals, enabling one to recognize specific odorants or tastants. Genetic differences in many of these receptors are known to play a role in how we perceive tastes and smells [12-15].

Results and discussion

Here, we report on a genome-wide association study (GWAS) of cilantro soapy-taste detection. Briefly, the GWAS was conducted in 14,604 unrelated participants of primarily European ancestry who responded to an online questionnaire asking whether they thought cilantro tasted like soap (Table 1). Two single-nucleotide polymorphisms (SNPs) were genome-wide significant ($p < 5 \times 10^{-8}$) in this population. One SNP, in a cluster of olfactory receptors, replicated in a non-overlapping group of 11,851 participants (again, unrelated and of primarily European ancestry) who reported whether they liked or disliked cilantro (see the 'Methods' section for full details). Figure 1 shows p values across the whole genome; Figure 2 shows p values near the most significant associations. A

quantile-quantile plot (Additional file 1) shows little ($\lambda = 1.007$) global inflation of p values. Index SNPs with p values under 10^{-6} are shown in Table 2 (along with replication p values); all SNPs with p values under 10^{-4} are shown in Additional file 2.

We found one significant association for cilantro soapy-taste that was confirmed in the cilantro preference population. The SNP rs72921001 ($p_{discovery} = 6.4 \times 10^{-9}$, odds ratio (OR) = 0.81, $p_{repl} = 0.0057$) lies on chromosome 11 within a cluster of eight olfactory receptor genes: OR2AG2, OR2AG1, OR6A2, OR10A5, OR10A2, OR10A4, OR2D2, and OR2D3. The C allele is associated with both detecting a soapy smell and disliking cilantro. Of the olfactory receptors encoded in this region, OR6A2 appears to be the most promising candidate underlying the association with cilantro odor detection. It is one of the most studied olfactory receptors (often as the homologous olfactory receptor I7 in rats) [16-19]. A wide range of odorants have been found to activate this receptor, all of which are aldehydes [17]. Among the unsaturated aldehydes, octanal binds best to rat I7 [18]; however, compounds ranging from heptanal to undecanal also bind to this receptor [17]. Several singly unsaturated n-aldehydes also show high affinity, including (E)-2-decenal [17]. These aldehydes include several of those playing a key role in cilantro aroma, such as decanal and (E)-2-decenal. Thus, this gene is particularly interesting as a candidate for cilantro odor detection. The index SNP is also in high LD ($r^2 > 0.9$) with three non-synonymous SNPs in OR10A2, namely rs3930075, rs10839631, and rs7926083 (H43R, H207R, and K258T, respectively). Thus, OR10A2 may also be a reasonable candidate gene in this region.

The second significant association, with rs78503206 ($p_{discovery} = 3.2 \times 10^{-8}$, OR = 0.68, $p_{repl} = 0.49$), lies in an intron of the gene SNX9 (sorting nexin-9; see Figure 2). SNX9 encodes a multifunctional protein involved in intracellular trafficking and membrane remodeling during endocytosis [20]. It has no known function in taste or smell and did not show association with liking cilantro in the replication population. This SNP is located about 80 kb upstream of SYNJ2, an inositol 5-phosphatase thought to be involved in membrane trafficking and signal

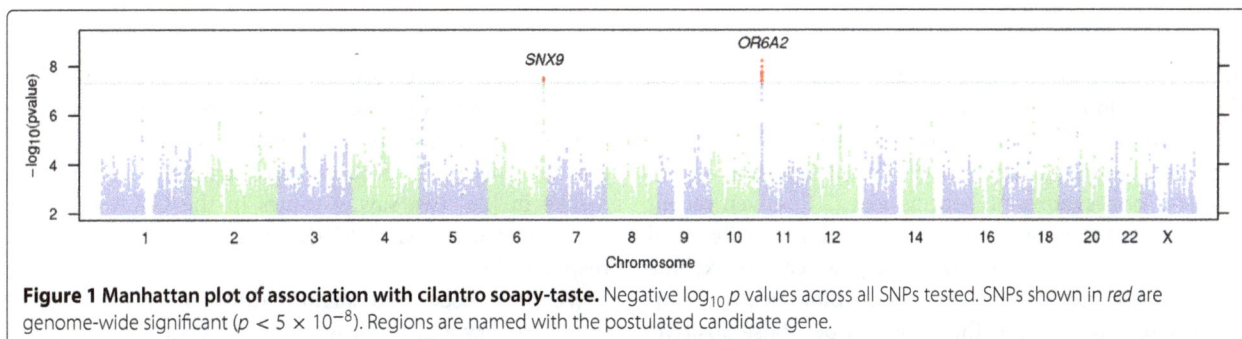

Figure 1 Manhattan plot of association with cilantro soapy-taste. Negative $\log_{10} p$ values across all SNPs tested. SNPs shown in *red* are genome-wide significant ($p < 5 \times 10^{-8}$). Regions are named with the postulated candidate gene.

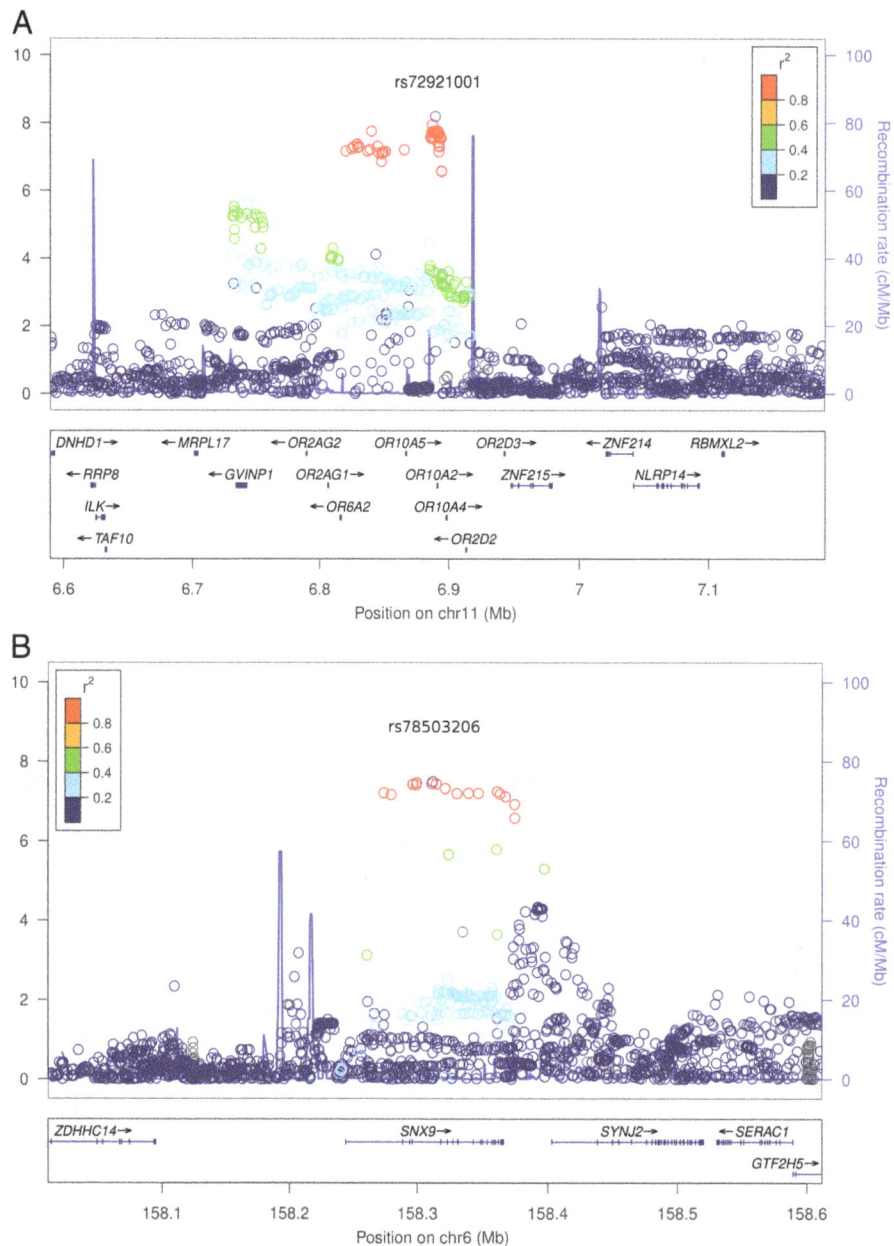

Figure 2 Associations with cilantro soapy-taste near rs72921001 **(A)** and rs78503206 **(B)**. Negative $\log_{10}$ p values for association (*left axis*) with recombination rate (*right axis*). *Colors* depict the squared correlation (r^2) of each SNP with the most associated SNP ((A) rs72921001 and (B) rs78503206, shown in *purple*). *Gray* indicates SNPs for which r^2 information was missing.

transduction pathways. In candidate gene studies, *SYNJ2* SNPs were found to be associated with agreeableness and symptoms of depression in the elderly [21] and with cognitive abilities [22]. In mice, a *Synj2* mutation causes recessive non-syndromic hearing loss [23]. Given recent evidence that the perception of flavor may be influenced by multiple sensory inputs (*cf.* [24,25]), we cannot exclude the *SYNJ2*-linked SNP as conveying a biologically meaningful association. While this SNP may be a false positive,

it could also be the case that this SNP is associated only with detecting a soapy smell in cilantro (and not in liking cilantro). In addition, we were unable to replicate the SNPs that were found to be nominally significant for cilantro dislike in [26] (we saw *p* values in the GWAS of 0.53, 0.41, and 0.53 for rs11988795, rs1524600, and rs10772397, respectively).

We have used two slightly different phenotypes in our discovery and replication, soapy-taste detection and

Table 2 Index SNPs for regions with $p < 10^{-6}$ for cilantro soapy-taste

SNP	Chromosome	Position	Gene	Allele	MAF	r^2	$p_{discovery}$	p_{repl}	OR (CI)
rs72921001	11	6,889,648	OR6A2	C/A	0.364	0.969	6.4×10^{-9}	0.0057	0.809 (0.753–0.870)
rs78503206	6	158,311,499	SNX9	C/T	0.077	0.980	3.2×10^{-8}	0.49	0.679 (0.588–0.784)
chr5:4883483	5	4,883,483	ADAMTS16	C/T	0.032	0.885	1.7×10^{-7}	0.51	0.526 (0.405–0.683)
rs7227945	18	4,251,279	DLGAP1/LOC642597	T/G	0.055	0.920	5.3×10^{-7}	0.96	1.447 (1.258–1.663)
rs6554267	4	56,158,891	KDR/SRD5A3	T/G	0.019	0.651	7.4×10^{-7}	0.85	1.975 (1.529–2.549)
rs13412810	2	192,420,461	MYO1B/OBFC2A	G/A	0.141	0.942	7.9×10^{-7}	0.78	0.770 (0.693–0.857)

The index SNP is defined as the SNP with the smallest p value within a region. The listed gene is our postulated candidate gene near the SNP. Alleles are listed as major/minor (in Europeans). MAF is the frequency of the minor allele in Europeans, and r^2 is the estimated imputation accuracy. $p_{discovery}$ and p_{repl} are the discovery and replication p values, respectively. The OR is the discovery odds ratio per copy of the minor allele (e.g., the A allele of rs72921001 is the allele associated with a lower risk of detecting a soapy taste).

cilantro preference, which are correlated ($r^2 \approx 0.33$). Detection of a soapy taste is reportedly one of the major reasons people seem to dislike cilantro. Despite having over 10,000 more people reporting cilantro preference, we have used soapy-taste detection as our primary phenotype because it is probably influenced by fewer environmental factors. Indeed, we see a stronger effect of rs72921001 on soapy-taste detection than on cilantro preference (OR of 0.81 versus 0.92). A GWAS on the replication set gave no genome-wide significant associations. SNPs with p values under 10^{-6} for this analysis are shown in Additional file 3.

We find significant differences by sex and ancestral population in soapy-taste detection (Tables 1 and 3). Women are more likely to detect a soapy taste (and to dislike cilantro) (OR for soapy-taste detection 1.36, $p = 2.5 \times 10^{-10}$; Table 1). African-Americans, Latinos, East Asians, and South Asians are all significantly less likely to detect a soapy taste compared to Europeans (ORs of 0.676, 0.637, 0.615, and 0.270, respectively, $p < 0.003$; see Table 3). Ashkenazi Jews and South Europeans did not show significant differences from Northern Europeans ($p = 0.84$ and 0.65, respectively). We tested the association between rs72921001 and soapy-taste detection

within each population. Aside from the European populations, there was only a significant association in the small South Asian group ($p = 0.0078$, OR = 0.18, 95% CI 0.053–0.64). This association is in the same direction as the association in Europeans. Note that the GWAS population in Table 1 is a subset of the 'Europe all' population in Table 3, filtered to remove relatives (see the 'Methods' section). While the differences in allele frequency across populations do not explain the differences in soapy-taste detection, our analysis does suggest that this SNP may affect soapy-taste detection in non-European populations as well.

We calculated the heritability for cilantro soapy-taste detection using the GCTA software [27]. We found a low heritability of 0.087 ($p = 0.08$, 95% CI −0.037 to 0.211). This estimate is a lower bound for the true heritability, as our estimate only takes into account heritability due to SNPs genotyped in this study. While this calculation does not exclude a heritability of zero, the existence of the association with rs72921001 does give a non-zero lower bound on the heritability. Despite the strength of the association of the SNP near OR6A2, it explains only about 0.5% of the variance in perceiving that cilantro tastes soapy. Our heritability estimate is lower than those given in a recent twin

Table 3 Cilantro soapy-taste by ancestry

Population	Not soapy (%)	Soapy (%)	Total	MAF	p value
Ashkenazi	634 (85.9%)	104 (14.1%)	738	0.355	0.56
South Europe	458 (86.6%)	71 (13.4%)	529	0.335	0.25
Europe all	13,213 (87.0%)	1,973 (13.0%)	15,186	0.373	1.23×10^{-8}
North Europe	11,794 (87.2%)	1,736 (12.8%)	13,530	0.376	1.17×10^{-8}
All	16,196 (87.6%)	2,299 (12.4%)	18,495	0.356	3.94×10^{-8}
African-American	545 (90.8%)	55 (9.2%)	600	0.224	0.87
Latino	820 (91.3%)	78 (8.7%)	898	0.350	0.29
East Asia	424 (91.6%)	39 (8.4%)	463	0.283	0.22
South Asia	322 (96.1%)	13 (3.9%)	335	0.371	0.0078

Number of people detecting a soapy taste by ancestry group, sorted from most to least soapy-taste detection. For reference, we have added the minor allele frequency of rs7107418 in each group. This SNP is a proxy for rs72921001 ($r^2 > 0.98$), with the minor G allele of rs7107418 corresponding to the minor A allele of rs72921001 (which is associated with less soapy tasting). The p value is the p value of association between soapy-taste and rs7107418 in each group.

study (0.38 for odor and 0.52 for flavor) [7]. This could be due to the differences in phenotypes measured between the two studies, or it could be possible that other genetic factors not detected here could influence cilantro preference. For example, there could be rare variants not typed in this study (possibly in partial linkage disequilibrium with rs72921001) that have a larger effect on cilantro preference. Such rare variants could cause the true heritability of this phenotype to be larger than we have calculated. For example, the heritability of height is estimated to be about 0.8; however, the heritability tagged by common SNPs is calculated at about 0.45 [26]. We note that there can be epigenetic modifiers of taste as well, for example, food preferences can even be transmitted to the fetus *in utero* through the mother's diet [24].

Survey responses, while very efficient for collecting large amounts of data, can only approximately measure the detection and/or perception of the chemicals in cilantro. This has implications for the interpretation of our results. For example, it is possible that the SNP rs72921001 could have a large effect on detection of a specific chemical in cilantro, but that the resulting effect on liking cilantro is much weaker, being modulated by environmental factors. For example, many people might initially dislike cilantro yet later come to appreciate it. This environmental component could also be the reason that our heritability estimates are low. It would thus be interesting to study the genetics of cilantro taste/odor perception in a group without prior exposure to cilantro to reduce the environmental effect, using more direct measures of cilantro perception (i.e., having the subjects actually taste and smell cilantro).

Conclusions
Through a GWAS, we have shown that a SNP, rs72921001, near a cluster of olfactory receptors is significantly associated with detecting a soapy taste to cilantro. One of the genes near this SNP encodes an olfactory receptor, OR6A2, that detects the aldehydes that may make cilantro smell soapy and thus is a compelling candidate gene for the detection of the cilantro odors that give cilantro its divisive flavor.

Availability of supporting data
We have shared full summary statistics for all SNPs with p values under 10^{-4} in Additional file 2. Due to privacy concerns, under our IRB protocol, we are unable to openly share statistics for all SNPs analyzed in the study.

Methods
Subjects
Participants were drawn from the customer base of 23andMe, Inc., a consumer genetics company. This cohort has been described in detail previously [15,28]. Participants provided informed consent and participated in the research online, under a protocol approved by an external AAHRPP-accredited IRB, Ethical and Independent Review Services (E&I Review).

Phenotype data collection
On the 23andMe website, participants contribute information through a combination of research surveys (longer, more formal questionnaires) and research 'snippets' (multiple-choice questions appearing as part of various 23andMe webpages). In this study, participants were asked two questions about cilantro via research snippets:

- 'Does fresh cilantro taste like soap to you?' (Yes/No/I'm not sure)
- 'Do you like the taste of fresh (not dried) cilantro?' (Yes/No/I'm not sure)

Among all 23andMe customers, 18,495 answered the first question (as either yes or no), 29,704 the second, and 15,751 both. Participants also reported their age. Sex and ancestry were determined on the basis of their genetic data. In both the GWAS set and the replication set, all participants were of European ancestry. In either group, no two shared more than 700 cM of DNA identical by descent (IBD, approximately the lower end of sharing between a pair of first cousins). In total, we were left with a set of 14,604 participants who answered the 'soapy' question for GWAS and 11,851 who answered only the taste preference question for a replication set. IBD was calculated using the methods described in [29]; the principal component analysis was performed as in [15]. To determine European and African-American ancestry, we used local-ancestry methods (as in [30]). Europeans had over 97% of their genome painted European, and African-Americans had at least 10% African and at most 10% Asian ancestry. Other groups were built using ancestry-informative markers trained on a subset of 23andMe customers who reported having four grandparents of a given ancestry.

Genotyping
Subjects were genotyped on one or more of three chips, two based on the Illumina HumanHap550+ BeadChip and the third based on the Illumina OmniExpress+ BeadChip (San Diego, CA, USA). The platforms contained 586,916, 584,942, and 1,008,948 SNPs. Totals of 291, 5,394, and 10,184 participants (for the GWAS population) were genotyped on the platforms, respectively. A total of 1,265 individuals were genotyped on multiple chips. For all participants, we imputed genotypes in batches of 8,000–10,000 using Beagle and Minimac [31-33] against the August 2010 release of the 1000 Genomes reference haplotypes [34], as described in [35].

A total of 11,914,767 SNPs were imputed. Of these, 7,356,559 met our thresholds of 0.001 minor allele frequency, average r^2 across batches of at least 0.5, and minimum r^2 across batches of at least 0.3. The minimum r^2 requirement was added to filter out SNPs that imputed less well in the batches consisting of the less dense platform. Positions and alleles are given relative to the positive strand of build 37 of the human genome.

Statistical analysis

For the GWAS, p values were calculated using a likelihood ratio test for the genotype term in the logistic regression model:

$$Y \sim G + \text{age} + \text{sex} + pc_1 + pc_2 + pc_3 + pc_4 + pc_5,$$

where Y is the vector of phenotypes (coded as 1 = thinks cilantro tastes soapy or 0 = does not), G is the vector of genotypes (coded as a dosage 0–2 for the estimated number of minor alleles present), and $pc_1, \ldots, pc_5$ are the projections onto the principal components. The same model was used for the replication, with the phenotype coded as 1 = dislikes cilantro or 0 = likes. We used the standard cutoff for genome-wide significance of 5×10^{-8} to correct for the multiple tests in the GWAS. ORs and p values for the differences in soapy-taste detection between sexes and population were calculated directly, without any covariates. Table 3 uses a proxy SNP for rs72921001, as our imputation was done only in Europeans, so we did not have data for rs72921001 in other populations.

For the heritability calculations, we used the GCTA software [27]. The calculations were done on genotyped SNPs only within a group of 13,628 unrelated Europeans. Unrelated filtering here was done using GCTA to remove individuals with estimated relatedness larger than 0.025. Thus, this group is slightly different from the GWAS set, as the GWAS set's relatedness filtering was done using IBD. We assumed a prevalence for soapy-taste detection of 0.13 for the transformation of heritability from the 0–1 scale to the liability scale. Otherwise, default options were used. We calculated heritability for autosomal and X chromosome SNPs separately; the estimates were 0.0869 (standard error 0.0634, p value 0.0805) for autosomal SNPs and 2×10^{-6} (standard error 0.010753, p value 0.5) for the X chromosome.

Additional files

Additional file 1: Quantile-quantile plot of association with cilantro soapy-taste. Observed p values versus theoretical p values under the null hypothesis of no association. The genomic control inflation factor for the study was 1.007 and is indicated by the red line; approximate 95% confidence intervals are given by the blue curves.

Additional file 2: All SNPs with $p < 10^{-4}$ for cilantro soapy-taste. Alleles are listed as major/minor. MAF is the frequency of the minor allele in Europeans, and r^2 is the estimated imputation accuracy. Positions and alleles are given relative to the positive strand of build 37 of the human genome. The gene column shows the position of the SNP in context of the nearest genes. The SNP position is within brackets, and the number of dashes gives approximate $\log_{10}$ distances.

Additional file 3: Index SNPs with $p < 10^{-6}$ for cilantro preference in the replication set. Results of a GWAS on cilantro preference in the replication set. Columns are as in Table 2.

Abbreviations

AAHRPP: Association for the Accreditation of Human Research Protection Programs; BCE: before common era; CI: confidence interval; GWAS: genome-wide association study; IRB: institutional review board; OR: odds ratio; OR2AG2, OR2AG1, OR6A2, OR10A5, OR10A2, OR10A4, OR2D2, OR2D3, members of olfactory receptor gene families 2, 6, and 10; SNP: single-nucleotide polymorphism; TAS1R/TAS2R: taste receptor gene families 1 and 2.

Competing interests

The authors of this paper are 23andMe employees and own stock options in the company.

Authors' contributions

NE, SW, CBD, AKK, JLM, DAH, UF, and JYT conceived and designed the experiments. NE analyzed the data and drafted the manuscript with contributions from all other authors. All authors read and approved the final manuscript.

Acknowledgements

We thank the customers of 23andMe for participating in this research and all the employees of 23andMe for contributing to the research.

References

1. Zohary D, Hopf M: *Domestication of Plants in the Old World: The Origin and Spread of Cultivated Plants in West Asia, Europe, and the Nile Valley.* New York: Oxford University Press; 2000.
2. Bostock J, Riley H: *The Natural History of Pliny. Volume 4 in Bohn's Classical Library.* London: H.G. Bohn; 1855.
3. Faas P: *Around the Roman Table.* Basingstoke: Palgrave Macmillan; 2002.
4. Gerard J: *The Herball or General Historie of Plants.* Amsterdam: Theatrum Orbis Terrarum; 1974. [1597 ed].
5. Leach H: **Rehabilitating the "stinking herbe": a case study of culinary prejudice.** *Gastronomica: The J Food Culture* 2001, **1**(2):10–15.
6. Mauer L, El-Sohemy A: **Prevalence of cilantro (Coriandrum sativum) disliking among different ethnocultural groups.** *Flavour* 2012, **1**:8.
7. Knaapila A, Hwang LD, Lysenko A, Duke FF, Fesi B, Khoshnevisan A, James RS, Wysocki CJ, Rhyu M, Tordoff MG, Bachmanov AA, Mura E, Nagai H, Reed DR: **Genetic analysis of chemosensory traits in human twins.** *Chem Senses* 2012, **37**(9):869–881.
8. Cadwallader K, Benitez D, Pojjanapimol S, Suriyaphan O, Singh T: **Characteristic aroma components of the cilantro mimics.** In *Natural Flavors and Fragrances. Volume 908.* Edited by Frey C, Rouseff RL. Washington: American Chemical Society; 2005:117–128.
9. Eyres G, Dufour JP, Hallifax G, Sotheeswaran S, Marriott PJ: **Identification of character-impact odorants in coriander and wild coriander leaves using gas chromatography-olfactometry (GCO) and comprehensive two-dimensional gas chromatography-time-of-flight mass spectrometry (GC x GC-TOFMS).** *J Sep Sci* 2005, **28**(9–10):1061–1074.
10. Li X, Staszewski L, Xu H, Durick K, Zoller M, Adler E: **Human receptors for sweet and umami taste.** *Proc Natl Acad Sci USA* 2002, **99**(7):4692–4696.
11. Chandrashekar J, Mueller KL, Hoon MA, Adler E, Feng L, Guo W, Zuker CS, Ryba NJ: **T2Rs function as bitter taste receptors.** *Cell* 2000, **100**(6):703–711.

12. Reed DR, Knaapila A: **Genetics of taste and smell: poisons and pleasures.** *Prog Mol Biol Transl Sci* 2010, **94**:213–240.

13. Kim UK, Jorgenson E, Coon H, Leppert M, Risch N, Drayna D: **Positional cloning of the human quantitative trait locus underlying taste sensitivity to phenylthiocarbamide.** *Science* 2003, **299**(5610):1221–1225.

14. Keller A, Zhuang H, Chi Q, Vosshall LB, Matsunami H: **Genetic variation in a human odorant receptor alters odour perception.** *Nature* 2007, **449**(7161):468–472.

15. Eriksson N, Macpherson JM, Tung JY, Hon LS, Naughton B, Saxonov S, Avey L, Wojcicki A, Pe'er I, Mountain J: **Web-based, participant-driven studies yield novel genetic associations for common traits.** *PLoS Genet* 2010, **6**:e1000993.

16. Kurland MD, Newcomer MB, Peterlin Z, Ryan K, Firestein S, Batista VS: **Discrimination of saturated aldehydes by the rat I7 olfactory receptor.** *Biochemistry* 2010, **49**(30):6302–6304.

17. Araneda RC, Kini AD, Firestein S: **The molecular receptive range of an odorant receptor.** *Nat Neurosci* 2000, **3**(12):1248–1255.

18. Krautwurst D, Yau KW, Reed RR: **Identification of ligands for olfactory receptors by functional expression of a receptor library.** *Cell* 1998, **95**(7):917–926.

19. Araneda RC, Peterlin Z, Zhang X, Chesler A, Firestein S: **A pharmacological profile of the aldehyde receptor repertoire in rat olfactory epithelium.** *J Physiol (Lond)* 2004, **555**(Pt 3):743–756.

20. Buajeeb W, Poomsawat S, Punyasingh J, Sanguansin S: **Expression of p16 in oral cancer and premalignant lesions.** *J Oral Pathol Med* 2009, **38**:104–108.

21. Luciano M, Lopez LM, de Moor MH, Harris SE, Davies G, Nutile T, Krueger RF, Esko T, Schlessinger D, Toshiko T, Derringer JL, Realo A, Hansell NK, Pergadia ML, Pesonen AK, Sanna S, Terracciano A, Madden PA, Penninx B, Spinhoven P, Hartman CA, Oostra BA, Janssens AC, Eriksson JG, Starr JM, Cannas A, Ferrucci L, Metspalu A, Wright MJ, Heath AC, et al.: **Longevity candidate genes and their association with personality traits in the elderly.** *Am J Med Genet B Neuropsychiatr Genet* 2012, **159B**(2):192–200.

22. Lopez LM, Harris SE, Luciano M, Liewald D, Davies G, Gow AJ, Tenesa A, Payton A, Ke X, Whalley LJ, Fox H, Haggerty P, Ollier W, Pickles A, Porteous DJ, Horan MA, Pendleton N, Starr JM, Deary IJ: **Evolutionary conserved longevity genes and human cognitive abilities in elderly cohorts.** *Eur J Hum Genet* 2012, **20**(3):341–347.

23. Manji SS, Williams LH, Miller KA, Ooms LM, Bahlo M, Mitchell CA, Dahl HH: **A mutation in synaptojanin 2 causes progressive hearing loss in the ENU-mutagenised mouse strain Mozart.** *PLoS ONE* 2011, **6**(3):e17607.

24. Bakalar N: **Sensory science: partners in flavour.** *Nature* 2012, **486**(7403):4–5.

25. Smith B: **Perspective: complexities of flavour.** *Nature* 2012, **486**(7403):S6.

26. Yang J, Benyamin B, McEvoy BP, Gordon S, Henders AK, Nyholt DR, Madden PA, Heath AC, Martin NG, Montgomery GW, Goddard ME, Visscher PM: **Common SNPs explain a large proportion of the heritability for human height.** *Nat Genet* 2010, **42**(7):565–569.

27. Yang J, Lee SH, Goddard ME, Visscher PM: **GCTA: a tool for genome-wide complex trait analysis.** *Am J Hum Genet* 2011, **88**:76–82.

28. Tung JY, Do CB, Hinds DA, Kiefer AK, Macpherson JM, Chowdry AB, Francke U, Naughton BT, Mountain JL, Wojcicki A, Eriksson N: **Efficient replication of over 180 genetic associations with self-reported medical data.** *PLoS ONE* 2011, **6**:e23473.

29. Henn B, Hon L, Macpherson JM, Eriksson N, Saxonov S, Pe'er I, Mountain JL: **Cryptic distant relatives are common in both isolated and cosmopolitan genetic samples.** *PLoS ONE* 2012, **7**(4):e34267.

30. Eriksson N, Tung JY, Kiefer AK, Hinds DA, Francke U, Mountain JL, Do CB: **Novel associations for hypothyroidism include known autoimmune risk loci.** *PLoS ONE* 2012, **7**(4):e34442.

31. Browning SR, Browning BL: **Rapid and accurate haplotype phasing and missing-data inference for whole-genome association studies by use of localized haplotype clustering.** *Am J Hum Genet* 2007, **81**:1084–1097.

32. Howie B, Fuchsberger C, Stephens M, Marchini J, Abecasis GR: **Fast and accurate genotype imputation in genome-wide association studies through pre-phasing.** *Nat Genet* 2012, **44**(8):955–959.

33. Abecasis G, Fuchsberger C: **minimac.** [http://genome.sph.umich.edu/wiki/minimac]

34. Altshuler D, Durbin RM, Abecasis GR, Bentley DR, Chakravarti A, Clark AG, Collins FS, De La Vega FM, Donnelly P, Egholm M, Flicek P, Gabriel SB, Gibbs RA, Knoppers BM, Lander ES, Lehrach H, Mardis ER, McVean GA, Nickerson DA, Peltonen L, Schafer AJ, Sherry ST, Wang J, Wilson R, Gibbs RA, Deiros D, Metzker M, Muzny D, Reid J, Wheeler D, et al.: **A map of human genome variation from population-scale sequencing.** *Nature* 2010, **467**:1061–1073.

35. Eriksson N, Benton GM, Do CB, Kiefer AK, Mountain JL, Hinds DA, Francke U, Tung JY: **Genetic variants associated with breast size also influence breast cancer risk.** *BMC Med Genet* 2012, **13**:53.

19

Plate shape and colour interact to influence taste and quality judgments

Peter C Stewart[*] and Erica Goss

Abstract

Background: Research has demonstrated that factors external to the food source can influence consumers' perceptions of food. Contextual factors including cutlery or tableware (for example, size and composition), the atmosphere (for example, noise levels and odours), and packaging (for example, shape and colour) have all been shown to influence the perceptual experience. Plateware has also been shown to influence taste perception since ratings of a dessert (strawberry mousse) were modified by plate colour but not by plate shape. In the current study, which used a 2 × 2 between-subjects design, the effect of plate colour (black versus white) and plate shape (round versus square) on taste perception is re-examined. Through sweetness, intensity, quality, and liking ratings of cheesecake, the current study extends the previous investigation to include an examination of the plate colour by plate shape interaction while using plates with more angular corners.

Results: Judgments made on simple elemental properties (sweetness and flavour intensity) and higher level compound property judgments (food quality or food liking) were shown to be differentially influenced by the interaction of plate colour and plate shape. Both elemental and compound property judgments were heightened by white round plates while compound judgments were also increased when food was presented on black square plates.

Conclusions: The results suggest that plate colour and shape influence taste perception but not in a straight-forward manner and instead the influence depends on the interaction of the two variables. Depending on which attribute of the perceptual experience is more important, knowledge of this interaction could be used advantageously by the culinary community.

Keywords: Taste perception, Plate, Shape, Colour, Sweetness, Flavour intensity, Food quality, Cross-modal, Preferences

Background

Most people would correctly say that taste is determined by more than just taste receptors on the tongue but they may be surprised by the extent to which this is true. Along with the gustatory response, it is commonly recognized that what a food smells like (olfactory cues), how it looks (visual cues), and how it feels in your mouth or hands (somatosensory cues) all influence the resulting perception of taste. However, it is not only the attributes of the food that influence taste perception, and instead, environmental/contextual factors also greatly impact the resultant perception of taste. The current study examined two factors external to the food source, plate shape and plate colour,

for their effect on perceptual judgments of sweetness, intensity, quality, and liking.

Research has shown that the manipulation of a variety of food-specific cue types will influence various aspects of taste perception. Olfactory cues have been shown to manipulate our perception of taste with a number of studies having shown an enhancement of sweetness ratings in response to pairing a food with odours usually associated with sweetness [1-3]. Additionally, in two experiments, Stevenson, Prescott, and Boakes [4] paired 20 different odours with sucrose taste solutions (E1) and citric acid taste solutions (E2). They found that odours with a strong learned association with sweetness (for example, caramel) enhanced sweetness ratings and suppressed sour ratings, thereby showing the importance of learning and memory in taste perception. In a more recent investigation, Djordjevic, Zatorre, and Jones-Gotman [5] also found an

* Correspondence: pstewart@grenfell.mun.ca
Psychology Program, Grenfell Campus, Memorial University of
Newfoundland, University Avenue, Corner Brook, NL A2H 6P9, Canada

odour-induced change in taste perception (OICTP) when they paired olfactory and gustatory stimuli. Specifically, the smell of strawberries led to a sweetness enhancement and a soy sauce odour increased ratings of saltiness.

The texture of food has also been shown to modify the perceptual experience [6-8]. Cook, Hollowood, Linforth, and Taylor [9] found that ratings of sweetness were significantly lessened in thicker solutions. Additionally, Walker and Prescott [10] showed that apple juice flavour was rated less sweet in a more viscous solution. Further, texture has been shown to directly and indirectly, via its influence on chewing parameters, influence flavour release (see Salles et al. [11] for a review).

Shape is closely related to texture and certain attributes of food seem to reliably share a synaesthetic relationship with shapes. Synaesthesia is a phenomenon whereby the stimulation of one sensory modality simultaneously produces an accompanying stimulation in a second sensory modality. Cytowic [12] discusses a synaesthete who tasted shapes. For this individual, an amateur chef who cooked according to the shape of the food rather than following any recipes, shapes had tastes and certain shapes tasted better than others. Although synaesthesia typically refers to a non-standard 'crossing' of the senses, it seems that when it comes to food we may all be synaesthetes to some extent. Research in non-synaesthetes has shown that most individuals have shape associations with certain food qualities. For example, wines are often described as having a rounded or pointed taste [13]. Particularly relevant to the current study, it has also been shown that sweetness is often associated with round shapes while bitterness and saltiness are associated with angular shapes [14,15]. Whether this flavour-shape association extends to environmental factors external to the food source is one focus of the current study.

Perhaps the most profound manipulation of the perceptual experience is driven by visual input. Specifically, it has been repeatedly shown that modifying the colour of a standard food influences the taste perception of that food [16-19]. Garber, Hyatt, and Starr [20] examined the effect of incongruent colours (and labelling) on taste perception of noncarbonated fruit drinks. By manipulating the colour of an orange-flavoured drink (orange versus purple versus clear) the researchers showed that participants were significantly less likely to correctly identify the flavour if the beverage was coloured purple or clear. They also reported an interaction between stimulus colour and package labelling for ratings of naturalness/expensiveness. This suggested that external factors, factors not directly related to the food itself, can influence perception of higher level cognitive judgments [20]. Whether this colour influence generalises to environmental stimuli, such as the colour of plate the food is presented on, is examined in the current investigation.

Recently, perception research has supported that the environment or context surrounding the food has been shown to be a significant modulator of taste perception [21]. Specific contextual variables include, but are not limited to, background music [22]; the size, weight, and composition of cutlery [23]; a product's packaging [24]; colour lighting [25]; and menu item naming [26].

An additional environmental variable that has come under recent scrutiny was plateware [27]. In two separate experiments, the researchers examined the potential influence of plate colour (E1: black versus white) and plate shape (E2: round versus square versus triangular) on ratings of sweetness, intensity, quality, and liking. All attribute ratings, except quality ratings, were shown to be significantly increased when the strawberry mousse was presented on white plates when compared to black plates. However, contrary to expectations, plate shape had no significant influence on attribute ratings. In addition to concluding that plate colour is a significant contributor to taste perception, the authors suggested that the jury was still out on plate shape since their plates may have been somewhat less angular than desired due to the plate corners being rounded. Further, Piqueras-Fiszman et al. [27] did not examine the interaction between plate colour and plate shape.

Like the Piqueras-Fiszman et al. studies, the majority of research into taste perception has taken a single attribute approach and only manipulated one environmental variable per experiment. It has been suggested that an important goal of future research should be to examine the simultaneous interaction of two or more environmental variables [21]. Continuing from the Piqueras-Fiszman et al. work, the current study replicated and extended the aforementioned findings using plates with more angular corners and by employing a factorial design. Results revealed significant interactions between plate shape and plate colour across each of the rated attributes.

Results

Aware that we had relatively small sample sizes in each experimental group (n = 12), we felt it important to first examine the experimental groups' composition for any pre-existing differences that may have confounded the results. We felt that these preliminary analyses were necessary since, as stated and referenced multiple times in the paper's introduction, many variables can exert a large influence on perception. All analyses were independent measures analyses and were tested against an alpha level of .05. Sample sizes, means, and standard deviations displayed as a function of gender and plate type can be seen in Table 1. Group composition did not significantly differ as a function of age or gender[a] suggesting our groups were well matched in terms of age and gender.

Table 1 Group characteristics and attribute descriptive statistics as a function of plate colour, plate shape, and participant gender (n = 48)

Variable	White round	White square	Black round	Black square
Male				
n (N = 27)	6	8	9	4
Age (years)	20.50 (2.20)	20.50 (2.20)	21.44 (5.48)	20.50 (3.70)
Sweetness (%)	77.50 (10.60)	55.04 (13.24)	49.46 (13.60)	63.31 (10.35)
Intensity (%)	77.87 (14.38)	38.37 (20.05)	47.49 (21.12)	50.81 (20.60)
Quality (%)	82.62 (21.78)	70.30 (20.50)	66.61 (18.98)	70.43 (18.95)
Liking of sample (%)	85.48 (24.38)	70.09 (24.62)	71.92 (12.42)	87.50 (8.34)
Hunger (%)	44.27 (16.71)	53.43 (22.24)	45.88 (19.08)	65.32 (26.75)
Liking in general (%)	85.93 (11.60)	59.81 (19.20)	71.15 (21.31)	77.96 (19.98)
Female				
n (N = 21)	6	4	3	8
Age (years)	23.67 (6.77)	20.75 (0.96)	19.67 (2.52)	27.63 (13.00)
Sweetness (%)	68.37 (18.27)	59.41 (7.04)	55.38 (16.83)	51.21 (17.24)
Intensity (%)	72.96 (21.44)	58.33 (20.03)	45.70 (21.28)	50.94 (24.37)
Quality (%)	93.01 (9.52)	68.95 (7.32)	61.82 (5.13)	87.37 (9.10)
Liking of sample (%)	87.99 (13.64)	77.28 (23.84)	76.52 (14.03)	86.63 (18.47)
Hunger (%)	30.01 (29.28)	56.32 (31.18)	42.65 (7.83)	38.64 (32.62)
Liking in general (%)	96.33 (4.69)	81.18 (16.87)	78.49 (7.11)	91.47 (14.00)
Total	N = 48			

As a result, these variables were dropped from any further analyses.

Using two separate 2 × 2 (colour × shape) ANOVAs, we tested if the groups differed in regard to state hunger levels and in their general regard for cheesecake. There were no significant main effects or interactions for either dependent variable, again suggesting relatively well matched groups.

Analyses of attribute ratings

Separate 2 (colour) × 2 (shape) independent measures ANOVAs were conducted with each attribute (sweetness, intensity, quality, and liking) as the dependent variable. All analyses were tested at a significance level of .05 with a Bonferroni correction applied.

Regarding sweetness ratings, there was a significant main effect of colour with samples on white plates being rated as sweeter than those on black plates, $F(1, 44) = 8.42$, $P < .05$, $\eta_p^2 = .16$, but there was no main effect of plate shape. Most importantly however, was the presence of a significant colour × shape interaction, $F(1, 44) = 6.70$, $P < .05$, $\eta_p^2 = .13$. This interaction suggested that round white plates were rated significantly sweeter than square white plates ($t(22) = 3.04$, $P < .05$) while there was no difference between the plate shapes for black plates (see Figure 1a).

For food intensity ratings, both the colour and shape main effects were significant; however, the significant interaction, $F(1,44) = 9.72$, $P < .05$, $\eta_p^2 = .18$, between the variables seemed to qualify the main effects. As with sweetness ratings, when compared to round black and square white plates, intensity ratings were significantly higher for round white plates, $t(22) = 4.12$, $P < .05$ and $t(22) = 4.01$, $P < .05$ respectively (see Figure 1b).

There were no significant main effects for either colour or shape on ratings of quality however, there was a significant interaction, $F(1, 44) = 13.30$, $P < .05$, $\eta_p^2 = .23$. Post hoc tests revealed that three of the four pairings were significant (round white versus round black: $t(22) = 3.28$; round white versus square white: $t(22) = 2.61$; round black versus square black: $t(22) = -2.55$) and the fourth pairing (square white versus square black: $t(22) = -1.84$, $P = .08$) approached significance (see Figure 1c).

Similar to quality ratings, there were no main effects of liking ratings but there was a significant interaction, $F(1, 44) = 7.29$, $P < .05$, $\eta_p^2 = .14$ (see Figure 1d). Post hoc tests revealed a significant colour effect for round plates ($t(22) = 2.10$) and a significant shape effect for black plates ($t(22) = -2.43$). The two other comparisons did not reach statistical significance (square white versus square black: $P = .09$; round white versus square white: $P = .12$).

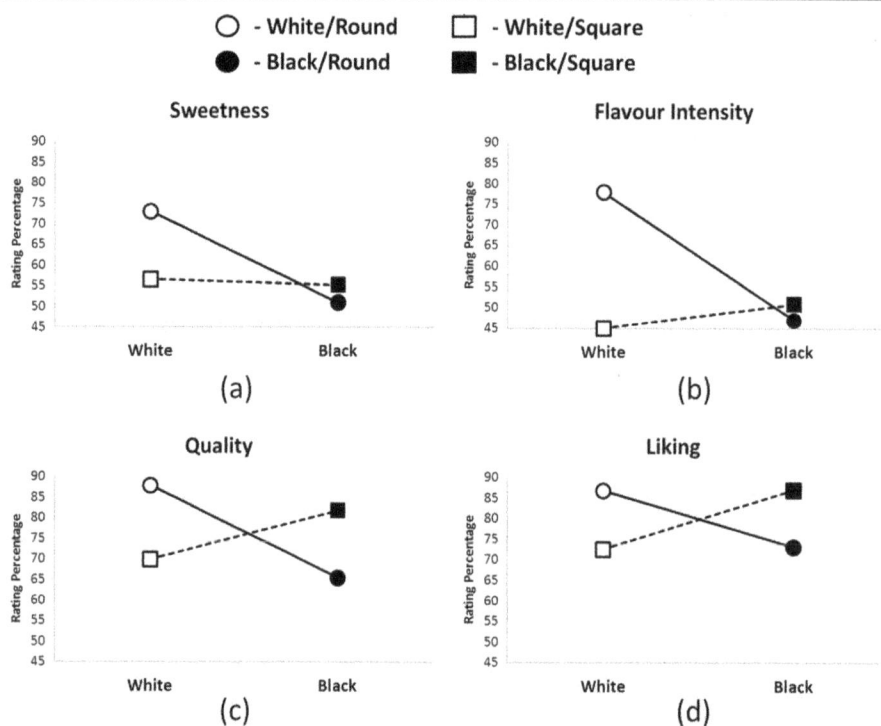

Figure 1 Significant plate colour by plate shape interactions on each of the attribute ratings. (a) Sweetness ratings: Food eaten on white round plates was rated significantly sweeter than that eaten on white square plates. (b) Intensity ratings: Food eaten from white round plates was rated significantly more intense than that eaten on white square plates. (c) Quality ratings: Food eaten on white round plates and black square plates was rated of significantly higher quality than that eaten on white square and black round plates. (d) Liking ratings: Food eaten on white round plates and black square plates was liked significantly more than that eaten on white square and black round plates. However, after considering participants' general regard for cheesecake, this interaction for liking ratings was negated.

The potential influences of hunger and general regard for cheesecake

Moskowitz *et al.* [28] showed that after a satiating intake of a glucose load, participants' ratings of pleasantness for subsequently presented sweet stimuli differed from participants who were sated from a standard breakfast or lunch. Although the current study did not gather very specific hunger-related information, to control for general hunger effects, we did collect a rating of state hunger. We also collected data pertaining to how much people generally liked cheesecake since it seemed likely that people who like a particular food will rate it differently than those who dislike the food or like it to a lesser extent. To examine these possibilities we conducted 2 × 2 analyses of covariance (ANCOVA) using ratings of hunger and general regard cheesecake as separate covariates for each of the four attributes (sweetness, liking, quality, and liking).

Neither self-reported levels of hunger nor levels of general cheesecake regard were significant covariates in analyses of sweetness or quality (P >.05). However, both hunger levels and cheesecake regard levels were significant covariates when intensity ratings were analysed. With that said, and importantly, the interaction between plate colour and plate shape on intensity ratings remained significant when accounting for hunger levels, F(1, 43) = 8.65, P <.05. When accounting for variance in intensity ratings due to general cheesecake regard, the interaction between plate shape and colour, F(1, 43) = 2.66, P = .11, was no longer significant although both main effects remained significant.

Interestingly, the strong interaction observed between plate colour and plate shape on liking ratings of the cheesecake sample was not influenced by the inclusion of hunger as a covariate but was completely nullified by the inclusion of participants' general regard for cheesecake as a covariate, F(1, 43) = .45, P co .51. This suggests that it is quite important to account for personal taste when analysing data of this sort.

Discussion

The current investigation examined the influence that plate shape and colour may have on the perceived sweetness, intensity, quality, and liking of a portion of cheesecake. It has been shown in the past that there is a white plate (over black) advantage but that plate shape did not significantly modulate attribute ratings [27]. Results of the current study, specifically the significant or near significant interactions across all attributes, suggest that both plate shape and plate colour are important but also that the relationship is anything but straightforward.

We have chosen to discuss these findings in terms of elemental versus compound judgments. By a compound judgment (for example, a judgment about quality or liking) we are referring to a higher level, gestalt judgment that is made by considering a number of individual elemental judgments (that is, judgments of sweetness, intensity, look, smell, *etcetera*) in order to come to what Delwiche [29] referred to as an emergent phenomenon. For example, when a quality judgment is made about a particular food, it seems likely that the final perception emerges from a combination of the look, smell, taste, context, and other elemental properties of the food. It is important to note that we are by no means suggesting that an elemental judgment is done on the basis of only one sensory modality; the background of this paper already suggested this is almost certainly not the case. We are rather saying that there are far fewer modalities involved with an elemental judgment than with a compound judgment. That is, people almost certainly judged sweetness by interpreting gustatory, visual, and olfactory input but it is less likely that quality or liking considerations factored into a sweetness rating. With that said, it is possible that there is an entirely nested reciprocal relationship between all these factors and future research could determine the validity of this possibility.

As can be seen in Figure 1(a,b), for elemental judgments (that is, sweetness and intensity) there was a clear white round plate enhancement suggesting that food served on this shape and colour plate leads to an increase in perceived sweetness and intensity. Although this needs to be extended to include different foods and plate attributes, this perceptual difference may have benefits to the food industry. We saw an approximate 20% increase in perceived sweetness and a 30% rating increase in intensity when food was served on a white round plate. Perhaps an unsweetened or less sweet dessert when served on a white round plate would be better received than if it was served on a plate of a different colour and shape. Anecdotally, diabetics often complain that the only desserts they can eat do not taste sweet enough or taste bad due to the artificial sweeteners used. Potentially, the combination of a low sugar dessert on a white round plate would lead an improvement in the overall perception. Conversely, a common foodstuff that may be viewed as being too sweet could be served on a square white plate or a black plate to possibly dampen the perceived sweetness.

The results show a somewhat different story for compound judgments (that is, quality and liking). Figure 1 (c,d) shows that although there is still a white round plate advantage for ratings of quality and liking, there is also an advantage for square black plates. Since the participants' general liking of cheesecake seemed to explain the interaction effect observed for liking ratings, we limit our further discussion to the quality interaction. We contend that the white round plate advantage in quality judgments is being driven by elemental judgments that are perceptually increased by white round plates. For example, in this study we have shown that perceived sweetness and intensity are increased by the presentation of food on white round plates and it seems possible, if not likely, that other factors that were not measured in this experiment also show this white round plate preference. Quality being a compound judgment would be the sum of these elemental judgments and the end result would be a more positive quality judgment for food presented on white round plates. We further contend that this is a likely explanation for the square black plate advantage also obtained for judgments of quality. According to the Ecological Valence Theory (EVT), in regard to colour, our preference for a particular colour is determined by a sum total of our past experience with that colour, experience not necessarily specific to the object currently being perceived [30]. Humans tend to prefer colours that summon implicitly positive cognitive associations while disliking colours that spur the opposite. It seems reasonable that this would also extend to and combine with shape, although we are aware of no studies directly assessing this. Perhaps square black objects and white round objects have positive connotations when it comes to assessments of quality. Although this notion has yet to be extended to food, Schloss, Strauss, and Palmer [31] found that pure black (or white) T-shirts, dress shirts, ties/scarves, and squares were preferred over any shade of grey and that black ties/scarves were preferred. They suggested that for many objects people preferred subdued colours rather than flashy colours and that this preference was likely determined from experience with the object and/or colour. It is important to also consider that black is often seen and used in advertising to denote sophistication, luxuriousness, elegance, and quality. However, these potential associations with blackness do not aid in much in the explanation of black *square* plates. With that said, we suggest that the black square plate advantage in quality judgments may be due to the sum total of experience driven preferences. It would be interesting for future research to try and tease apart this plate effect using the EVT methodologies.

It is also possible that something as simple as the degree of familiarity is influencing participant attribute ratings. It seems likely that of the four plate types, white round plates are the most familiar and black square plates are the most novel. Sheau-Fen, Sun-May, and Yu-Ghee [32] found that higher levels of familiarity were positively related with higher levels of perceived quality for a number of store brand consumer items. An opposing finding, yet actually complimentary for our purposes, is that consumer items that were judged to be unique or novel due to minor manipulations are judged as more desirable than the typical item [33]. Furthermore, Bornstein [34] concluded that although it is generally reproductively advantageous to prefer the typical to the novel it is at times advantageous (that is, during child development) to favour novel stimuli. Marrying these ideas together, the enhanced ratings observed for the food eaten from the white plates may be the result of familiarity while the obtained quality and liking enhancements for the black square plates may be due to novelty effects.

It is also known that the order in which participants complete questionnaire items can influence their answers to the questions. Malhotra [35] showed that ratings on simple questions, questions that are not challenging and take little effort, were susceptible to order effects. Since all participants in this study completed the attribute ratings in the same order (that is, sweetness followed by intensity followed by quality followed by liking), it is possible that each subsequent rating was influenced by the previous leaving sweetness ratings the only purely independent rating of the four. Without replicating the experiment we cannot know what effect this would have on the current data. However, because this was a between-subjects design and participants only viewed one plate type, there is no reason to predict that any potential order effect would be different for the black square group compared to any other group but future investigations may benefit from an order counterbalancing or randomization of rating items.

Although the square and round plates were of identical widths, the surface area of the round plates (572 cm^2) was less than the square plates (729 cm^2). Since the dimensions of our cake sample were consistent and we did find a white *round* plate enhancement across all ratings and a black *square* enhancement for quality and liking ratings, it is possible that both were surface area effects. Our data cannot rule out this possibility. Although a previous study by Rolls *et al.* [36] showed that food consumption was not different when served on plates of different sizes, this finding is not specific to taste judgments.

Finally, as previously stated, the Piqueras-Fiszman *et al.* [27] study reported no effect of plate shape on ratings of a strawberry mousse whereas the current study reports a new finding, a significant plate colour by plate shape interaction for all attributes. We have already mentioned a number of possible explanations for the pattern of results observed but it is important to address why a shape influence is observed in this study but not the previous [27]. Three reasons stand out for this. First,

the authors of the previously mentioned study suggested that their square plates were rounded at the corners leading to a softening of any possible shape influences. We used square plates that had much sharper points at the corners. Perhaps the effect of shape would have been observed in the Piqueras-Fiszman *et al.* study if plates similar to our own were used [27]. Second, the manifestation of a shape effect in the current study could be because the food sample we used was different. The multisensory influences discussed may be food specific and where the previous authors used strawberry mousse, we used cheesecake. This food stuff difference alone may be enough to explain the shape influence observed here but it could also be a function of the shape of the food, the colour of the food, or some interaction of these variables. Lyman [37] suggested that taste perception may be influenced by figure-ground (that is, food-plate) contrasts. That is, perceptions of taste may vary between foods of different colours, especially when served on different colour plates. Parallel possibilities may also exist for shape variables. As such, the differences in findings between Piqueras-Fiszman *et al.* [27] and the current study might be explained by these contrast differences. The different results may be due to their mousse samples being a different colour (reddish) and presented in a different shape (half sphere or pyramidal) than our cheesecake sample (yellowish and cylindrical). Future research might systematically vary the food (or plate) colour/shape to investigate the influence of simultaneous contrast on taste perception. Finally, a third possibility is that the interactions we observed in our study may have been observed in the Piqueras-Fiszman *et al.* study had they been evaluated [27].

Conclusions

The results of this experiment suggest that the influence of plate colour and shape on taste perception is, not surprisingly, more complex than expected. There appears to be a substantial white round plate advantage although, black square plates have their place also. It seems that basic judgments (that is, sweetness or intensity) are enhanced by white round plates while more complex judgments (that is, quality or liking) are enhanced by both white round and black square plates. We suggest that this may be due to specific learned associations or some sort of familiarity/novelty effect.

Regardless of what is driving this effect, the knowledge that plate shape and plate colour do interact to influence taste perceptions is important to the culinary industry. Chefs certainly want their food to taste a certain way. Knowing that food presented on a white round plate will be perceived as sweeter, for example, would allow them to modify the sweetness levels of their product such that the desired level of *perceived* sweetness is achieved.

Methods

Participants

A total of 48 individuals (27 males and 21 females), ranging in age from 16 to 54 years old ($M = 22.23$, $SD = 6.68$), volunteered to participate in the study. Genders did not differ in age, $t(46) = -1.70$, $P > .05$. All participants voluntarily stopped by two tables set up in the university cafeteria and initially completed a standard informed consent process which additionally involved reporting any food- or ingredient-related allergies. An ingredient list was made available to participants but, for safety reasons, any participant who reported a food-related allergy was thanked for their interest but not permitted to take part in the study. This study received ethics approval from the Grenfell Campus, Memorial University Research Ethics Board.

Materials

President's Choice (PC™) pre-packaged Original New York style cheesecake was used as the food stimulus. It was presented in a cylindrical fashion, with each portion cut to the same size (6.5 cm in diameter) via a biscuit cutter. Black and white porcelain plates of circular and square shapes were used along with tablecloths, napkins, and plastic spoons. Although the overall surface area was greater for the square plates, keeping the diameter/width of the round and square plates equal (27 cm) was thought to be more important (see Figure 2). With that said, the proportion of surface area covered by food has been shown to play a role in perception and should potentially be controlled for in future studies [38].

A questionnaire, adapted from Piqueras-Fiszman et al. [27], was used to gather information on participants' perceptions of the cheesecake sampled. Responses to questions regarding sweetness, intensity, quality, and liking were made on unstructured 10 cm scales, each anchored with 'Not at all sweet', 'Not at all intense', 'Very low quality', and 'Extremely dislike' on the left ends, and 'Very sweet', 'Very intense', 'Very high quality', and 'Extremely like' on the right ends. Participants were required to place an 'X' on each scale wherever they thought best represented their experience of the food sampled. Two further questions assessed participants' hunger at the time of the study and their general regard for cheesecake. Responses to these were also made on unstructured scales but were anchored with 'Extremely hungry' and 'Extremely dislike' on the left, and 'Extremely full' and 'Extremely like' on the right. Participant age and gender were also recorded.

Procedure

One round table was used to present the questionnaires and copies of the informed consent form while the second table, which was covered in a transparent plastic

Figure 2 The four plate types.

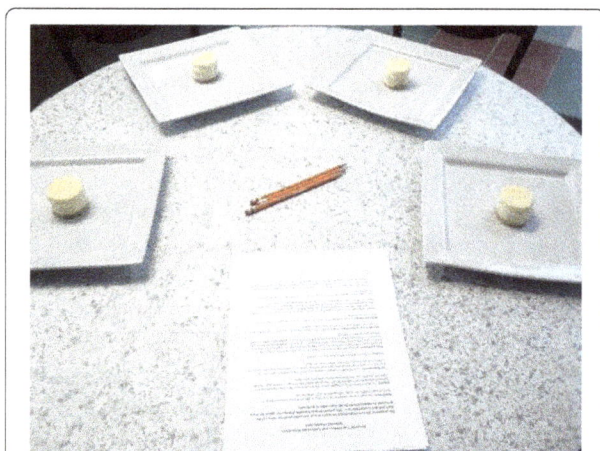

Figure 3 Table set-up for a trial of plates from one experimental condition (white square). No participant ever saw more than one type of plate.

tablecloth, presented the samples to be tasted. Data were collected in blocks with each block consisting of four plates corresponding to one of the four conditions. Each block was repeated three times, in a pseudo-random order, for a total of 12 trials per plate condition. The four plates were spaced equally apart on the half of the table opposite of the researcher and the circular portion of cheesecake was situated in the centre of each plate (see Figure 3).

Volunteer participants were asked whether they possessed any food-related allergies and if they did not, they were invited to begin the experiment by completing an informed consent process. The true nature of the study was withheld from participants until the debriefing session and instead participants were told they would take part in a taste test. At no point during their participation were subjects aware that this was a taste perception study comparing plate colour and shape. Participants were then seated at the table in front of one of the four plates and were instructed to not touch the plate and to refrain from conversing with other participants. As the cheesecake had a layer of graham crumbs as its lowest layer, with the rest of the cake composition being uniform, the participants were instructed to taste one spoonful from the *top* of the sample using the spoon provided. Participants then immediately completed the questionnaire and were then permitted to finish their cheesecake. Once the questionnaire was completed, participants were given a second copy of the informed consent form to keep for their own records, and a debriefing form explaining the true purpose of the study was provided. Any questions they had were answered and they were told how/when they could obtain the group results of the experiment. To avoid biasing any other potential participants, all participants were asked

to not discuss, for a period of three days, the true nature of the study with incoming participants or anyone else who may participate.

Endnote

[a]A 2 × 2 analysis of variance (ANOVA), with plate shape and plate colour as factors and age as the dependent variable, revealed no significant interaction or main effects. Further, age was not significantly correlated with any dependent variable. Both analyses suggested that the groups were well matched for age, and that age, at least within the age range measured here, seemed to not be predictive of judgments of sweetness, intensity, quality, or liking. We also examined if the groups differed in composition with regard to gender. Two chi-square tests, one for each plate colour, examined any relationship between plate shape and gender. There was no significant relationship observed for white plates suggesting similar gender compositions across plate shape conditions. However, for black plates a significant relationship was found between plate shape and gender, $\chi2 = 4.20$, $P < .05$, with significantly more females in the black square condition compared to the black circle condition. To further assess if gender was influencing the attribute ratings we again conducted ANCOVAs, using gender as a covariate, for each of the four attributes. Gender was not a significant covariate for any of the analyses ($P > .05$), suggesting that the gender difference in group compositions did not confound the obtained results.

Competing interests
The authors declare that they have no conflicts of interest, financially or otherwise.

Authors' contributions
PS conceived of the study, participated equally with EG in the study design and data analysis, and drafted/revised the manuscript. EG completed the data collection and aided in the initial stages of manuscript preparation. All authors have read and approved the final manuscript.

Acknowledgements
The authors appreciate the support of Grenfell Campus, Memorial University of Newfoundland for the funding used to purchase study materials and for allotting space in the university cafeteria. The authors also wish to thank all participants who volunteered to complete the study and the anonymous reviewers who aided in manuscript preparation.

References
1. Frank RA, Byram J: **Taste–smell interactions are tastant and odorant dependent.** *Chem Senses* 1988, 13:445–455.
2. Cliff M, Noble AC: **Time-intensity evaluation of sweetness and fruitiness and their interaction in a model solution.** *J Food Sci* 1990, 55:450–454.
3. Clark CC, Lawless HT: **Limiting response alternatives in time-intensity scaling: an examination of the halo dumping effect.** *Chem Senses* 1994, 19:583–594.
4. Stevenson RJ, Prescott J, Boakes RA: **Confusing tastes and smells: How odours can influence the perception of sweet and sour tastes.** *Chem Senses* 1999, 24:627–635.

5. Djordjevic J, Zatorre RJ, Jones-Gotman M: **Odor-induced changes in taste perception.** *Exp Brain Res* 2004, **159**:405–408.

6. Pangborn RM, Trabue IM, Szczesniak AS: **Effect of hydrocolloids on oral viscosity and basic taste intensities.** *J Texture Stud* 1973, **4**:224–241.

7. Okajima K, Spence C: *Effects of visual food texture on taste perception.* ; 2011. http://i-perception.perceptionweb.com/journal/I/article/ic966.

8. Saint-Eve A, Deleris I, Panouille M, Dakowski F, Cordelle S, Schlich P, Souchon I: **How texture influences aroma and taste perception over time in candies.** *Chem Percept* 2011, **4**:32–41.

9. Cook DJ, Hollowood TA, Linforth RST, Taylor AJ: **Perception of taste intensity in solutions of random-coil polysaccharides above and below c*.** *Food Qual Prefer* 2002, **13**:473–480.

10. Walker S, Prescott J: **The influence of solution viscosity and different viscosifying agents on apple juice flavor.** *J Sensory Stud* 2000, **15**:285–307.

11. Salles C, Chadnon M-C, Feron G, Guichard E, Laboure H, Morzel M, Semon E, Tarrega A, Yven C: **In-mouth mechanisms leading to flavour release and perception.** *Critic Rev Food Sci Nutr* 2011, **51**:67–90.

12. Cytowic RE: *The Man Who Tasted Shapes: A Bizarre Medical Mystery Offers Revolutionary Insights Into Reasoning, Emotion, and Consciousness.* New York: Putnam; 2003.

13. Lehrer A: *Wine & Conversation.* 2nd edition. Oxford: Oxford University Press; 2009.

14. Ngo MK, Misra R, Spence C: **Assessing the shapes and speech sounds that people associate with chocolate samples varying in cocoa content.** *Food Qual Prefer* 2011, **22**:567–572.

15. Ngo MK, Spence C: **Assessing the shapes and speech sounds that consumers associate with different kinds of chocolate.** *J Sensory Stud* 2011, **26**:421–428.

16. Moir HC: **Some observations on the appreciation of flavor in food stuffs.** *Chem Indust* 1936, **55**:145–148.

17. Heckler SE, Childers TL: **The role of expectancy and relevancy in memory for verbal and visual information: What is incongruency?** *J Consume Res* 1992, **18**:475–492.

18. Stillman JA: **Color influences flavor identification in fruit-flavored beverages.** *J Food Sci* 1993, **58**:810–812.

19. Dubose CN, Cardello AV, Maller O: **Effects of colorants and flavorants on identification of perceived flavor intensity, and hedonic quality of fruit-flavored beverages and cakes.** *J Food Sci* 1980, **45**:1393–1399.

20. Garber LL, Hyatt EM, Starr RG: **The effects of food color on perceived flavor.** *J Market Theor Prac* 2000, **8**:59–72.

21. Spence C, Harrar V, Piqueras-Fiszman B: **Assessing the impact of the tableware and other contextual variables on multisensory flavour perception.** *Flavour* 2012, **1**:12.

22. Crisinel A, Cosser S, King S, Jones R, Petrie J, Spence C: **A bittersweet symphony: systematically modulating the taste of food by changing the sonic properties of the soundtrack playing in the background.** *Food Qual Prefer* 2012, **24**:201–204.

23. Piqueras-Fiszman B, Laughlin Z, Miodownik M, Spence C: **Tasting spoons: assessing the impact of the material of the spoon on the taste of the food.** *Food Qual Prefer* 2012, **24**:24–29.

24. Spence C, Ngo M: **Assessing the shape symbolism of the taste, flavour, and texture of foods and beverages.** *Flavour* 2012, **1**:12.

25. Oberfeld D, Hecht H, Allendorf U, Wickelmaier F: **Ambient lighting modifies the flavour of wine.** *J Sensory Stud* 2009, **24**:797–832.

26. Irmak C, Vallen B, Robinson SR: **The impact of product name on dieters' and nondieters' food evaluations and consumption.** *J Consume Res* 2011, **38**:390–405.

27. Piqueras-Fiszman B, Alcaide J, Roura E, Spence C: **Is it the plate or is it the food? Assessing the influence of the color (black or white) and shape of the plate on the perception of the food placed on it.** *Food Qual Prefer* 2012, **24**:205–208.

28. Moskowitz HR, Kumraiah V, Sharma KN, Jacobs HL, Sharma SD: **Effect of hunger, satiety and glucose load upon taste intensity and taste hedonics.** *Physiol Behav* 1976, **16**:471–475.

29. Delwiche JF: **You eat with your eyes first.** *Physiol Behav* 2012, **107**:502–504.

30. Palmer SE, Schloss KB: **An ecological valence theory of human color preference.** *Proc Nat Acad Sci* 2010, **107**:8877–8882.

31. Schloss KB, Strauss ED, Palmer SE: **Object color preferences.** *Color Res Appl* 2012, **38**:393–411.

32. Sheau-Fen Y, Sun-May L, Yu-Ghee W: **Store brand proneness: effects of perceived risks, quality and familiarity.** *Austr Market J* 2012, **20**:48–58.

33. Carpenter GS, Glazer R, Nakamoto K: **Meaningful brands from meaningless differentiation: the dependence on irrelevant attributes.** *J Market Res* 1994, **31**:339–350.

34. Bornstein R: **Exposure and affect: overview and meta-analysis of research, 1968–1987.** *Psycholog Bull* 1989, **106**:265–289.

35. Malhotra N: **Order effects in complex and simple tasks.** *Pub Opin Quart* 2009, **73**:180–198.

36. Rolls BJ, Roe LS, Halverson KH, Meengs JS: **Using a smaller plate did not reduce energy intake at meals.** *Appetite* 2007, **49**:652–660.

37. Lyman B: *A Psychology of Food: More Than a Matter of Taste.* New York: Avi, van Nostrand Reinhold; 1989.

38. Zampollo F, Wansink B, Kniffin KM, Shimizu M, Omori A: **Looks good enough to eat: How food plating preferences differ across cultures and continents.** *Cross-Cult Res* 2012, **46**:31–49.

20

Taste and appetite

Per Møller

Abstract

In this short paper, I discuss two interpretations of the implications of food reward for healthy eating. It is often argued that foods that are palatable and provide sensory pleasure lead to overeating. I discuss an example of an experiment that claims to demonstrate this, to many people, intuitively reasonable result. I point out a number of assumptions about reward and eating behaviour underlying this sort of thinking and ask whether overeating might not instead, to a large extent, result from avoiding reward and sensory satisfaction. Four different experimental results that support the suggestion that 'quality can replace quantity' are briefly reviewed.

Keywords: Taste, Appetite, Overeating, Reward, Sensory pleasure

Humans eat foods, not nutrients. Homeostatic appetite mechanisms based on nutrients are therefore not sufficient to explain human food behaviour. Also, if homeostatic mechanisms were the only determinants of food intake, the recent problems of overeating and obesity would be hard to explain. Other control systems of ingestive behaviour and energy balance have therefore been identified [1]. These systems deal mainly with motivational, cognitive and emotional aspects of eating behaviour. Rewards derived from eating figure strongly in these extensive neural networks. Sensory pleasure from the taste of foods is therefore a major determinant of food intake.

Eating is initiated when a state of hunger is reached, but under most circumstances, not just any food will do; usually, people experience hunger for particular foods under particular circumstances.

Since foods provide reward [2], it is important to understand the processes of hedonic eating [3,4] and in particular, how these processes interact with homeostatic mechanisms controlling energy balance [1].

In this paper, I will discuss two interpretations of the implications of food reward for healthy eating.

Pleasure comes in different disguises: as the immediate sensation of wanting and liking a food when it is eaten or as a longer lasting feeling of well-being after a meal. Berridge and his coworkers have proposed a model of reward based on liking, wanting and learning [2,5]. Liking has been studied very much, despite its inability

Correspondence: pem@life.ku.dk
Department of Food Science, University of Copenhagen, Rolighedsvej 30, 1958 Frederiksberg, Denmark

to predict very much of people's food behaviour [6]. Motivational processes of wanting and desire seem to change more during a meal and to be better able to predict behaviour [7]. Obviously, pleasure derived from a meal also depends on expectations *prior* to eating it and on bodily and mental satisfactions and well-being experienced *after* a meal. These are problems that are virtually untouched by scientific investigation. We need to devise new methods of quantifying pleasure and satisfaction. These methods will probably have to rely on measurements of different types of memory and on measurements of interoceptive states [8].

Optimally, the foods we eat should be perceived as appetitive, not just as filling. Will high gastronomic quality of foods consumed on a daily basis leads to overeating, thereby exacerbating problems of overweight and obesity? This view has indeed surfaced in certain scientific circles [9-11]. It might, to some, seem almost self-evident, but to others, like myself, not at all so. From highly unscientific introspection and conversations with friends and colleagues about these matters, it seems that most of us eat far less of high-quality Parmesan cheese when it is offered, than of cheap, not so tasty hard cheeses. The same applies to wines and chocolate and all other types of food. Very few people can eat a whole 100 g bar of Valrhona chocolate in one go but easily perform this feat with chocolate of a lesser quality. From a more epidemiological point of view, one would wonder why the obesity problem in France is less severe than in other affluent countries with foods and meals generally of a lower quality than those served in France [12]. Many scientists have argued that increased pleasure and variety lead to overeating. There is probably little doubt that sensory-specific satieties guide us to eat meals which contain different tastes and textures and this is one of the nature's tricks to help us eat diets balanced in macro- and micro-nutrients without needing to know anything whatsoever about nutrition science [4,13,14]. On the other hand, experiments with real meals under ecologically valid circumstances, as opposed to the often very artificial arrangements and foods subjects face in laboratories, suggest that 'liking' *per se* does not predict when a meal ends [6]. Nevertheless, many workers claim to have demonstrated that pleasure and high variety are important factors for overeating. One example of this kind of thinking is demonstrated in a recent paper by Epstein and coworkers [9].

Epstein et al. randomly assigned 16 obese and 16 non-obese women (aged 20–50 years) to receive a macaroni and cheese meal presented 5 times, either daily for 1 week or once a week for 5 weeks. They also claim to have measured 'habituation' to the food stimuli. Habituation to a stimulus is an expression of the decrement in behavioural and physiologic responses to a stimulus, often observed when repeatedly presenting the same stimulus over and over again. Habituation is an attentional effect that does not involve sensory adaptation/fatigue or motor fatigue. Epstein et al. interpret it differently, describing habituation as a form of learning. Referring to previous work by themselves and others that investigated short-term habituation in their use of the term, the question they ask in this paper is whether there is such a thing as 'long-term' habituation to food.

Whether or not any *effects observed on intake* are causally related to 'habituation' is interesting, but not crucial to potential applications of results like these in the design of meal schemes. The results of the experiment showed that for both obese and non-obese women, daily presentation of a bland food resulted in faster habituation and less energy intake than did once-weekly presentation of the bland food. The smaller energy intake in the once-a-day condition was not *very* significant and might very well have resulted from a serious design flaw of the experiment. Nevertheless, the authors conclude that, if you are offered the same (bland) meal on 5 consecutive days, you will consume less of that particular meal on day 5 than you will on the fifth encounter if you are only offered the (particular) meal once a week for 5 weeks. This result led Epstein et al. to suggest that '*reducing variety may be an important component of interventions for obesity. Habituation may provide a mechanism for the effects of variety on energy intake, such that within-session habituation during a meal can lead to reduced intakes with reduced variety of foods*' [9].

This experiment was considered so important by the editors of the American Journal of Clinical Nutrition that it prompted an editorial written by Nicole M Avena and Mark S Gold [10]. Avena and Gold are fascinated by this work and write.... '*The findings of Epstein et al. provide support and guidance in developing dietary advice, such as the suggestion that people try to eat the same food each day, in which case habituation may develop that would reduce the likelihood of overeating and subsequent obesity*'.

And further... '*Thus, the work of Epstein et al. is important to consider in contemplating and designing meal plans in our variety-rich environment. Clearly, school-lunch planners and public health officials should note that diversity in the menu is not necessarily a virtue, and in fact it may be associated with promoting excess food intake and increased body mass index*'.... In summary, it is suggested that we should '*try to eat the same food each day*' and the call is out for '*school-lunch planners and public health officials*' to note these results.

These writings represent one interpretation of the implications of food reward for healthy eating. It basically claims that unless we severely limit rewards obtained from eating, we run the risk that the obesity epidemic will become even worse than it already is.

Eating food when hungry is obviously rewarding. This makes evolutionary sense. Since eating is necessary for survival, the signals needed to initiate the process of eating must be strong. But it does not follow logically that because initiating signals are strong, people will continue eating beyond satiation and sensory satisfaction. The argument rests on an assumption that the desire for reward is unlimited. This might indeed be the case in certain pathological states, but that it is generally the case is an assumption. In other rewarding human activities, it is well known that 'refractory periods' are necessary to fully enjoy the activity.

As noted above, people often consume substantially less of a food that provides more sensory pleasure than they do of a blander version of the food. That is, the more sensory rewarding a food is, the less people tend to eat of it. If this is the case, sensory satisfaction could promote healthier eating rather than the opposite. I will briefly discuss four sets of data suggesting that this might actually be the case.

The question can be phrased as whether 'quality' can replace 'quantity'.

The striatum is an area in the reward circuit in the brain, which has been implicated in many types of rewarding behaviours. Dopamine is an important neurotransmitter for the functioning of the striatum. Wang and coworkers [15] used positron emission tomography (PET) to measure the availability of dopamine receptors in the striatum in obese individuals and found an inverse relationship between BMI and availability of dopamine receptors. Since dopamine modulates reward circuits, this result suggests that dopamine deficiency in obese individuals may perpetuate pathological eating as a means to compensate for decreased activation of these circuits. That is, eating is driven by reward and continues until enough reward has been obtained. Under the assumption that well-tasting/high sensory quality foods provide more reward per energy unit than bland foods, this result supports the hypothesis that 'quality can replace quantity'.

In an experiment on the effects of trigeminal stimulation (hot spices) on hunger and satiety, HH Reisfelt and I came across a result which is relevant for the present discussion [16]. The subjects in the experiment attended the laboratory twice. On one of the visits, they were served an ordinary industrially manufactured tomato soup and were asked to report on hunger and satiety feelings, as well as on liking and wanting (and other measures which are not important in this context). During the other visit, they were served the same base soup, but this time, we had spiced the soup with chili.

We found that *satiety increased faster* when subjects ate the soup spiced with chili. Also, wanting of more of the spiced soup decreased faster over time than wanting of the base soup, even though wanting of the spiced soup was higher initially. The faster satiation and decrease in wanting when eating the spiced soup might conceal a wish to stop eating caused by a lower appreciation of the spiced soup than of the ordinary soup. We found, however, the opposite effect. The subjects liked better the spiced soup that satiated them faster. That is, eating a more rewarding food does not imply that normal subjects will eat more of it.

In a paper entitled 'eating what you like induces a stronger decrease in wanting to eat' [17], Lemmens et al. demonstrated just that effect with a randomized crossover design. In this experiment, the subjects came to the laboratory twice. During one visit, they were served a portion of chocolate mousse and during the other visit, a portion of cottage cheese. Caloric content was the same in both servings, and the subjects' hunger feelings were the same on the two visits. Chocolate mousse was liked more than the cottage cheese. By means of an image-based method, wanting for a large number of different foods was measured before and after intake of the foods. Lemmens et al. found that wanting dropped significantly for most food categories after intake of the chocolate mousse whereas this was not the case after eating the cottage cheese, which was liked less than the chocolate mousse. This result suggests that it is not a good idea to limit intake of liked foods in order to limit overall intake, under the assumption that people will tend to eat more of a food the more they want it.

Pelchat and coworkers [18] investigated brain activity using functional magnetic resonance imaging (fMRI) in people who had eaten two different diets for 1.5 days prior to the experiment. One group ate a monotonous diet, vanilla-flavoured Boost, whereas the other group ate a normal diet. The subjects in the normal diet group were also given two cans of Boost to familiarize themselves with it. Information about favorite foods was collected from all the subjects. After the 1.5 days of eating a normal diet or a monotonous diet, the subjects were scanned while they were told to imagine the sensory properties of a number of their favorite foods as well as of the Boost. The monotonous diet group showed greater activation to the craved or liked foods than to the monotonous Boost. Craving-related activations were detected in the hippocampus, insula and caudate. These areas have previously been reported to be involved in drug craving. Interestingly, no such differences were found for the normal diet group.

This result suggests that eating a monotonous diet induces stronger food cravings of liked foods that are often energy-dense. Under most circumstances, this will lead to a larger energy intake.

A better understanding of reward-related implications for healthy eating is, of course, not sufficient to fully understand and help prevent inappropriate eating behaviour. Habit and preference formation [19-21] and especially designing schemes where children (and other people) come to appreciate foods which are low in energy content is also important as is more research into self-regulation [22,23].

Before we understand these different basic scientific problems better, scientists should probably be a little less cocky in handing out advice to political decision makers.

Competing interests
The author declares that he has no competing interests.

References
1. Berthoud H, Morrison C: **The brain, appetite, and obesity.** *Annu Rev Psychol* 2008, 2008(59):55–92.
2. Berridge K: **Food reward: brain substrates of wanting and liking.** *Neurosci Biobehav Rev* 1996, 20(1):1–25.
3. Kahneman D, Diener E, Schwarz N (Eds): *Well-Being: The Foundations of Hedonic Psychology.* New York: Russell Sage; 1999.
4. Sørensen L, Møller P, Flint A, Martens M, Raben A: **Effect of sensory perception of foods on appetite and food intake: a review of studies on humans.** *Int J Obesity* 2003, 27:1152–1166.
5. Berridge KC, Robinson TE: **Parsing reward.** *Trends Neurosci* 2003, 26:507–513.
6. Reinbach H, Martinussen T, Møller P: **Effects of hot spices on energy intake, appetite and sensory specific desires in humans.** *Food Quality Pref* 2010, 21:655–661.
7. Olsen A, Ritz C, Hartvig D, Møller P: **Comparison of sensory specific satiety and sensory specific desires to eat in children and adults.** *Appetite* 2011, 57:6–13.
8. Craig AD: **How do you feel? Interoception: the sense of the physiological condition of the body.** *Nat Rev Neurosci* 2002, 3:655–666.
9. Epstein LH, Carr KA, Cavanaugh MD, Paluch RA, Bouton ME: **Long-term habituation to food in obese and nonobese women.** *Am J Clin Nutr* 2011, 94:371–376.
10. Avena NM, Gold MS: **Variety and hyperpalatability: are they promoting addictive overeating?** *Am J Clin Nutr* 2011, 94:367–368.

11. Møller P, Köster EP: **Variety and overeating: comments on long-term habituation to food.** *Am J Clin Nutr* 2012, **95**:981.

12. Rozin P, Kabnick K, Pete E, Fischler C, Shields C: **The ecology of eating: smaller portion sizes in France than in the United States help explain the French paradox.** *Psychol Sci* 2003, **14**(5):450–454.

13. Rolls BJ, Rolls ET, Rowe EA, Sweeney K: **Sensory specific satiety in man.** *Physiol Behav* 1981, **27**:137–142.

14. Rolls BJ, Rowe EA, Rolls ET: **How sensory properties of foods affect human feeding-behavior.** *Physiol Behav* 1982, **29**:409–417.

15. Wang GJ, Volkow ND, Logan J, Pappas NR, Wong CT, Zhu W, Netusll N, Fowler JS: **Brain dopamine and obesity.** *Lancet* 2001, **357**:354–357.

16. Møller P: **Gastrophysics in the brain and body.** *Flavour* 2013, **2**:8.

17. Lemmens S, Schoffelen P, Wouters L, Born J, Martens M, Rutters F, Westerterp-Plantenga M: **Eating what you like induces a stronger decrease of 'wanting' to eat.** *Physiol Behav* 2009, **98**:318–325.

18. Pelchat ML, Johnson A, Chan R, Valdez V, Ragland JD: **Images of desire: food-craving activation during fMRI.** *NeuroImage* 2004, **23**:1486–1493.

19. Schaal B, Marlier L, Soussignan R: **Human fetuses learn odours from their pregnant mother's diet.** *Chem Senses* 2000, **25**:729–737.

20. Hausner H, Nicklaus S, Issanchou S, Mølgaard C, Møller P: **Breastfeeding facilitates acceptance of a novel dietary flavour compound.** *Clin Nutr* 2010, **29**:141–148.

21. Hausner H, Olsen A, Møller P: **Mere exposure and flavour-flavour learning increase 2–3 year-old children's acceptance of a novel vegetable.** *Appetite* 2012, **58**:1152–1159.

22. de Ridder DTD, Lensvelt-Mulders G, Finkenauer C, Stok FM, Baumeister RF: **Taking stock of selfcontrol: a meta-analysis of how trait self-control relates to a wide range of behaviors.** *Pers Soc Psychol Rev* 2012, **16**:76–99.

23. Hofmann W, Schmeichel BJ, Baddeley A: **Executive functions and self-regulation.** *Trends Cogn Sci* 2012, **16**:174–180.

21

Assessing the influence of the color of the plate on the perception of a complex food in a restaurant setting

Betina Piqueras-Fiszman[1*], Agnes Giboreau[2] and Charles Spence[1]

Abstract

Background: Nowadays, more and more importance is given to how restaurant dishes are visually presented. With regard to the color of the plate, several recent studies have demonstrated that identical foods served on plates (or in containers) of different colors are often perceived differently at both the sensorial and hedonic levels. However, to date, these effects have not been tested in an ecologically valid setting with a range of more complex foods in order to assess the generalizability of the findings. The aims of the present study were to test the extent to which the color of the plate may influence the gustatory and hedonic experiences of a complex food. Specifically, we investigated diners' perception of three desserts served on a white or black plate in a between-participants experimental design in a real restaurant setting.

Results: The results demonstrated that the color of the plate exerted a significant influence on people's perception of the food, but that this effect varied as a function of the type of dessert served. The effects cannot be explained only in terms of color contrast. Color-flavor associations, for example, black with intense chocolate flavor, or even sophisticated chocolate, could have an impact too. Interestingly, the perceptual pattern for each dessert was constant for each plate used; that is, for all of the attributes rated, the higher scores were obtained with the same plate, for all of the desserts.

Conclusions: These results confirm the importance of the color of the plate (or background color) on people's expectation and perception of food, even in realistic and less controlled conditions, such as that of a restaurant.

Keywords: Food perception, Color influence, Flavor intensity, Liking, Consumer studies, Restaurant setting, Ecologically valid contexts, External validity

Background

Food and drink is normally consumed from certain containers/receptacles, such as plates, bowls, cups, glasses or product packaging (for example, drinks cans or plastic yoghurt pots). Great importance is often given to how the food or drink is visually presented [1-3], both in supermarkets and restaurants. Researchers are interested in examining the extent to which food and drink can be made to look more appealing and appetizing [4]. There is also interest in trying to modulate consumers' impressions of serving size and intake by modifying factors other than the food itself [5,6]. It is becoming increasingly clear that

our perception of food is affected not only by the various sensory properties of the food, but also by our expectations about it and other contextual factors [7,8].

With regard to the influence of the appearance of food on people's perception of its flavor, it is important to note that the color of a food or beverage often dominates over other sources of information regarding the flavor [9-13].

Focusing on the culinary context, several recent studies have demonstrated that the color of the surroundings in which food is presented (for example, cups, plates, tablecloth and even ambient lighting) has an effect on both consumers' perception of the food as well as the amount that they serve themselves and, very likely, consume. Recent studies have also provided evidence documenting the impact of the color of food containers on taste and flavor perception [14-17].

* Correspondence: Betina.piqueras-fiszman@psy.ox.ac.uk
[1]Department of Experimental Psychology, University of Oxford, South Parks Road, Oxford OX1 3UD, UK
Full list of author information is available at the end of the article

Regarding the effects of the color of the container on serving size, van Ittersum and Wansink [18] (study 2) found that those participants in a high color contrast condition served 9.8% (P <0.01) more than the target serving size on a larger plate, and 13.5% (P <0.01) less than the target serving size on a smaller plate. Meanwhile, in another study, van Ittersum and Wansink [18] (study 5) tested the effect of color contrast between the food and the plate on serving sizes in a realistic serving situation. Their results revealed that participants in the low color contrast condition (white pasta sauce on a white plate or red pasta sauce on a red plate) served themselves significantly (P <0.01) more pasta than participants in the high color contrast condition (white pasta sauce on a red plate or red pasta sauce on a white plate). These two studies can potentially be framed in terms of the Delboeuf illusion, the name given to the illusion whereby a central circle appears smaller when surrounded by a much larger concentric circle than when surrounded by a circle that is only slightly larger [19,20]. It has been shown that this illusion is enhanced by color contrast, and it could therefore provide a possible explanation for why and how plate size can influence people's serving behavior in a variety of real life situations.

With regard to the contrast effect between the color of the food and the color of the plate (that is, the color of the dish, cup, and so on), there are also mechanisms that may explain the perception of certain 'illusory' or more saturated colors. For instance, the orange of a carrot is intensified if it is served on a blue (as compared to a white) plate due to the phenomenon of simultaneous contrast [21]. If the color of the plate (or background) affects the way in which people perceive the color of the food [21-23] and the color of the food is known to affect the perception of flavor, then the color of the plate (and any contrast effects that it elicits) could be expected to influence the perceived properties of the food (for example, the flavor intensity). Piqueras-Fiszman et al. [24] studied this effect under laboratory conditions in a within-participants experiment using white and black plates. The results revealed that a strawberry mousse of homogeneous color and texture tasted significantly more intense and sweeter, and was liked more, when served on a white plate as compared to when served on a black plate. However, although these results are interesting from a purely sensory/psychological perspective, they have not been tested in an ecologically valid context and with more complex foods (that is, foods varying in color, texture, and so on).

Under more realistic conditions, such as that represented by dining in a restaurant setting, diners do not try the same food served from different plates; and, in addition, diners may consume different drinks in the meantime (such as soft drinks, wine or water), consume the meal at a variable pace and converse. Consequently, less attention may be paid to elements, such as where the food is presented. Moreover, in restaurants and bars, food is not presented in exactly the same way to all of the diners, as happens in the majority of controlled laboratory studies. Therefore, the previous effects obtained might not be as evident (if present at all) when assessed in a real life situation. On the other hand, consumers are often not aware of the environmental factors that may greatly influence their food choices or consumption behaviors [25].

Many authors have already demonstrated the effects of the physical testing environment on food acceptability and choice [26-29]. In addition, a great deal of research has been carried out recently on the context effect [30-32]. While the majority of these studies have focused on the effect of the context on hedonic ratings and food choice, demonstrating in many cases that the food is liked more in a 'real life' context (although the results are not consistent across all of the studies that have been published), the focus of this study is slightly different.

The present study was designed to investigate the influence of the color of the plate on the visual and gustatory experiences of different foods, as observed in previous research [24]. However, in contrast to previous research, the study reported here was conducted in a real restaurant (and eating situation) with three different desserts (made from various elements of different colors, flavors and textures) in order to assess the extent to which the cross-modal perceptual effects found in laboratory settings can be generalized to naturalistic testing conditions. Specifically, our research questions were: Which color of plate will make each dessert seem more appetizing? Which color of plate will the color intensity of the desserts appear enhanced? Will the perception of certain sensory attributes (for example, flavor intensity) be affected by the visual perception (for example, color intensity)? Will flavor-liking ratings be affected by the appearance-liking ratings? Additionally, will there be a consistent pattern of results between the different color of plates used, and between the visual and taste-related attributes assessed?

Results

All of the diners ate nearly all of the food served, and all diners finished the dessert (as reported by the restaurant manager). This information helps to match the conditions across participants.

The effect of the color of the plate on pre-tasting attributes

Appetizing rating

The color of the plate exerted a significant effect on consumers' appetizing appraisal of the desserts (P <0.0001), and the interaction between the color of the plate and the dessert was also significant (P <0.0001). This effect was only observed for Dessert A, which was perceived as significantly more appetizing when served on the white plate as compared to the black plate (mean (M) = 7.7 versus 5.0, P <0.0001). As shown in Figure 1A, Dessert C was also perceived as more appetizing when served on the white plate (although the difference was small, M = 7.4 versus 6.8), while Dessert B was rated as slightly more appetizing on the black plate (M = 6.8 versus 7.5, on the white and black plate, respectively). In addition, the interaction of the meal session with the color of the plate was also significant (P <0.0001). Diners rated the desserts served on the white plate for the lunch sessions as more appetizing (across the desserts, M = 7.6 versus 5.8, P <0.001), while for dinner, diners rated the desserts as similarly appetizing, regardless of the color of the plate (Figure 1B).

Liking of the overall presentation rating

Perhaps unsurprisingly, the results of liking of the overall presentation ratings followed exactly the same pattern as the appetizing ratings. According to the results of the analysis of variance (ANOVA), the dessert and the color of the plate exerted a significant effect on consumers' liking ratings of the overall presentation of the dish (P <0.0001 and P <0.01, respectively). However, this effect was not observed for all the desserts, as indicated by the significant interaction between the plate and the type of dessert (P <0.0001). In fact, only Dessert A was liked significantly more when served on the white plate than on the black plate (M = 6.9 versus 4.7, P <0.0001; Figure 2A). The results also highlighted a tendency for Dessert C to be liked more on the white plate (M = 7.3 versus 6.8), whereas the opposite results were observed for Dessert B (M = 5.2

versus 6.9); although these latter two differences between plate colors were not significant.

Regarding the meal session, its interaction with the color of the plate was significant (P <0.001). For dinner, the appearance of the dessert was similarly liked when served on either plate (M = 6.7 versus 6.8), while for lunch, serving the dessert on a white plate had a positive impact on consumers' liking of the presentation (M = 7.3 versus 5.9, P <0.001; Figure 2B). However, this was not the case for all of the desserts, as suggested by the interaction between the color of the plate and the type of dessert (P <0.0001).

Color intensity rating

Regarding the perceived color intensity of the desserts, significant effects were observed for the color of the plates, the type of dessert and their interaction (P <0.05, P <0.001 and P <0.01, respectively). In general, taking the three desserts as a group, their color was perceived as more intense when served on the white plate (M = 6.6 versus 6.0). However, at an individual level, the intensity of the color of the desserts was not significantly different when any of the desserts were served on either the white or black plate, although for Dessert C this difference reached marginal levels of significance (M = 7.4 on the white plate versus 6.5 on the black plate, P = 0.54; Figure 3A). Regarding the meal session, although it was not a significant factor (nor was its interaction with any other factors), the previous pattern of results observed in the other ratings was still observed (interaction with the dessert, P = 0.06), suggesting that the color intensity of the desserts (A and C) was enhanced on the white plate, only for the lunch session (M = 6.7 versus 5.7, P = 0.055; Figure 3B).

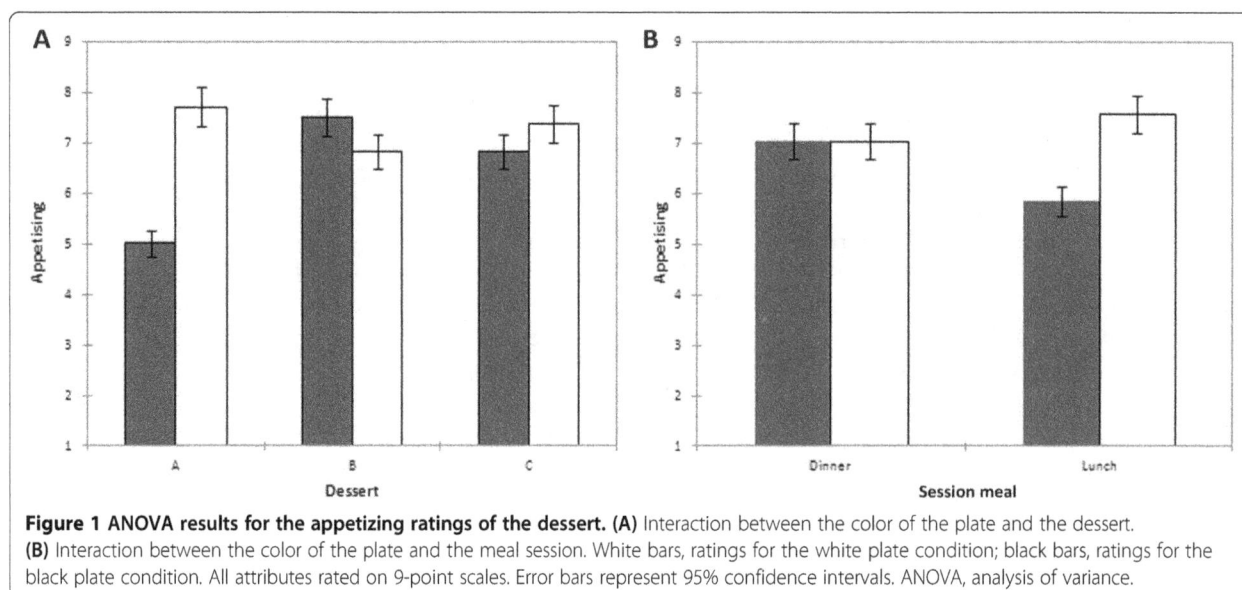

Figure 1 ANOVA results for the appetizing ratings of the dessert. (A) Interaction between the color of the plate and the dessert. **(B)** Interaction between the color of the plate and the meal session. White bars, ratings for the white plate condition; black bars, ratings for the black plate condition. All attributes rated on 9-point scales. Error bars represent 95% confidence intervals. ANOVA, analysis of variance.

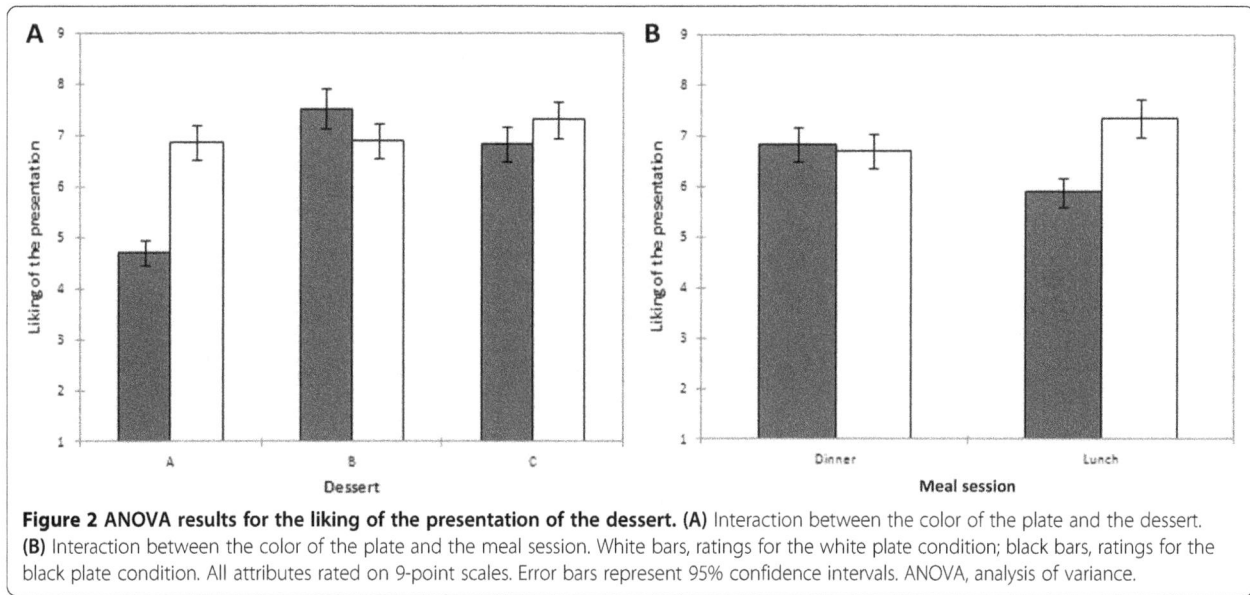

Figure 2 ANOVA results for the liking of the presentation of the dessert. (A) Interaction between the color of the plate and the dessert. **(B)** Interaction between the color of the plate and the meal session. White bars, ratings for the white plate condition; black bars, ratings for the black plate condition. All attributes rated on 9-point scales. Error bars represent 95% confidence intervals. ANOVA, analysis of variance.

The effect of color of the plate on sensory attributes

The results presented so far demonstrate an effect of the color of the plate on the attributes, which were mainly based on the appearance of the presentation of the desserts (prior to tasting). The following section describes the effect of the color of the plate on the oral perception of the dessert.

Flavor intensity

The flavor intensity results were similar to the results for visual attributes, but no main effects of the color of the plate were observed. The flavor intensity was only affected by the type of dessert and its interaction with the

color of the plate (P <0.001 and P <0.01, respectively). Only the flavor of Dessert B (the dessert with darker brown-colored tones, which presented less of a contrast with the black plate) was perceived as nearly significantly more intense (6.5 versus 7.3, P = 0.08) when presented on the black plate (Figure 4). This suggests that the results cannot be explained by a possible contrast effect, but rather by other factors. It could be the case that presenting this dessert on a black background elicited associations with stronger intensity products (for example, chocolate/coffee products with dark-brown or black packaging are usually associated with a stronger flavor or a more sophisticated chocolate/coffee product).

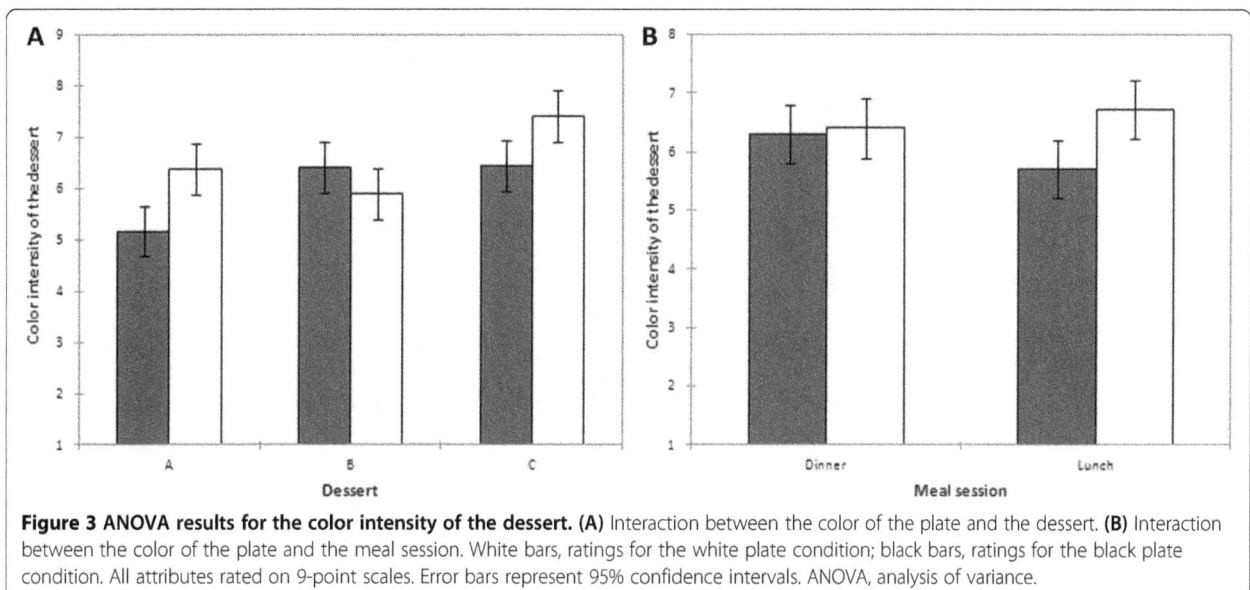

Figure 3 ANOVA results for the color intensity of the dessert. (A) Interaction between the color of the plate and the dessert. **(B)** Interaction between the color of the plate and the meal session. White bars, ratings for the white plate condition; black bars, ratings for the black plate condition. All attributes rated on 9-point scales. Error bars represent 95% confidence intervals. ANOVA, analysis of variance.

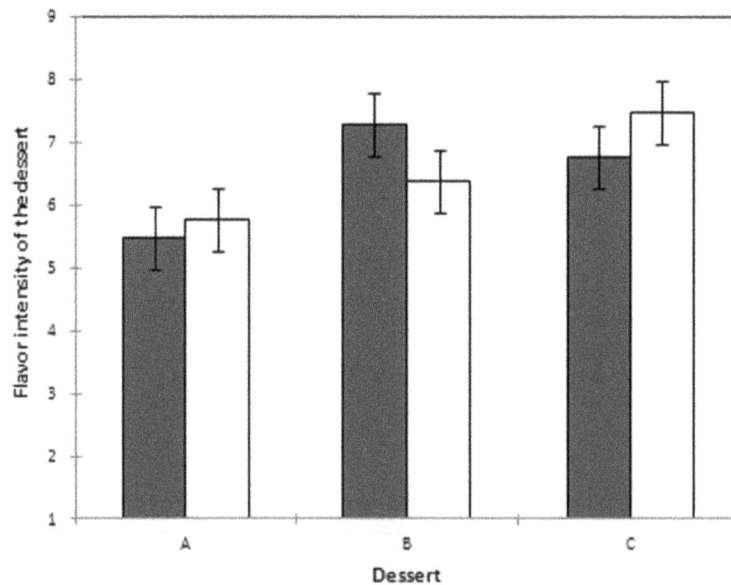

Figure 4 ANOVA results for the flavor intensity of the dessert. Interaction between the color of the plate and the dessert. White bars, ratings for the white plate condition; black bars, ratings for the black plate condition. All attributes rated on 9-point scales. Error bars represent 95% confidence intervals. ANOVA, analysis of variance.

Sweetness intensity

The type of dessert was the only attribute to have a significant effect on the perceived sweetness ($P < 0.0001$), which meant that Dessert B was rated as sweeter than Desserts A and C, regardless of the color of the plate they were served on (Figure 5). On average, Dessert B was rated slightly (0.5 point) sweeter when served on the black plate than the white plate, and the opposite was observed for the other two desserts.

It is interesting to highlight that the pattern of results was maintained. The sweetness intensity of Dessert B scored higher on the black plate, and the contrary was observed for Desserts A and C, for all the attributes, even if not reaching significant levels. When the regression results

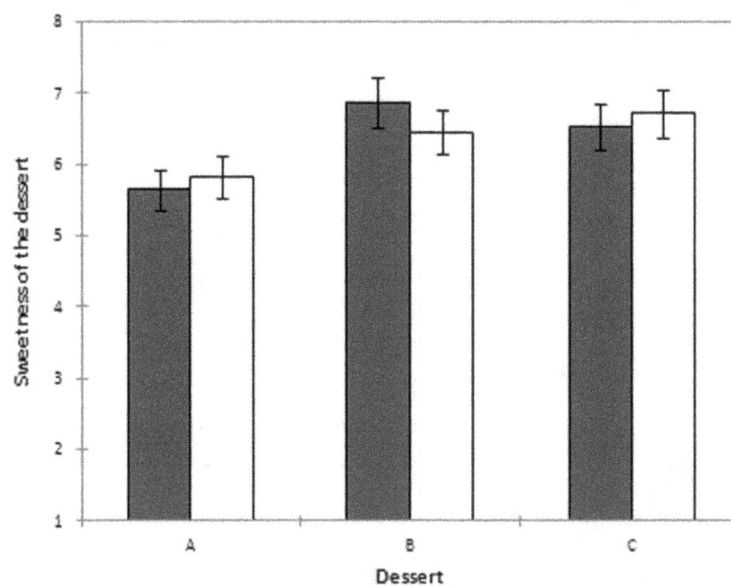

Figure 5 ANOVA results for the sweetness intensity of the dessert. Interaction between the color of the plate and the dessert. White bars, ratings for the white plate condition; black bars, ratings for the black plate condition. All attributes rated on 9-point scales. Error bars represent 95% confidence intervals. ANOVA, analysis of variance.

between the reported color intensity, flavor intensity and sweetness intensity of the desserts were examined (Table 1), the participants' ratings of color intensity were highly correlated to the results reported for flavor intensity (for the white and black plates, R = 0.539, r^2 = 0.36 and R = 0.603, r^2 = 0.29, respectively) but not to the sweetness ratings, as the low correlation coefficients indicate (R = 0.272, r^2 = 0.07 and R = 0.238, r^2 = 0.06, respectively). Therefore, it could be argued that when the participants perceived the color of the dessert as more intense, their perception of the flavor of the food served was also altered in a similar manner. However, as described earlier, the perceived (or reported) color intensity also varied as a function of the type of dessert, indicating that the observed correlations also depended on which dessert was served on which plate.

Overall liking

Regarding the overall liking, which was assessed after tasting the dessert, only the interaction effect between the color of the plate and the dessert was significant. Figure 6 reveals that Desserts A and C resulted in very similar scores for both plates, while essentially the reverse pattern was obtained for Dessert B. The latter was liked significantly more on the black plate than on the white plate (M = 7.6 versus 6.5, P <0.05) and marginal differences were observed for Dessert C, which was liked slightly more on the white plate (M = 7.6 versus 6.8, P = 0.09).

The effect that the hedonic attributes rated visually (that is, appetizing and liking of the presentation) on the final overall liking were highly positively correlated, as would be expected (Table 2). The regression results highlight that the participants' ratings of the final overall liking were highly correlated with the results reported for appetizing ratings (for the white and black plates, R = 0.483, r^2 = 0.23 and R = 0.649, r^2 = 0.42, respectively) and for liking of the presentation (R = 0.524, r^2 = 0.27 and R = 0.472, r^2 = 0.23, respectively). However, depending on the dessert, the results differed to a great extent. Figures 1 and 2 (left panels) show that the ratings given to Dessert A (on a white and black plate) differed significantly for both appetizing and liking of the presentation, while the only dessert which was perceived significantly different in terms of its overall liking from one plate to the other was Dessert B. Therefore, while a correlation was observed, it cannot be entirely explained in terms of a halo effect in the ratings (Figure 6).

Table 1 Correlations (Pearson coefficient) between reported color intensity, flavor intensity and sweetness intensity, by color of plate

Plate color	Flavor intensity	Sweetness intensity
Black	0.603[b]	0.205[a]
White	0.539[b]	0.272[a]

[a]Significant effect at P <0.05; [b]Significant effect at P <0.0001.

Discussion

The present study was designed to investigate whether or not the color (either black or white) of the plate would exert a significant influence on how appealing and appetizing three different desserts were rated (prior to consumption), and if the perception of these visual attributes affected the perceived flavor, sweetness intensity and final overall liking of diners in a restaurant setting. While other researchers have focused on the impact of visual cues and the presentation of food on consumers' acceptance and preference [1,2,7], including the estimation of portion size [18], to date the effect of an extrinsic factor, such as the color of the plate on visual and oral perception of food, has not been studied in any depth [24]. Importantly, this study was performed in an entirely ecologically valid setting (a real restaurant), under natural conditions (that is, using a between-participants design, with participants able to interact and consume their meal at their own pace) and with three different desserts (served in as similar a manner as possible).

The results demonstrate that the color of the plate affected consumers' perception mainly for the attributes based on visual judgments (liking of the presentation, appetizing and color intensity of the dessert). One limitation to be taken into account when interpreting the results is that although the ratings given for the visual attributes were positively correlated, it could be that there was a halo effect (that is, ratings of appetizing may have overridden ratings of the color intensity and overall liking of the presentation). Although balancing the order of presentation of these attributes in the questionnaire could have ruled out this order effect, it is likely that the diners would have rated these three related attributes in a similar way, regardless of the order of presentation of the questionnaire items.

The attributes of the desserts based on chemosensory qualities of the food stimuli (that is, flavor intensity and sweetness) were affected mainly by the type of dessert served, but the extent to which these attributes were affected depended on the plate (background color) as well. Figures 2-6 demonstrate that Desserts A and C always received higher scores when presented on the white plate, while the opposite was observed for Dessert B. This consistency across the attributes suggests that, by itself, the color of the plate has the potential to elicit a positive appraisal, and even enhance certain sensory qualities of foods. However, this effect depends to a great extent on the food presented, together with many other parameters, such as the context, material of the plate, and congruity between these factors and the food itself [7,31]. Once again, the similarity in the response patterns among the attributes could have been due to a halo effect between the attributes rated.

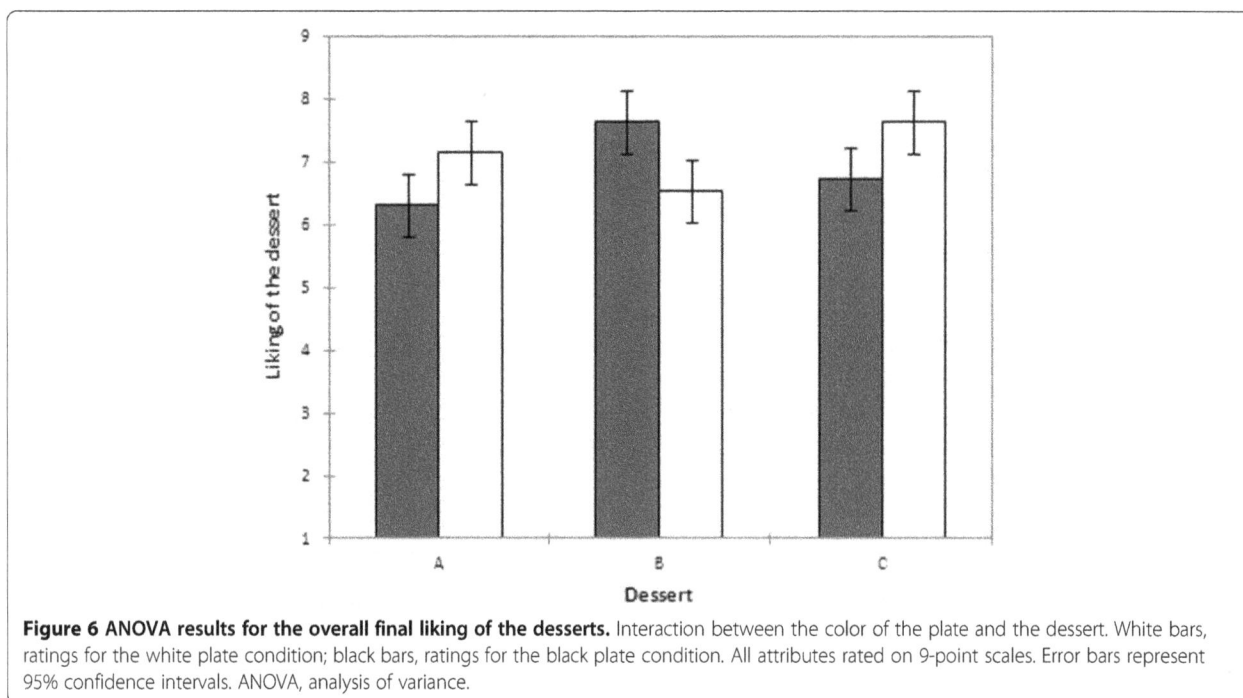

Figure 6 ANOVA results for the overall final liking of the desserts. Interaction between the color of the plate and the dessert. White bars, ratings for the white plate condition; black bars, ratings for the black plate condition. All attributes rated on 9-point scales. Error bars represent 95% confidence intervals. ANOVA, analysis of variance.

In this study, given the complexity of the visual appearance of the desserts (that is, with layers and decorations of different colors and tastes), the results cannot be explained solely in terms of color contrast between the dessert and the plate, since it was Dessert B, the dessert with darker brown-colored tones, that received higher scores when served on the black plate. This result highlights the complex reasons why consumers prefer certain food products and even perceive some attributes as enhanced when served from one container, or background, rather than another [17]. The reason for the Dessert B result could be that the black background elicited a more intense flavor (that is, it evoked this expectation), which resulted in higher attractiveness and appetizing ratings. However, flavor intensity was not significantly affected.

However, the relationship between the visual and taste-related attributes is of greater interest. Regarding color perception, the flavor intensity patterns observed for all the desserts (Figure 4) were similar to the perceived color intensity of the desserts (Figure 3), and the two attributes were highly correlated. Certain sensory dimensions (such as intensity) are processed by several different modalities.

Table 2 Correlations (Pearson coefficient) between reported ratings of appetizing, liking of the presentation (pre-tasting attributes) and final overall liking

Plate color	Appetizing	Liking of the presentation
Black	0.649[a]	0.472[a]
White	0.483[a]	0.524[a]

[a]Significant effect at $P < 0.0001$.

Therefore, it would not be surprising if consumers implicitly perceived a food product as more intense when its color(s) is perceived as such [33].

However, it could be that the desserts simply 'looked' better to diners when served on a certain colored plate. Thus, as part of a halo effect, the rest of the ratings (sensory and hedonic) may also have been enhanced [34]. In any case, it has already been widely demonstrated that the visual appearance of food (presented in a certain container, as it usually reaches the consumer or diner) can affect both sensory discriminative and hedonic appraisal [8,17,35], whether as part of a halo effect or as a genuine perceptual effect.

Regarding the lunch and dinner sessions, the lunch sessions had natural lighting conditions, while artificial lighting was used for the dinner sessions. These are the usual conditions in which diners eat at a restaurant; therefore, we decided to keep them as natural and realistic as possible. It is important to highlight that the effect of the session (only interaction effects with the color of the plate and only in ratings of visual attributes, as described in Results) could be due to the differences in illumination conditions (note that the order of presentation of the plates was counterbalanced between the sessions).

The results regarding the changes in the sensory attributes of the desserts as a function of the plate color are not as strong as reported in the Piqueras-Fiszman *et al.* study [24]. This is understandable, since in the context of a real restaurant, the stimuli cannot be perfectly matched for all consumers and the conditions for all of the diners cannot be kept identical (that is, diners may or may not

Week		Wed	Thurs	Friday
1	Lunch	--	White	Black
	Dinner	White	Black	White
2	Lunch	--	Black	White
	Dinner	Black	White	Black
		Dessert A	Dessert B	Dessert C

Figure 7 Outline of the 2-week study procedure. Dessert A was a *fraisier*, Dessert B was a *fraicheur* of raspberry and vanilla, and Dessert C consisted of a *vacherin* glacé with vanilla, raspberry and basil.

drink while eating, certain diners may eat more rapidly than others, diners may be distracted while eating, and so on) [30]. Consequently, this makes observations hard to analyze and interpret because testing conditions cannot be controlled as strictly as in laboratory conditions and there could be many other intervening factors [36,37]. In addition, diners engaged in conversation are less likely to pay attention to the presentation of a food (in particular, to the plate). These observations contrast with testing under laboratory conditions, where participants are tested individually and are more focused on the task. However, on the other hand, these results are more ecologically valid and although no generalization can be put forward to suggest which background color is better in terms of flavor for a given food (dessert), the results nevertheless highlight that the color of the plate can exert a significant impact on consumers' expectations and appraisal of the food, prior and after consumption, despite the complex uncontrolled conditions, which makes obtaining significant results even more challenging.

Therefore, the results could be particularly relevant in situations where plated dishes are presented to diners prior to tasting, such as restaurant websites, social media, food events, pictorial menus and buffets.

Conclusions

The present study showed that the color of the plate affected consumers' perception mainly for the attributes based on visual appraisal (liking the presentation of the dish, how appetizing the dessert looked and the color intensity of the dessert). The attributes of the desserts, such as the flavor and sweetness intensity, were affected mainly by the type of dessert served, but the extent to which these attributes were affected depended on the plate (background color) as well. Some of the results can be explained in terms of color contrast; however, the associations that consumers can hold for certain colors and flavors (regarding the intensity dimension) can play an important role too. Therefore, these findings contribute to the emerging literature on how extrinsic variables can influence food perception, highlighting that the impact is dependent on the specific food evaluated and that results in real life conditions can be slightly diminished as compared to laboratory conditions. Certainly, more research is needed to confirm the validity and robustness of such results. Nevertheless, chefs can capitalize on these findings and further exploit the characteristics of the plates in order to discover potential new ways to systematically enhance expectations, perception and experience of food, apart from modifying the ingredients and decoration of the food, and mostly in situations where the plated food is showcased in advance prior to consumption.

Methods

The study was carried out at the experimental restaurant of the Institut Paul Bocuse, Lyon, France, which works as a fully operational restaurant (that is, people book, dine and pay for their meals). Three different desserts were served either on white or black plates according to the balanced between-participants experimental design shown in Figure 7. Each dessert was prepared on a different day of the week (Wednesday, Thursday or Friday), while presented on a white or black plate either during the lunch or dinner services, balanced across 2 weeks from 6 June 2012 to 15 June 2012.

Figure 8 The three desserts. From left to right: Dessert A (*fraisier*), Dessert B (*fraicheur* of raspberry and vanilla) and Dessert C (*vacherin* glacé with vanilla, raspberry and basil), presented on white or black plates.

Table 3 Participant information by day and session

Session	Wednesday (males/females)	Thursday (males/females)	Friday (males/females)
Lunch	N/A	19/32	18/34
Dinner	22/26	25/25	27/25
Total (n)	48 (22/26)	101 (44/57)	104 (45/59)
Age (years)	Mean = 45; SD = 12	Mean = 42; SD = 13	Mean = 44; SD = 16

Food stimuli

The dessert part of the meal was chosen as the stimulus for this study, since it is normally served in a portion and is therefore easier to ensure that diners consume it in the same manner, compared to other courses comprised of several different elements. There was flexibility in terms of the choice of dessert used (that is, we could discuss and decide with the chefs the types of desserts). Berry-based desserts were chosen, since they were reasonably consistent with the Piqueras-Fiszman *et al.* study [24] and were served as part of the menu in the weeks when the study was conducted (Figure 7). Dessert A was a *fraisier* (main colors: yellow, white and red), Dessert B was a *fraicheur* of raspberry and vanilla (main colors: light brown, white and red), and Dessert C consisted of a *vacherin* glacé with vanilla, raspberry and basil (main colors light pink, white and cream) (Figure 8). Importantly, although ideally the desserts should have been presented (and decorated) identically, this was practically impossible to control in this realistic setting. Therefore, they were served in as similar a manner as possible. In addition, a closed fixed course menu (no possibility to choose) was kept constant throughout the 2 weeks in which the study was conducted in order to rule out any bias that could have been introduced by different dishes tasted before the desserts. The courses served prior to the desserts were presented on an assortment of tableware, that is, some starters were served in a transparent bowl, while some main courses were served on a slate plate or a traditional white plate (but were kept constant and the same for all the diners).

Plates

White and black plates were ordered especially for use in this study, since they are the most commonly used colors in restaurant settings, as observed from several tableware catalogues for restaurants and hospitality. The plates were of the same shape and size (rectangular, 30 × 26 cm), but the white plate had a glossy finish, while the black plate had a matt finish. Although, strictly-speaking, black and white are not considered as colors, they will be referred to as such for ease of exposition here.

Participants

A total of 253 diners participated in the study and nearly all of them were French. Of these participants, 142 (56%) were female (M age = 43.4 years; SD = 13.8 years). No particular recruitment process was followed. The participants simply consisted of people who had chosen to book a table for lunch or dinner at the restaurant, and taking part in the study was completely voluntary. Detailed information concerning the participants' age and gender distribution by day and session is shown in Table 3.

Procedure

The participants were welcomed to the experimental restaurant (Figure 9) and it was explained to them that a quick questionnaire would be delivered at the end of the

Figure 9 Experimental restaurant.

meal in order to know what they thought about the dessert. This was worded so that the participants would most likely consider that the aim of the study was to give feedback to the trainee chefs. Participants were asked to read the questionnaire first and then proceed with the dessert. A5-size pencil-and-paper questionnaires and pens were delivered with the dessert.

In order to examine whether the color of the plate influenced the liking of the presentation and how appetizing the desserts appeared, the following questions were asked: 1) 'How appetizing is the dessert?' (visually); and 2) 'How much do you like the appearance of the dish overall?'. Since desserts with different colors as stimuli make it complex to test any possible contrast effect, a question regarding the diners' perception of the intensity of the color of the dessert was included: 3) 'How intense is the color of the dessert?'. Then, in order to check whether the appearance attributes were related to gustatory perception, once participants had tasted the dessert, questions were asked about the intensity of the flavor (overall) and of sweetness (the main taste of these desserts): 4) 'How intense is the flavor of the dessert?'; and 5) 'How intense is the sweetness of the dessert?'. In addition, liking was also asked regarding the overall perception of the dessert: 6) 'How much did you like the dessert?'. Basic demographic questions were also included in the questionnaire.

The participants were asked to rate each question on a 9-point scale labeled at anchors with 'not at all' and 'very much'. Questions 1 to 3 were completed prior to tasting, while the remaining questions were completed during or at the end, to keep the situation as natural for the participants as possible.

After the meal, the questionnaires were collected, the diners paid for their meal as in any normal restaurant and they were thanked for taking part in the study.

The human study reported in this article was performed in accordance with the ethical standards of the Declaration of Helsinki (1964) and approved by the ethical committee of the Institut Paul Bocuse. All procedures were carried out with the adequate understanding and written consent of the participants.

Data analysis

In order to determine whether the color of the plates exerted a significant effect on the six attributes in question, a three-way ANOVA was performed on the data to examine the meal session (lunch or dinner), the dessert (A, B or C), the plate (white or black) and the interactions as explanatory variables. In addition, two regression analyses were performed to explore the correlation between the ratings of the color intensity, flavor intensity and sweetness intensity, and the correlation between the

ratings for appetizing, liking of the presentation (appearance attributes) and overall final liking (post-tasting).

When the effects were significant, honestly significant differences were calculated using Tukey's test. Differences were considered significant when $P \leq 0.05$. Statistical analyses were performed using XLSTAT 2011 (Addinsoft, NY, USA).

Abbreviations
ANOVA: analysis of variance; M: mean; SD: standard deviation.

Competing interests
The authors declare that they have no competing interests.

Authors' contributions
BPF, AG and CS conceived the idea of the study and designed the details. BPF conducted the study and analyzed the data. BPF and CS wrote the manuscript. BPF, AG and CS read and approved the final manuscript.

Acknowledgements
The authors would like to thank the Ministerio de Educación, Spain, for the grant awarded to BPF. In addition, the authors would like to thank Camille Schwarz, Caroline Jacquier, Rémy Mondon and Alain Dauvergne from Institut Paul Bocuse for their valuable help.

Author details
[1]Department of Experimental Psychology, University of Oxford, South Parks Road, Oxford OX1 3UD, UK. [2]Institut Paul Bocuse, Château du Vivier, BP 25-69131, Écully Cedex, France.

References
1. Mielby LH, Kildegaard H, Gabrielsen G, Edelenbos M, Thybo AK: **Adolescent and adult visual preferences for pictures of fruit and vegetable mixes – Effect of complexity.** *Food Qual Prefer* 2012, **26:**188–195.
2. Hurling R, Shepherd R: **Eating with your eyes: effect of appearance on expectations of liking.** *Appetite* 2003, **41:**167–174.
3. Zellner DA, Siemers E, Teran V, Conroy R, Lankford M, Agrafiotis A, Ambrose L, Locher P: **Neatness counts. How plating affects liking for the taste of food.** *Appetite* 2011, **57:**642–648.
4. Zellner DA, Lankford M, Ambrose L, Locher P: **Art on the plate: Effect of balance and color on attractiveness of, willingness to try and liking for food.** *Food Qual Prefer* 2010, **21:**575–578.
5. Marchiori D, Corneille O, Klein O: **Container size influences snack food intake independently of portion size.** *Appetite* 2012, **58:**814–817.
6. van Kleef E, Shimizu M, Wansink B: **Just a bite: Considerably smaller snack portions satisfy delayed hunger and craving.** *Food Qual Prefer* 2013, **27:**96–100.
7. Zellner DA: **Contextual influences on liking and preference.** *Appetite* 2007, **49:**679–682.
8. Spence C, Harrar V, Piqueras-Fiszman B: **Assessing the impact of the tableware and other contextual variables on multisensory flavour perception.** *Flavour* 2012, **1:**7.
9. Shankar MU, Levitan CA, Spence C: **Grape expectations: the role of cognitive influences in color-flavor interactions.** *Conscious Cogn* 2010, **19:**380–390.
10. Spence C: **The color of wine – Part 1.** *The World of Fine Wine* 2010, **28:**122–129.
11. Verhagen JV, Engelen L: **The neurocognitive bases of human multimodal food perception: sensory integration.** *Neurosci Biobehav Rev* 2006, **30:**613–650.
12. Spence C, Levitan C, Shankar MU, Zampini M: **Does food color influence taste and flavor perception in humans?** *Chem Percept* 2010, **3:**68–84.
13. Stevenson RJ: *The Psychology of Flavour.* Oxford: Oxford University Press; 2009.
14. Guéguen N: **The effect of glass color on the evaluation of a beverage's thirst-quenching quality.** *Curr Psychol Lett Brain Behav Cogn* 2003, **11:**1–6.
15. Harrar V, Piqueras-Fiszman B, Spence C: **There's more to taste in a coloured bowl.** *Perception* 2011, **40:**880–882.

16. Piqueras-Fiszman B, Spence C: The influence of the color of the cup on consumers' perception of a hot beverage. *J Sensory Stud* 2012, **27**:324–331.

17. Schifferstein HNJ: The drinking experience: Cup or content? *Food Qual Prefer* 2009, **20**:268–276.

18. Van Ittersum K, Wansink B: Plate size and color suggestibility: The Delboeuf illusion's bias on serving and eating behavior. *J Consum Res* 2012, **39**:215–228.

19. Weintraub DJ, Cooper LA: Coming of age with the Delboeuf illusion: Brightness contrast, cognition, and perceptual development. *Developmental Psychol* 1972, **6**:187–197.

20. Weintraub DJ, Schneck MK: Fragments of Delboeuf and Ebbinghaus illusions: contour/context explorations of misjudged circle size. *Percept Psychophys* 1986, **40**:147–158.

21. Hutchings JB: *Food Colour and Appearance.* Glasgow: Blackie Academic and Professional; 1994.

22. Ekroll V, Faul F, Niederée R: The peculiar nature of simultaneous colour contrast in uniform surrounds. *Vision Res* 2004, **44**:1765–1786.

23. Lyman B: *A Psychology of Food: More than a Matter of Taste.* New York: Van Nostrand Reinhold; 1989.

24. Piqueras-Fiszman B, Alcaide J, Roura E, Spence C: Is it the plate or is it the food? Assessing the influence of the color (black or white) and shape of the plate on the perception of the food placed on it. *Food Qual Prefer* 2012, **24**:205–208.

25. Stroebele N, De Castro JM: Effect of ambience on food intake and food choice. *Nutrition* 2004, **20**:821–838.

26. Bell R, Meiselman HL, Pierson BJ, Reeve WG: Effects of adding an Italian theme to a restaurant on perceived ethnicity, acceptability, and selection of foods. *Appetite* 1994, **22**:11–24.

27. de Graaf C, Cardello AV, Kramer FM, Lesher LL, Meiselman HL, Schutz HG: A comparison between liking ratings obtained under laboratory and field conditions: The role of choice. *Appetite* 2005, **44**:15–22.

28. Meiselman HL: Experiencing food products within a physical and social context. In *Product Experience.* Edited by Schifferstein HNJ, Hekkert P. San Diego, CA: Elsevier; 2008:559–580.

29. Rolls BJ, Shide DJ: Both naturalistic and laboratory-based studies contribute to the understanding of human eating behavior. *Appetite* 1992, **19**:76–77.

30. Petit C, Sieffermann JM: Testing consumer preferences for iced-coffee: Does the drinking environment have any influence? *Food Qual Prefer* 2007, **18**:161–172.

31. Hein KA, Hamid N, Jaeger SR, Delahunty CM: Application of a written scenario to evoke a consumption context in a laboratory setting: Effects on hedonic ratings. *Food Qual Prefer* 2010, **21**:410–416.

32. Hein KA, Hamid N, Jaeger SR, Delahunty CM: Effects of evoked consumption contexts on hedonic ratings: A case study with two fruit beverages. *Food Qual Prefer* 2012, **26**:35–44.

33. Zellner DA, Durlach P: Effect of color on expected and experienced refreshment, intensity, and liking of beverages. *Am J Psychol* 2003, **116**:633–647.

34. Churchill A, Meyners M, Griffiths L, Bailey P: The cross-modal effect of fragrance in shampoo: Modifying the perceived feel of both product and hair during and after washing. *Food Qual Prefer* 2009, **20**:320–328.

35. Ariely D: *Predictably Irrational: The Hidden Forces that Shape our Decisions.* New York: Harper Collins; 2008.

36. Meiselman HL, Johnson JL, Reeve W, Crouch JE: Demonstrations of the influence of the eating environment on food acceptance. *Appetite* 2000, **35**:231–237.

37. Spence C, Piqueras-Fiszman B: *The Perfect Meal: The Multisensory Science of Food and Dining.* Oxford: Wiley-Blackwell. in press.

22

Assessing the shape symbolism of the taste, flavour, and texture of foods and beverages

Charles Spence* and Mary Kim Ngo

Abstract

Consumers reliably match a variety of tastes (bitterness, sweetness, and sourness), oral-somatosensory attributes (carbonation, oral texture, and mouth-feel), and flavours to abstract shapes varying in their angularity. For example, they typically match more rounded forms such as circles with sweet tastes and more angular shapes such as triangles and stars with bitter and/or carbonated foods and beverages. Here, we suggest that such shape symbolic associations could be, and in some cases already are being, incorporated into the labelling and/or packaging of food and beverage products in order to subconsciously set up specific sensory expectations in the minds of consumers. Given that consumers normally prefer those food and beverage products that meet their sensory expectations, as compared to those that give rise to a 'disconfirmation of expectation', we believe that the targeted use of such shape symbolism may provide a means for companies to gain a competitive advantage in the marketplace. Here, we review the latest research documenting a variety of examples of shape symbolism in the food and beverage sector. We also highlight a number of the explanations for such effects that have been put forward over the years. Finally, we summarise the latest evidence demonstrating that the shapes a consumer sees on the label and even the shape of the packaging in which the product is served can all impact on a consumer's sensory-discriminative and hedonic responses to food and beverage products.

Keywords: Shape symbolism, Taste, Flavour, Oral somatosensation, Food and beverage packaging, Sensory expectations

Introduction

Looking through the annals of marketing history, one finds occasional mention of the existence of a variety of cross-modal correspondences between abstract shapes and basic tastes. Generally speaking, cross-modal correspondences can be defined as a tendency for an individual to match (or associate) a feature, or attribute, in one sensory modality with a sensory feature, or attribute, in another sensory modality, no matter whether physically present or merely imagined [1,2]. So, for example, back in 1971, Ernst Dichter [3] reported that consumers typically matched more rounded shapes (for example, circles) with foods and beverages having a prominent sweet taste, while matching bitter-tasting foodstuffs with more angular shapes (such as stars and triangles) instead. Dichter's colleague, Louis Cheskin, also discussed such

shape-taste associations. He was particularly interested in how they could be utilised by marketers in the design of product packaging to set up the right, or appropriate, expectations in the mind of the consumer (for example, Cheskin [4]). A number of beverage products already appear to use angular red shapes, frequently a star, but sometimes also a triangle or pyramid, in order to subconsciously signal to the consumer that the contents of the packaging are carbonated, and/or bitter-tasting. Why else, one might ask, do so many brands of sparkling water adorn their product packaging with such symbols (Figure 1A, B, and C)? Well, it turns out that these shapes may be more than purely symbolic.

On the one hand, it could be argued that the triangular shape on the front of the Apollinaris bottle (Figure 1B) is actually meant to symbolise a mountain. One could also point to the fact that stars are commonly used to recognize prize-winning, high quality, and/or success in whatever they are associated with. Hence, one might think that this is the reason why such shapes appear so

* Correspondence: charles.spence@psy.ox.ac.uk
Department of Experimental Psychology, Crossmodal Research Laboratory, University of Oxford, South Parks Road, Oxford OX1 3UD, UK

Figure 1 Angular red shapes are a prominent attribute of the packaging/logo of many sparkling beverages, including (A) San Pellegrino, (B) Apollinaris, and (C) Saskia, and angularity is also an integral part of several brands of beer, including (D) Heineken, (E) Bass and Co., and (F) Sapporo.

frequently on the labelling of bottled waters. However, the key point to note here is that such angular shapes seem to appear much more frequently on the front (not to mention the sides, back, and top) of carbonated waters than on the packaging of still waters. Should this latter observation turn out to be statistically robust, it would suggest that some form of shape symbolic relation between angularity and carbonation may be at play over and above any other semantic meanings that might be associated with such recognizable angular forms.

Heineken, one of the major producers of beer, another carbonated alcoholic beverage, has also incorporated the red star as an integral part of the company's logo (Figure 1D). In this case, however, it turns out that the shape-symbolic correspondence between carbonation and angularity was not dreamt up by Madison Avenue's marketing guru, as Cheskin [4] was sometimes described. It is, in fact, an old Dutch brewer's symbol that originated from a prize winning/high-quality assessment in the brewer's early days ([5]). Similarly, one of the first

trademarks to be registered in the UK back in 1876 was the angular red triangle on Bass and Co.'s 'Pale Ale' [6] (Figure 1E). Sapporo, in Japan, has also incorporated a prominent star (originally red, though they recently changed the colour to a sandy yellow) on the front of their logo (Figure 1F). And, according to the company's Italian website (www.sanpellegrino.com [7]), the red stars that appear up to seven times on some bottles of San Pellegrino sparkling mineral water have been an integral part of the design of this highly successful product's packaging for more than a century now. Although these examples are admittedly anecdotal, when taken together, they do at least suggest that the successful use of shape symbolism in packaging design may help to give products a slight but significant, not to mention long-lasting, competitive advantage in the marketplace [8].

Although such non-verbal signals (for example, abstract forms) normally go unnoticed by the consumer's conscious mind, that does not necessarily mean that such cues are not picked-up and utilized by the unconscious mind [9]. Indeed, a quick walk down the aisles of the local supermarket provides sufficient evidence to suggest that shape symbolism effects such as those just described are not uncommonly used by marketers and graphic designers, either deliberately or otherwise. The reason for this may be to convey nonverbal information to the consumer about the sensory properties (for example, carbonation, bitterness, sweetness, and so on) of packaged food and beverage items. One explanation for such patterns in the marketplace is that they provide evidence of 'intelligent design' on the part of the creative teams involved. On the other hand, however, it could also be argued, albeit rather speculatively at this stage, that the use of such shape symbolism in connection with a number of popular and long-lasting food and beverage products/brands may actually reflect nothing more that the 'survival of the fittest'; that is, product packaging, or brand logos, whose shape symbolic associations do not match the sensory attributes of the product itself are less likely to survive for long in the highly competitive marketplace than those that do. Thus, those examples where the designer or design team intuitively happened to get it 'right' (for example, the prominent red stars associated with the packaging of San Pellegrino sparkling water), or at least to come up with a 'consensual' design [10], may simply end up staying on the shelves for longer, that is, selling better. Of course, if this suggestion were to be correct, then one might speculate that researchers and marketers may well find inspiration regarding the existence of heretofore undiscovered (or at least undocumented) shape symbolic correspondences simply by looking in the marketplace for any statistical regularities in the imagery (for example, along the angular versus rounded continuum) associated with different tastes, oral-somatosensory attributes, and/or flavours in food and/or beverage products.

In what follows, we start by briefly highlighting the origins of laboratory research on shape symbolism. We then go on to summarise the data from a number of recent studies, both published and unpublished, in which cross-modal correspondences between shapes (or angularity/curvilinearity) and the sensory attributes of a variety of different food and beverage items have been documented. We review the latest evidence from those studies that have demonstrated that the shapes that consumers are exposed to in a food or beverage consumption context can actually change their responses to a variety of foodstuffs. Finally, we outline a number of explanations for such cross-modal correspondences that have been put forward over the years. Note here that relevant shape information can be conveyed by, or associated with, any abstract imagery present on the label or logo, the shape of the label, or even the very shape of the product packaging itself [9,11-17].

Laboratory research on shape symbolism

The phenomenon of shape symbolism was originally documented by psychologists and linguists back at the end of the 1920s [18,19], though note that at that time it was referred to as sound (rather than shape) symbolism. Sound symbolism is the name given to the idea that certain speech sounds may signify (or be associated with) certain attributes in that which they refer to [2,20]. While research on the topic of sound symbolism goes back at least as far as Plato [21], the dominant view, until recently, seems to have been that the link between the signifier and the sign is arbitrary [22]. However, while this is often the case, a growing body of research now suggests the existence of a number of sound symbolism effects in language [20,23]. Shape symbolism refers to the idea that people associate certain attributes (in this case, attributes of food and beverage products) with specific shapes, typically varying in their size and or angularity/degree of curvature. Both sound and shape symbolism can therefore be seen as specific forms of cross-modal correspondence [2]. It was the linguist Edgar Sapir [19] who first reported that if asked to assign each of the meaningless names 'Mil' and 'Mal' to a large and a small table, then the majority of people say that the word 'Mal' was more appropriate for the larger table (thus demonstrating a cross-modal correspondence between the size of objects and specific speech sounds). Subsequently, the association between the 'i' sound and smallness has been demonstrated in many different cultures/languages across the world, for example, Newman [24] (though see Diffloth [25] for a dissenting voice).

Perhaps more relevant when it comes to the use of shape symbolism on product packaging are the findings reported by the German psychologist Wolfgang Köhler [18]. He noted that when given the meaningless names 'Baluma' and 'Takete', the majority of people match the 'rounded' sounds of 'Baluma' with the more organic shape, and the 'harsher' sound of the plosive stops in 'Takete' with the angular star-like shape instead (Figure 2). When he later changed 'Baluma' to 'Maluma', to avoid any unwanted potential associations with the word balloon, he obtained exactly the same pattern of results [see Köhler [26], Nielsen and Randall [27], Ramachandran and Hubbard [28], and Bremner et al. [29]]. Until very recently, no one seems to have given much thought to the possibility that cross-modal correspondences might also exist between shapes and tastes or flavours (though see Fónagy [30,31], for a solitary exception). The absence of serious scientific interest in the existence of cross-modal correspondences between shapes and tastes, the oral-somatosensory attributes of foodstuffs, and flavours contrasts with the already-mentioned (albeit unsubstantiated) suggestions that have occasionally appeared in the

Figure 2 Which of these shapes is called 'Baluma;' (or 'Maluma' or 'Bouba') and which is called 'Takete' (or 'Kiki' [25]). The majority of people around the world pair the more organic amoeba-like shape with the 'rounded' speech sounds of 'Baluma'.

marketing/packaging press over the last 50 years or so (for example, see Dichter [3], Cheskin [4], and McNeal [32]).

Assessing the shape symbolism of tastes and flavours

A recently published series of experiments has, though, now started to document the existence of a number of robust and easy-to-demonstrate crossmodal correspondences between shapes and the various sensory properties of a variety of real food and beverage items (see references [7,33-38]). So, for example, it has been shown that people tend to match sweet-tasting foods with organic (and rounded) shapes while matching more angular shapes with bitter- and sour-tasting foodstuffs (see Table 1 for a summary of what is currently known about shape symbolism in the food and beverage sector). Carbonation, as in a sparkling beverage, is also matched by consumers with more angular shapes (such as a star). These cross-modal associations, or correspondences, can

Table 1 Summary of shape symbolic associations with food and beverage products

Angular shapes	Organic shapes	Study
Cranberry jam	Blueberry jam	Gallace *et al.* [33]
Mint chocolate	Brie	
Salt and vinegar crisps	Regular crisps	
Dark chocolate	Milk chocolate	Ngo *et al.* [34]
Sparkling water	Still water	Spence and Gallace [17]
Maltesers	Rolos chocolates	
Cranberry juice	Brie	
Sparkling water	Still water	Ngo *et al.* [37]
Dark chocolate	Milk chocolate	Ngo and Spence [34]
Solid mint chocolate	Mint fondant	
Tunworth cheese	Lancashire Cheddar	Spence *et al.* [38] (unpublished work)
Keen's Cheddar	Stawley goat's cheese	
	Leipäjuusto cheese	
Bitter tastes	Sweet tastes	Dichter [3]

normally be demonstrated in relatively small, and easy to conduct, studies. Such experiments typically involve testing no more than 10 to 20 participants, each giving a single response to each food or beverage on an anchored Visual Analogue Scale (VAS).

While the majority of research assessing the shape symbolism of tastes and flavours has been conducted with actual consumers sampling real food and beverage products, we have recently demonstrated that at least certain of these results can also be obtained simply by instructing people to imagine the taste of particular foods and/or beverages. Thus far, mentally imagining the taste of a food has been shown to work for relatively simple, or stark, food comparisons, such as that between still and sparkling water, or between milk and dark chocolate. What is more, under such circumstances, Internet-based data collection appears to be just as reliable at highlighting the cross-modal correspondences between angularity and oral-somatosensory texture as more traditional laboratory-based data collection techniques [35,37]. Taken together, these simple techniques are now allowing academic researchers, not to mention sensory marketers, to rapidly assess just how many cross-modal correspondences there are between shapes, primarily varying in terms of their angularity, and the tastes, flavours, and oral-somatosensory attributes of foodstuffs [9].

Below, we briefly summarize some of the key results to have emerged from the latest research on the topic of shape symbolism in the food and beverage sector. In one study, Ngo *et al.* [34] provided evidence of the existence of a cross-modal correspondence between the angularity of visually presented shapes and the bitterness of chocolate samples. The participants in this study tasted three samples of commercially produced chocolate that varied in their cocoa content: one 30% cocoa milk chocolate, another 70% cocoa dark chocolate, and the third 90% cocoa dark chocolate. After tasting each sample, the participants were instructed to place a mark at the appropriate point along an analogue line scale anchored at one end by an organic shape and at the other end by an angular shape. The results revealed that as the cocoa content of the chocolate samples increased, so the participants tended to mark a point on the scale that was closer to the angular-shaped end of the line (Figure 3). Note that in order to control for any bias that participants might have toward responding on one side of the scale *vs.* the other (a phenomenon known as pseudoneglect – shown when neurologically normal individuals systematically mis-bisect the mid-point of a line shown visually [39]), Ngo *et al.* sometimes presented the more angular anchor on the left of the scale and at other times on the right of the scale instead. Crucially, the relative positioning of the shapes on the left *vs.* right of the scale does not seem to have a noticeable effect on the pattern of results observed in this kind of study.

A follow-up study [35] both confirmed these findings and additionally revealed that consumers consistently matched milk and mint fondant chocolates (50% cocoa dark chocolate with a smooth and creamy mint filling) with organic shapes, while matching dark (70% cocoa) and solid mint (50%) chocolates with angular shapes instead. Note that the taste/flavours of the mint fondant and solid mint were essentially the same. The key distinction between the two foodstuffs was in terms of the texture, with mint fondant being creamy and soft, and solid mint having a somewhat harder texture. The results of this latter study therefore highlighted the key role that oral-somatosensory texture can play in determining the cross-modal correspondences that consumers exhibit for everyday foodstuffs such as chocolate.

Elsewhere, researchers have demonstrated that consumers match the oral-somatosensory properties of Maltesers (honeycomb-centred milk chocolate produced by Nestlé) with more angular shapes while matching Rolos (Nestlé) and Caramel Nibbles (Cadbury; both soft caramel filled milk chocolates) with more organic shapes instead [7]. Once again, given that the chocolate coating of these products was very similar, the shape symbolic relationship would appear to have been driven by the crunchy honeycomb versus soft caramel centres of these chocolate products (that is, by the oral-somatosensory texture of the products).

Deroy and Valentine [36] recently examined cross-modal correspondences for several different kinds of beer (6.4% alcohol Adelscott, 5.5% alcohol 1664 Blanche, and 4.8% alcohol Bitburger) and various 2D and 3D shapes. The participants in this study were given a total of 34 shapes from which to choose (Figure 4). The results demonstrated that the Adelscott beer was considered sweeter-tasting than the other beers and was incidentally matched with more rounded, voluminous shapes, while participants judged the Bitburger to be more bitter-tasting and matched it with angular shapes instead. It is, though, worth noting here that the results of the two studies that have used a wide variety of possible shapes [36,40] have not, as yet, provided any particularly clear results; nor, we would argue, have they provided any especially strong justification for using so many possible stimuli in future shape symbolism research in the food and beverage sector. Hence, for the moment at least, we would be tempted to advocate the use of the simpler VAS contrasting an organic/rounded shape on the one hand with an angular

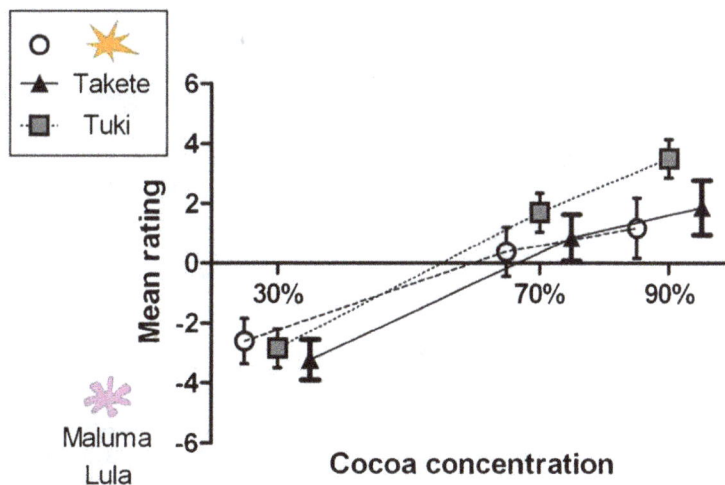

Figure 3 Results from a study conducted by Ngo et al. [34] demonstrating both sound and shape symbolism effects with increasing bitterness (for example, as the cocoa content increases in the chocolate) being matched with more angular shapes and words starting with a harder plosive sound ('t' in this case). The units on the Y-axis indicate the distance (in cm) along the 10-cm line scale, with 0 reflecting the centre point. Figure adapted and reprinted from Ngo et al. [34], Figure 2, with permission.

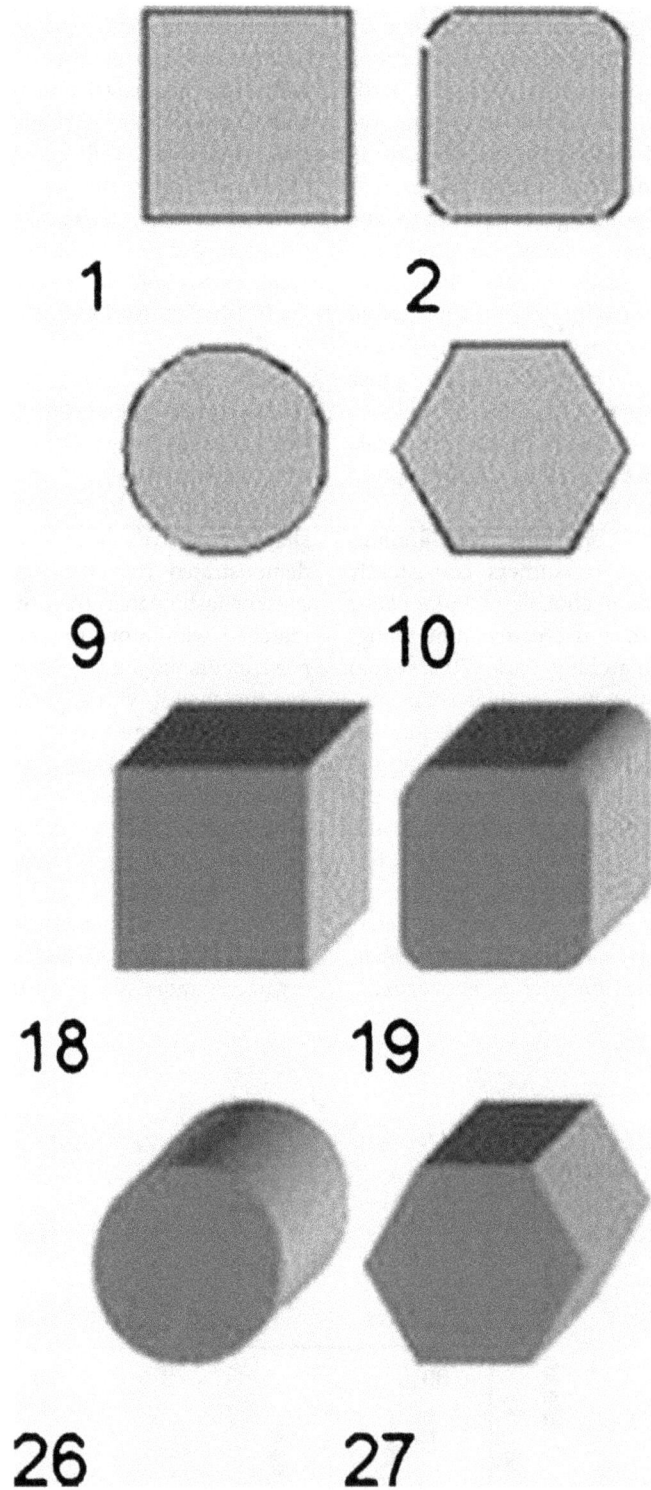

Figure 4 Figure showing a sample from the 34 different two- and three-dimensional shapes (based on a range of shapes originally introduced by Cytowic and Wood [40]) used in Deroy and Valentin's [36] recently published study in order to assess the properties of three different beers (Bitburger, Adelscott, and 1664 Blanche). Modified and reprinted from Deroy and Valentin [36], Figure 2, with permission.

shape on the other, at least until we understand more about the key drivers underlying shape symbolism effects in the food and beverage sector.

Recently, Spence *et al.* [38] extended the exploration of shape and sound symbolism to the more complex flavour profiles associated with various aged cheeses. These researchers examined each separate component of the flavour Gestalt of the cheese by having participants separately rate the gustatory, olfactory, and oralsomatosensory attributes on the angular *vs.* organic shape response scale (Figure 5). The strong and pungent flavours of Tunworth, a Camembert-style cheese, and Keen's Cheddar were matched by participants with the angular shape. By contrast, the milder flavours of Stawley goat's cheese and Lancashire Cheddar were rated as being more organic in shape. The results suggested that participants based their responses primarily on the taste/flavour (note that although participants were asked to report on taste, normal consumers often confuse taste and smell) rather than on the smell or texture of the cheeses that they were evaluating. Similar results were also obtained from cheese experts (for example, cheese mongers) and from typical store consumers. As such, the shape symbolism approach to describing cheeses may hold the potential to provide a more accessible means for individuals, experts and consumers alike, to talk about the qualities of a product such as cheese. A follow-up study conducted recently on a group of Finnish participants demonstrated that Leipäjuusto, a mild cheese that tends to 'squeak' against the teeth when a consumer bites into it, was consistently matched with the organic shape. The latter results once again suggest that

taste was the leading contributor to participants' shape symbolic correspondences.

It is worth noting in passing here that many people often describe cheeses as having a 'sharp' taste (see Williams [41] and Spence *et al.* [38]). In this case, the synaesthetic use of the adjective sharp, which is normally only used to refer to the tactile attribute of an object (according to the dictionary definition, the term 'sharp' refers to something with a thin cutting edge or a sharp point), appears to be used by people to describe the noticeable acidity that is a prominent feature of many cheeses. This is especially true of British Cheddar cheeses. Wine aficionados have also been known to refer to certain pleasant wines as well rounded [42,43]. Here, though, it is rather less clear whether the shape descriptor 'rounded' is being used in a literal sense, or rather just as a colloquial (or metaphorical) means of expressing the generally pleasing and well-balanced nature of the particular example of fermented grape juice under consideration.

Using shape symbolism to set consumer expectations in the food and beverage sector

It is our contention that the appropriate use of shape symbolism on product packaging, especially in the food and beverage sector, can be used to help set up the right sensory expectations in the minds of consumers. It has been suggested that the colours and shapes on product packaging can be just as, if not more, important than any text/descriptions in terms of setting up, consciously or otherwise, expectations about a product's likely sensory qualities [44,45]. This may be especially true given how

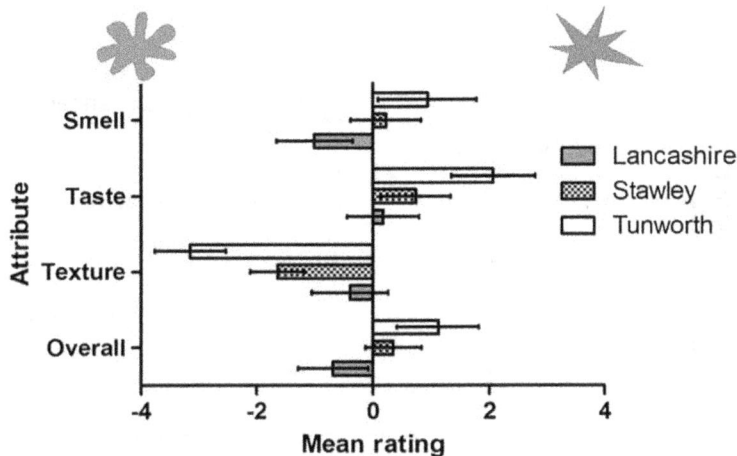

Figure 5 Figure showing the results of recent research in which participants had to rate the individual attributes of a range of aged cheeses (aroma, taste/flavour, oral-somatosensory texture, and overall experience) using response scales anchored by the organic and angular shapes. The results show that the taste element of the overall flavour drove participants' cross-modal associations. Ideally, the positioning of the anchors for these scales (for example, on the left or right of the page) should be randomized in order to control for any left-right biasing effects [39].

little attention consumers give to certain kinds of labels (for example, back labels on wine bottles [46-48]). The theory here is that under the majority of conditions, molecular gastronomy restaurants excepted [49], consumers prefer a product if its taste, aroma, flavour, and/or oral-somatosensory attributes match their prior sensory expectations [50-51]. Marketers and graphic designers may, then, be well advised to try and set up the appropriate expectations through any shape displayed on the packaging (Figure 6). In the section that follows, we will see how viewing shapes (for example, as on packaging) might be expected to influence the actual taste of the product consumed. Relevant here are the results of a survey conducted a few years ago by AC Nielson Co. on fast-turnover consumer products. The suggestion was that more than 50% of new product failures can be attributed to there being some sort of mismatch, or unintended interaction, between the product and its packaging [6].

Explaining the existence of cross-modal correspondences between shapes and the sensory attributes of foods and beverage products?

According to a large body of older empirical research using the semantic differential technique [53], the meaning of concepts/words can be mapped by assessing people's responses to a small number of scales that are anchored by pairs of polar opposite terms, normally adjectives (for example, good *vs.* bad, old *vs.* young, and so on). The most popular (discriminative) dimensions to be extracted from the presentation of such scales to consumers were activity, evaluation, and potency. Note that

evaluation refers to the pleasantness/unpleasantness of the stimuli. The idea is that people may simply use the angular/rounded response scale as a proxy to indicate how pleasant they find the taste, flavour, and/or oral-somatosensory texture of a food or beverage product. Relevant here is the fact that people tend to prefer rounded over more angular shapes [54]. Certainly there are grounds for suggesting that mapping pleasantness might provide part of the answer to the underlying explanation for the existence of shape symbolism in the food/beverage sector. Note that angularity, bitterness, and carbonation are all initially disliked as stimuli [55,56]. The common meanings associated with different kinds of sensory stimuli [57-60] may therefore provide a possible explanation for at least certain of the shape symbolism effects that have been documented in the food and beverage sector to date. The affective value, or pleasantness, of sensory stimuli therefore provides one way in which the 'meaning' of very different types of sensory stimuli could be matched. However, that said, the latest evidence assessing the cross-modal correspondences that people have with dark chocolate, a foodstuff that some people find very pleasant while others rate as unpleasant, suggests that pleasantness is unlikely to provide the whole story here [61].

It is also worth noting that the term sparkling, often used in English-speaking countries to refer to carbonated water, may also lead to specific associations with stars that sparkle. However, while this more semantic explanation of shape symbolic matching works when it comes to trying to explain the cross-modal association between the star-

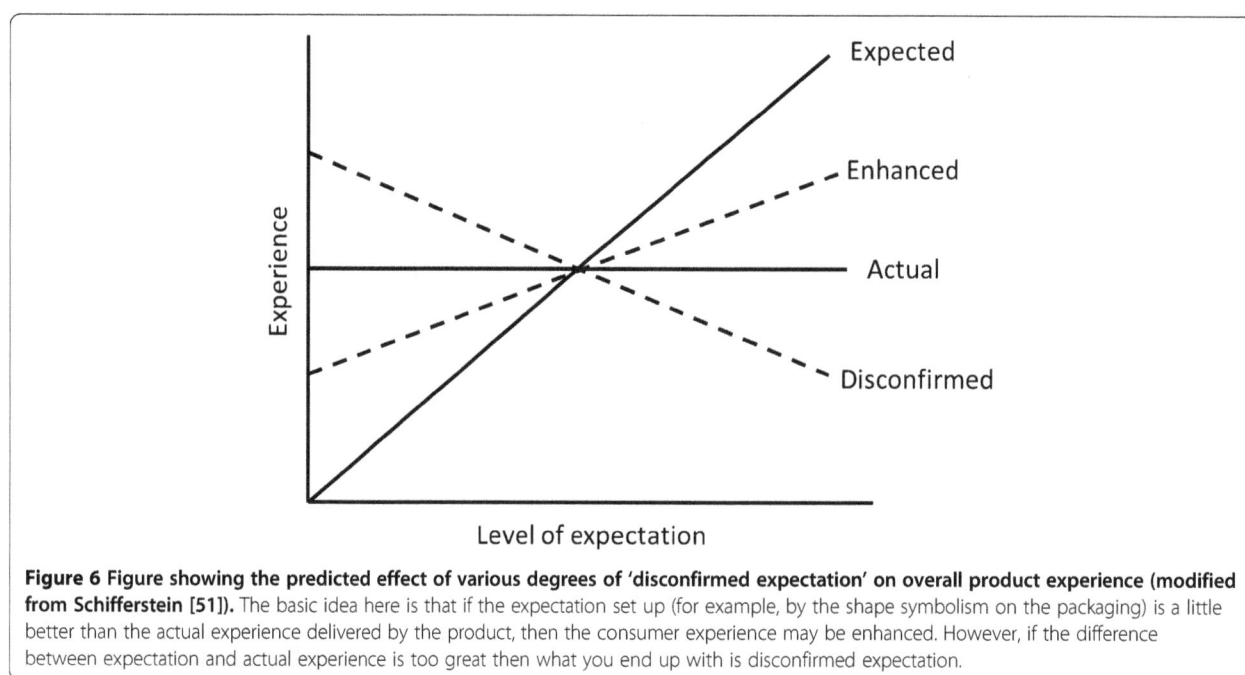

Figure 6 Figure showing the predicted effect of various degrees of 'disconfirmed expectation' on overall product experience (modified from Schifferstein [51]). The basic idea here is that if the expectation set up (for example, by the shape symbolism on the packaging) is a little better than the actual experience delivered by the product, then the consumer experience may be enhanced. However, if the difference between expectation and actual experience is too great then what you end up with is disconfirmed expectation.

shape and carbonated water, the fact that participants show an equally strong tendency to match carbonation with a triangle as with a star, argues against this being the most appropriate explanation for this particular cross-modal correspondence. Thus, it seems to be something about angularity, *per se*, rather than any concrete meaning that can be attached to the abstract shapes embodying that angularity that is doing the work here.

In terms of shape, McNeal [32] has suggested (although without providing any evidence) that triangles constitute lively shapes and hence are good for sparkling beverages. At least intuitively, sparkling beverages, acidity, and crunchy textures in food would all seem to be associated with activity. By contrast, still water, sweet tastes, and softer textures, as in the case of soft caramel, would seem to be more passive kinds of stimuli instead. Hence, it could be argued that the similar meaning associated with shapes and flavour experiences associated with foods and beverages might underlie at least certain examples of the sound symbolic mappings that have been documented to date (for example, the mapping between angular shapes and carbonation). One might also wonder whether some sort of metaphorical mapping can also be put forward to help explain shape symbolic mappings in the food and beverage sector [30,31,62-65].

Given that angularity on product packaging does indeed appear to be linked to certain carbonated beverages (specifically carbonated water and beers/lagers), it becomes difficult to ascertain whether consumers, when asked to match an angular *vs.* rounded shape with carbonation in sparkling water, say, are basing their response on the statistical correlations present in the marketplace, or on some deeper (possibly metaphorical) association between angularity and taste/flavour. In order to try and address this question, Bremner *et al.* [29] recently conducted a study of shape symbolism with the remote Himba from Kaokoland in Namibia. This group have no written language, nor access to supermarkets/advertising, and so on. These researchers asked whether this group would exhibit the same kinds of cross-modal correspondences as the Western participants tested in the majority of the studies reported to date. Interestingly, while this group demonstrated the basic bouba/kiki effect [18,28], they showed no significant association when it came to assessing the angular *vs.* rounded shapes to samples of sparkling *vs.* still water. What is more, when it came to assessing three samples of chocolate varying in their cocoa content, the Himba actually showed the opposite pattern of results to Western participants. That is, they actually matched bitter tastes with rounded forms. These results therefore demonstrate that at least certain shape symbolic associations between tastes, flavours, and the oral-somatosensory attributes of food and beverage products are culture specific. These results are also consistent with the suggestion that

Western participants' shape symbolic responses to the tastes, flavours, and oral-somatosensory attributes of foods may be learnt (no matter whether this is done explicitly or implicitly) from the statistics of the marketplace.

Finally, in the future, it will be important to probe whether the amoeba-like and star-like shapes used in many of the studies on shape symbolism are necessarily the most appropriate for research in this area. These shapes have been inherited, more or less unchanged, from Köhler's [18] early work [7,28]. One might therefore ask, especially in light of certain comments from the marketers [3], whether these shapes are any more (or less) appropriate than other shapes, such as a circle or triangle, say [59]. Or, in other words, is it anything more than an accident of history that researchers do not use a circle and a triangle or square instead of the amoeba-like shape and the star? One might also ask what the relevant attribute of the shapes that are matched to the oral-somatosensory attributes of food really is. Research using more complex arrays of shapes may help to provide an answer here [see Cytowic and Wood [40] and Hanson-Vaux *et al.* [66]]. As we have seen already, one possibility is that people may match stimuli across the modalities in terms of their pleasantness [53,61,67].

In summary, then, a number of different plausible explanations for the existence of reliable cross-modal associations between the angularity of shapes on the one hand, and the taste and flavour of food and beverage products on the other, can be postulated. While several explanations have gained some degree of empirical support, it is important to note that there may simply not be a single explanation for all such cross-modal matching effects [1,2].

Does shape symbolism influence the sensory-discriminative properties of food and beverage items?

Although there is currently little information directly relevant to this question, those studies that have been published do indeed suggest that shape symbolic sensation transference [4,16] can occur. The participants in an as yet unpublished study by Gal *et al.* [17] first had to judge which of three simultaneously presented (and similarly sized) shapes occupied the largest surface area. Next, they had to taste a piece of Cheddar cheese. Those participants who had evaluated three angular shapes rated the cheese as tasting significantly sharper (by around 7%) than another group of participants who had just evaluated three organic shapes instead. While a relatively large number (>200) of participants had to be tested in this study in order to generate a result that was only just statistically significant, the potential implications of these findings in terms of the use of shape symbolism in/on product packaging could well be worth pursuing. Think only of the much larger number of 'participants' who are likely to interact with any moderately successful product launched into the marketplace than were tested by Gal *et al.* Meanwhile, Becker *et al.* [12] have recently shown that the curvature of a product's packaging can also affect a consumer's rating of the taste of, in this case, yoghurt, even when the shape of the container was only seen on a computer monitor while the yoghurt itself was consumed from a tray.

The participants in a study by Seo *et al.* [68] had to match each of eight food odours (guava, honey melon, mint, Parmesan cheese, pepper, truffle, vanilla, and violet) with one of 19 different abstract symbols (Figure 7), varying in shape from those that were more organic to those that were more angular. The participants reliably matched certain shapes to particular odours: as Seo *et al.* put it, *The odors generally regarded as being pleasant (e.g., vanilla, banana, violet, honey melon, and mint) were paired with circle- or curve-shaped symbols, whereas, the odors judged generally as being unpleasant (e.g., Parmesan cheese, truffle, and pepper) were paired with square- or angular-shaped symbols.* In the second part of this study, Seo and colleagues documented significant differences in odour pleasantness and intensity ratings for two of the odours, those smelling of violet and of Parmesan cheese, when they were presented together with an abstract shape that had been judged as being congruent than when they were presented with an incongruent shape, or else with no visual stimulus at all. In particular, the odour of violet was rated as being more intense and more pleasant while Parmesan cheese odour was rated as being less pleasant (when presented with a congruent as compared to an incongruent symbol). Thus, the pairing of a congruent shape with the odour gave rise to a more extreme perceptual response in terms of intensity and pleasantness. That is, those odours that were considered pleasant to begin with became even more pleasant and more intense when presented with the congruent symbol, while those odours that were individually rated as unpleasant became even more unpleasant with the presentation of the congruent symbol.

Participants' brain responses were also measured using olfactory event-related potentials (ERPs). Using this technique, Seo *et al.* [68] were able to demonstrate that this particular cross-modal correspondence between abstract visual shapes and odours was influencing the neural responses of participants at a relatively early stage of information processing. In particular, the N1 component had both a higher amplitude and a shorter latency on trials that were cross-modally congruent as compared to those that were cross-modally incongruent in terms of their shape symbolism [69,70]. Given that these cross-modal interactions occurred so soon after stimulus onset, Seo *et al.*'s results suggest that the presentation of the abstract shape was having an effect on participants' perception of the odour, rather than just on their subsequent rating of that odour experience [71]. These results involving expectations set by the use of shape symbolism are consistent with other research demonstrating the impact of other ways of inducing product expectations on product perception [72].

Conclusions

The last few years have seen rapid growth in our understanding of, and evidence for, cross-modal correspondences between shapes and tastes, flavours, and the oral-somatosensory attributes of food and beverage products. Visual inspection of the products currently available

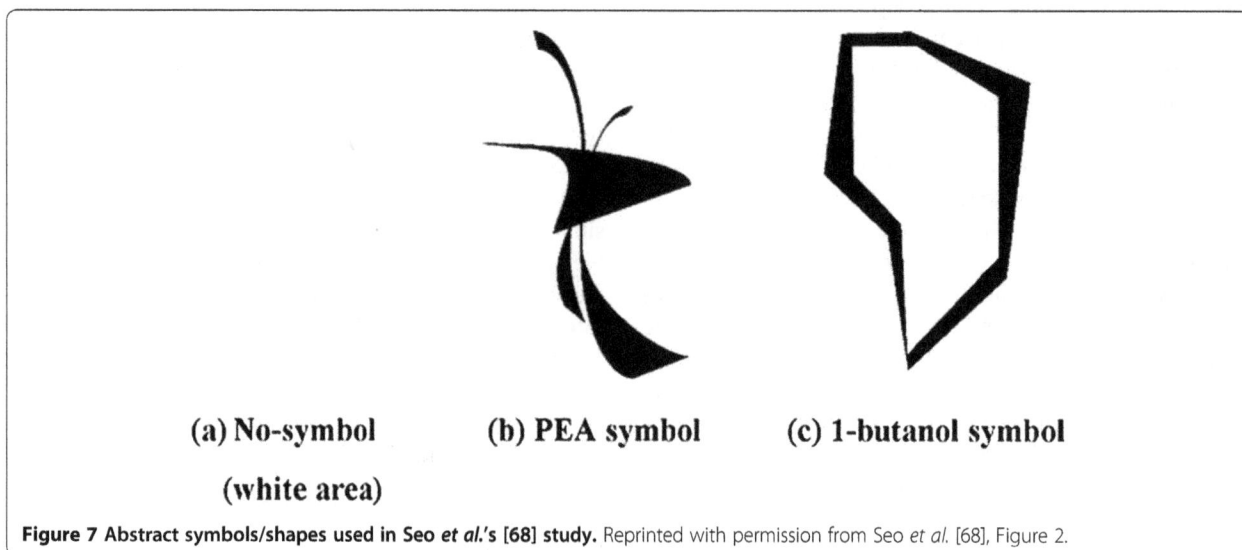

(a) No-symbol
(white area)

(b) PEA symbol

(c) 1-butanol symbol

Figure 7 Abstract symbols/shapes used in Seo *et al.*'s [68] study. Reprinted with permission from Seo *et al.* [68], Figure 2.

in a typical supermarket suggests that these shape correspondences have, on occasion, already been used. Though, as we saw earlier, whether this is as a result of intuition on the part of the marketers or merely an example of 'survival of the fittest' in the marketplace is currently hard to say. (Of course, it could also be the case that when a particular shape correspondence is (un)successful on the market that it will shape the future intuitions of the marketers.) That said, the research reviewed here clearly demonstrates how widespread and easy to assess such cross-modal correspondences are, even in the food and beverage sector. The effects reported here also appear to be fairly robust, given that very similar cross-modal matching results have been obtained when the same foodstuffs were assessed in various different studies. Shape symbolism can extend from everything from the shape of labels to the shape of the containers themselves [12,16]. Having demonstrated an extensive array of shape symbolism effects in the food and beverage sector in Western participants, researchers will obviously need to assess how widespread such effects are in different cultures/markets in the future (Bremner *et al.* [29] and Heinrich *et al.* [73]).

The available evidence provides some limited support for the suggestion that the expectations that are set up by shape can influence people's product perception [12,17,68]. At present, though, the effects of capitalizing on cross-modal correspondences between shape and taste/flavour for commercial applications definitely needs longer-term follow-up studies in order to demonstrate the potential implications of shape symbolism to marketing in the food and beverage sector. In the future, it will also be crucial to demonstrate that the use of such shape symbolism impacts on consumers' choice between products in more realistic environments, such as the supermarket aisle. In terms of the future of this kind of research, to the extent that we taste what we expect to taste, both in terms of the sensory discriminative and hedonic properties of foods and beverages, one could imagine using shape symbolism in order to enhance sweetness or saltiness, say, when the actual formulation of various popular brands has been modified in order to meet the latest health targets. This normally calls for a reduction of these tasty ingredients, such as salt, sugar, carbonic acid (H_2CO_3), and fat, a reduction that often leads to complaints from brand-loyal consumers [74]. It will also be interesting to investigate whether any shape symbolic expectations can be maximised by making sure that, where possible, the shape symbolic associations of the shape of the packaging, label, and even logo convey a congruent message. Here, one might already think of the angular bottle and label shape used on the Listerine bottle versus the rounded bottle and label used for Scope, both brands of mouthwash [75]. What is certainly clear is that there is growing interest in this area from academic researchers [76-78].

In conclusion, we believe that the appropriate utilization of shape symbolism in the marketplace may have a number of advantages when compared to other more traditional marketing techniques [9]: On the one hand, shape symbolism effects are more likely to work in the global marketplace, since they are typically unaffected by language [that said, it currently remains a question for future research to determine which cross-modal correspondences are universal and which are more culture specific (see Diffloth [25] and Bremner *et al.* [29])]; on the other hand, the use of shape symbolism on product packaging also seems to have the advantage that it can set up sensory expectations in the mind of the consumer whether or not they happen to be paying attention to the meaning of the shapes that they are exposed to.

Competing interests

The authors declare that they have no competing interests.

Authors' contributions

CS and MN both contributed to the writing of this review. Both authors also read and approved the final version of the manuscript.

Acknowledgements

MN would like to thank The Clarendon Fund, Oxford University, for funding her PhD research. The two authors contributed equally to the writing and research involved in this article.

References

1. Parise CV, Spence C: **Audiovisual crossmodal correspondences**. In *Oxford Handbook of Synaesthesia*. Edited by Simner J, Hubbard E. Oxford: Oxford University Press; in press.
2. Spence C: **Crossmodal correspondences: A tutorial review**. *Atten Percept Psychophys* 2011, **73**:971–995.
3. Dichter E: **The strategy of selling with packaging**. *Package Eng Mag* 1971, July:16a–16c.
4. Cheskin L: *Secrets of marketing success: An expert's view on the science and art of persuasive selling.* New York: Trident Press; 1967.
5. **The Dutch touch** [http://175proof.com/triedandtested/the-dutch-touch/].
6. Sacharow S: *The package as a marketing tool.* Radnor, PA: Chilton Book Company; 1982.
7. Spence C, Gallace A: **Tasting shapes and words**. *Food Qual Prefer* 2011, **22**:290–295.
8. Miller C: **The shape of things: beverages sport new packaging to stand out from the crowd**. *Market News* 1994, **28**:1–2.
9. Spence C: **Managing sensory expectations concerning products and brands: capitalizing on the potential of sound and shape symbolism**. *J Consum Psychol* 2012, **22**:37–54.
10. Koriat A: **Subjective confidence in one's answers: the consensuality principle**. *J Exp Psychol Learn* 2008, **34**:945–959.
11. Ares G, Deliza R: **Studying the influence of package shape and colour on consumer expectations of milk desserts using word association and conjoint analysis**. *Food Qual Prefer* 2010, **21**:930–937.
12. Becker L, Van Rompay TJL, Schifferstein HNJ, Galetzka M: **Tough package, strong taste: the influence of packaging design on taste impressions and product evaluations**. *Food Qual Prefer* 2011, **22**:17–23.
13. Overbeeke CJ, Peters ME: **The taste of desserts' packages**. *Percept Motor Skill* 1991, **73**:575–580.
14. Smets GJF, Overbeeke CJ: **Expressing tastes in packages**. *Design Stud* 1995, **16**:349–365.
15. Wang RWY, Sun CH: *Analysis of interrelations between bottle shape and food taste.* Paper presented at the 2006 Design Research Society, International Conference in Lisbon (IADE); [http://www.iade.pt/drs2006/wonderground/proceedings/fullpapers/DRS2006_0054.pdf].

16. Spence C, Piqueras-Fiszman B: **The multisensory packaging of beverages**. In *Food packaging: Procedures, management and trends*. Edited by Kontominas MG. Hauppauge. NY: Nova Publishers; in press.

17. Gal D, Wheeler SC, Shiv B: *Cross-modal influences on gustatory perception*.

18. Köhler W: *Gestalt psychology*. New York: Liveright; 1929.

19. Sapir E: **A study in phonetic symbolism**. *J Exp Psychol* 1929, **12**:225–239.

20. Hinton L, Nichols J: *Ohala JJ (Eds): Sound symbolism*. Cambridge: Cambridge University Press; 1994.

21. Jowett B: *Plato's Cratylus with introduction*. Rockville, MD: Serenity Publishers; 2009.

22. Saussure F: In *Course in general linguistics*. Edited by Bally C, Sechehaye A. IL: Open Court; 1983. Translated by Harris R. La Salle.

23. Robson D: **Language's missing link**. *New Sci* 2011, **16** July:30–33.

24. Newman SS: **Further experiments in phonetic symbolism**. *Am J Psychol* 1933, **45**:53–75.

25. Diffloth G: **i: big, a: small**. In *Sound symbolism*. Edited by Hinton L, Nichols J, Ohala JJ. Cambridge: Cambridge University Press; 1994:107–114.

26. Köhler W: *Gestalt psychology: An introduction to new concepts in modern psychology*. New York: Liveright Publication; 1947.

27. Nielsen A, Rendall D: **The sound of round: evaluating the sound-symbolic role of consonants in the classic Takete-Maluma phenomenon**. *Can J Exp Psychol* 2011, **65**:115–124.

28. Ramachandran VS, Hubbard EM: **Synaesthesia - A window into perception, thought and language**. *J Conscious Stud* 2001, **8**:3–34.

29. Bremner A, Caparos S, Davidoff J, de Fockert J, Linnell K, Spence C: **Bouba and Kiki in Namibia? Western shape-symbolism does not extend to taste in a remote population**. *Cognition*, . in press.

30. Fónagy I: *Die Metaphern in der Phonetik [The metaphors in phonetics]*. The Hague: Mouton and Co.; 1963.

31. Fónagy I: **A research instrument**. In *Languages within language: An evolutive approach*. Edited by Fónagy I. Amsterdam: John Benjamins; 2001:337–357.

32. McNeal JU: *Consumer behaviour: An integrative approach*. Boston: Little, Brown and Company; 1982.

33. Gallace A, Boschin E, Spence C: **On the taste of "Bouba" and "Kiki": an exploration of word-food associations in neurologically normal respondents**. *Cogn Neurosci* 2011, **2**:34–46.

34. Ngo MK, Misra R, Spence C: **Assessing the shapes and speech sounds that people associate with chocolate samples varying in cocoa content**. *Food Qual Prefer* 2011, **22**:567–572.

35. Ngo MK, Spence C: **Assessing the shapes and speech sounds that people associate with different kinds of chocolate**. *J Sens Stud* 2011, **26**:421–428.

36. Deroy O, Valentin D: **Tasting shapes: investigating cross-modal correspondences**. *Chemosens Percept* 2011, **4**:80–90.

37. Ngo MK, Piqueras-Fiszman B, Spence C: **On the colour and shape of still and sparkling water: implications for product packaging**. *Food Qual Prefer* 2012, **24**:260–268.

38. Jewell G, McCourt ME: **Pseudoneglect: A review and meta-analysis of performance factors in line bisection tasks**. *Neuropsychologia* 2000, **38**:93–110.

39. Spence C, Ngo MK, Percival B, Smith B: **Crossmodal correspondences: Assessing shape symbolism for cheese**. *Food Qual Prefer* 2013, **28**:206–212.

40. Cytowic RE, Wood FB: **Synaesthesia II: psychophysical relations in the synaesthesia of geometrically shaped taste and colored hearing**. *Brain Cognition* 1982, **1**:36–49.

41. Williams JM: **Synesthetic adjectives: a possible law of semantic change**. *Language* 1976, **52**:461–478.

42. Lehrer A: *Wine & conversation*. 2nd edition. Oxford: Oxford University Press; 2009.

43. Peynaud E: *The taste of wine: the art and science of wine appreciation*. London: Macdonald & Co; 1987. Translated by Schuster M.

44. Singh S: **Impact of color on marketing**. *Manage Decis* 2006, **44**:783–789.

45. Swientek B: *Uncanny developments*. *Beverage Ind* 2001, **92**:38–39.

46. Charters S, Lockshin L, Unwin T: **Consumer responses to wine bottle back labels**. *J Wine Res* 1999, **10**:183–195.

47. Bender MM, Derby BM: **Prevalence of reading nutrition and ingredient information on food labels among adult Americans: 1982–1988**. *J Nutr Educ* 1992, **24**:292–297.

48. Wansink B, Sonka ST, Hasler CM: **Front-label health claims: When less is more**. *Food Policy* 2004, **29**:659–667.

49. Piqueras-Fiszman B, Spence C: **Sensory incongruity in the food and beverage sector: art, science, and its commercialization**. *Petits Propos Culinaires*, **95**:74–118.

50. Deliza R, MacFie HJH: **The generation of sensory expectation by external cues and its effect on sensory perception and hedonic ratings: a review**. *J Sens Stud* 1997, **2**:103–128.

51. Schifferstein HNJ: **Effects of product beliefs on product perception and liking**. In *Food, people and society: a European perspective of consumers' food choices*. Edited by Frewer L, Risvik E, Schifferstein H. Berlin: Springer Verlag; 2001:73–96.

52. Yeomans M, Chambers L, Blumenthal H, Blake A: **The role of expectancy in sensory and hedonic evaluation: the case of smoked salmon ice-cream**. *Food Qual Prefer* 2008, **19**:565–573.

53. Osgood CE, Suci GJ, Tannenbaum PH: *The measurement of meaning*. Urbana: University of Illinois Press; 1957.

54. Bar M, Neta M: **Humans prefer curved visual objects**. *Psych Sci* 2006, **17**:645–648.

55. Spence C: **Multi-sensory integration & the psychophysics of flavour perception**. In *Food oral processing – Fundamentals of eating and sensory perception*. Edited by Chen J, Engelen L. Oxford: Blackwell Publishing; 2012:203–219.

56. Steiner JE: **Innate, discriminative human facial expressions to taste and smell stimulation**. *Annals NY Acad Sci* 1974, **237**:229–233.

57. Bozzi P, Flores D'Arcais G: **Ricerca sperimentale sui rapporti intermodali fra qualità espressive [Experimental research on the intermodal relationships between expressive qualities]**. *Arch Psicol Neurol Psichiatr* 1967, **28**:377–420.

58. O'Boyle MW, Tarte RD: **Implications for phonetic symbolism: the relationship between pure tones and geometric figures**. *J Psycholinguist Res* 1980, **9**:535–544.

59. Hevner K: **Experimental studies of the affective value of colors and lines**. *J Appl Psychol* 1935, **19**:385–398.

60. Poffenberger AT, Barrows BE: **The feeling value of lines**. *J Appl Psychol* 1924, **8**:187–205.

61. Crisinel A-S, Spence C: **The impact of pleasantness ratings on crossmodal associations between food samples and musical notes**. *Food Qual Prefer* 2011, **24**:136–140.

62. Deroy O: *Tastes and shapes: Synaesthesia or metaphor.*: CenSes, University of London; Presentation given at The Nature of Taste, between Science and Aesthetics meeting; December 3 2009.

63. Marks LE: **Metaphor and the unity of the senses**. In *Sensory science theory and applications in foods*. Edited by Lawless HT, Klein BP, Lawless HT, Klein BP. New York: Marcel Dekker; 1991:185–205.

64. Marks LE, Hammeal RJ, Bornstein MH: **Perceiving similarity and comprehending metaphor**. *Monogr Soc Res Child* 1987, **52**:1–102.

65. Liu CH, Kennedy JM: **Form symbolism, analogy, and metaphor**. *Psychon B Rev* 1997, **4**:546–551.

66. Maurer D, Pathman T, Mondloch CJ: **The shape of boubas: sound-shape correspondences in toddlers and adults**. *Dev Sci* 2006, **9**:316–322.

67. Seo H-S, Arshamian A, Schemmer K, Scheer I, Sander T, Ritter G, Hummel T: **Cross-modal integration between odors and abstract symbols**. *Neurosci Lett* 2010, **478**:175–178.

68. Hanson-Vaux G, Crisinel AS, Spence C: **Smelling shapes: Crossmodal correspondences between odors and shapes**. *Chem Senses*, .

69. Bien N, ten Oever S, Goebel R, Sack AT: **The sound of size: Crossmodal binding in pitch-size synesthesia: A combined TMS, EEG, and psychophysics study**. *NeuroImage* 2012, **59**:663–672.

70. Kovic V, Plunkett K, Westermann G: **The shape of words in the brain**. *Cognition* 2009, **114**:19–28.

71. Bar M, Neta M: **Visual elements of subjective preference modulate amygdala activation**. *Neuropsychologia* 2007, **45**:2191–2200.

72. Woods AT, Lloyd DM, Kuenzel J, Poliakoff E, Dijksterhuis GB, Thomas A: **Expected taste intensity affects response to sweet drinks in primary taste cortex**. *Neuroreport* 2011, **22**:365–369.

73. Henrich J, Heine SJ, Norenzayan A: **The weirdest people in the world?** *Behav Brain Sci* 2010, **33**:61–135.

74. Tozer J: **The great HP Sauce revolt: online fury over salt cut that 'ruins the taste**. [http://www.dailymail.co.uk/news/article-2036349/HP-Sauce-revolt-Online-fury-salt-cut-ruins-taste.html].

75. Parise CV, Spence C: **Assessing the associations between brand packaging and brand attributes using an indirect performance measure.** *Food Qual Prefer* 2012, **24**:17–23.

76. Westerman SJ, Gardner PH, Sutherland EJ, White T, Jordan K, Watts D, Wells S: **Product design: Preference for angular versus rounded design elements.** *Psychol Market* 2012, **29**:595–605.

77. Westerman SJ, Sutherland EJ, Gardner PH, Baig N, Critchley C, Hickey S, *et al*: **The design of consumer packaging: Effects of manipulations of shape, orientation, and alignment of graphical forms on consumers' assessments.** *Food Qual Pref* 2013, **27**:8–17.

78. Stutts CA, Torres A: **Taste interacts with sound symbolism.** *North Amer J Psychol* 2012, **14**:1. Downloaded from http://www.freepatentsonline.com/article/North-American-Journal-Psychology/281111805.html.

23

Molecular gastronomy is a scientific discipline, and note by note cuisine is the next culinary trend

Hervé This[1,2]

Abstract

For the past two decades, there has been much confusion about molecular gastronomy. This confusion has arisen because people ignore that the word gastronomy does not mean cuisine, it means knowledge about food. Similar to 'molecular biology', molecular gastronomy is a scientific discipline that looks for the mechanisms of phenomena occurring during dish preparation and consumption. As with any other scientific discipline, it can have many applications. One of the first was 'molecular cuisine' but since 1994, 'note by note cuisine' has also been promoted. The latter involves preparing dishes using pure compounds, or more practically mixture of compounds obtained by fractioning plant or animal tissues, instead of using these tissues themselves. Note by note cuisine raises issues in various fields: science, technology, nutrition, physiology, toxicology and politics.

Keywords: Cuisine, Molecular cuisine, Molecular gastronomy, Note by note cuisine

Molecular gastronomy

In 1988, a new scientific discipline, molecular gastronomy, was defined as 'looking for the mechanisms of phenomena occurring during dish preparation and consumption' [1,2]. This new definition presented the opportunity to discuss the precise content of molecular gastronomy and its relationship with other existing fields of science.

There has always been much confusion between science and technology when it comes to food, including over exactly what food is. Dictionaries give the definition: 'any substance that can give to living beings the elements necessary for their growth or for their preservation' [3]. However, one has to recognize that human beings very seldom eat non-transformed tissues or natural products; raw materials are transformed so that chemical and physical changes determine the final composition of all food as well as its 'bioactivity', a term which we propose to describe the sensory effects, nutritional value, eventual toxic effects, and so on, of the various compounds released by food systems [4].

During food preparation, plant or animal tissues are at least washed and cut, and most food are thermally processed. For example, even for a simple carrot salad, which requires no thermal processing, there is a big difference between the raw product in the field and what is consumed - that is, grated carrots on a plate: this is because cutting the tissue triggers enzymatic reactions [5] and because compounds get transferred between the dressing and the plant tissue [6]. This analysis leads to the conclusion that reagents and products of 'culinary transformations' (transformations performed in the kitchen) should not both be called food. The specific transformation occurring from the raw materials to the final prepared dish is worth studying, both for scientific and technological reasons.

Making the difference between science and technology clear is particularly important for molecular gastronomy because of the confusion between science and cooking (see, for example [7]), and because the public unduly fears 'chemistry in the plate' (this will prove important for the future development of note by note cuisine, described below). The French chemist, Antoine-Laurent de Lavoisier, was right when, in his article about meat

Correspondence: herve.this@agroparistech.fr
[1]INRA, UMR 1145, Group of Molecular Gastronomy, 16 Rue Claude Bernard, Paris 75005, France
[2]AgroParisTech, UFR de Chimie Analytique, 16 Rue Claude Bernard, Paris 75005, France

stocks [8], he wrote that the two fields of science and technology 'do not meet'. The French chemist and biologist Louis Pasteur, who was so successful in both science and technology, was also fiercely opposed to the expression applied sciences [9]. We believe that there is an important difference between the science and the technology of food transformations [10]. The latter is not comparable to the scientific study of the phenomena occurring during dish preparation and consumption, that is, molecular gastronomy. Let us add that phrases such as 'culinary science' or 'science of cooking' [11] are wrong, strictly speaking, unless science here means knowledge in general. Also, there are no 'scientific chefs', contrary to what is too often encountered in the media [12]. Of course the discipline of molecular gastronomy had precedent [13], and many chemical or physical phenomena occurring during dish preparation and consumption were studied before 1988 [14]. However, food science in the 1980s neglected culinary processes. For example, textbooks such as the classic *Food Chemistry* contained almost nothing on culinary transformations (this is still the case in the most recent edition) [15], with less than 0.5% of the important chapter on meat dedicated to culinary transformations (for example, meat shrinkage during heating because of collagen denaturation); most of the chapter described raw meat composition and structure, or industrial products (sausages, meat extracts, and so on). The same textbook contained nothing about the effect of thermal processing on wine, despite the widespread use of cooked wine in culinary activity (48% of French classical sauces contain wine; J Henne, MB France, K Belkhir and HT, submitted work).

The complexity of culinary transformations and the general lack of funding by the food industry for studies outside of its field were probably responsible for food science drifting slowly toward the science of ingredients and food technology, neglecting the phenomena that occur when cooking cassoulet, goulash, hollandaise sauce, and so on. It was considered an eccentricity when a paper on béarnaise sauce was published in a scientific journal in the 1970s [16]. This lack of interest in culinary transformations is why the late Nicholas Kurti (1908 to 1998) [17], former professor of physics in Oxford, and I decided in March 1988 that a new discipline had to be introduced [2].

The situation at that time in food science was more or less the same as it had been for molecular biology some decades before. The term molecular biology was first used by Warren Weaver in 1938 to describe certain programs funded by the Rockefeller Foundation, where it simply meant the application of techniques developed in the physical sciences to investigate life processes [18]. The first scientific practitioner to call his work 'molecular biology' was William Astbury, who used the term

before 1950 to mean the study of structures, functions, and genesis of biological molecules [19]. What Kurti and I had in mind was more or less the same, but concerning another field of knowledge, so the name molecular and physical gastronomy was chosen.

The choice of 'gastronomy' in this title was obvious: it does not mean haute cuisine but rather 'intelligent knowledge of whatever concerns man's nourishment' [20]. When Kurti died in 1998, the name was abbreviated to molecular gastronomy, and Kurti's name was given to the international meetings of the discipline.

The appeal of this new field was and remains scientifically clear: if one wants to discover new phenomena, the exploration of a new field is a safe bet, as there is plenty of easy exploration. As always, when new knowledge is produced, it is possible to make technological applications. Since 2000, innovations based on molecular gastronomy have been introduced almost every month (frequently, names of famous chemists of the past are given to new kinds of dishes) [21].

However, the initial program of this discipline was inappropriate because it mixed science and technology. In 2000 [22], it was realized that any recipe has three parts: a technically useless part, a 'definition', and 'technical added information'. This last term describes information that is not absolutely necessary to make the dish (old wives tales, proverbs, tips, methods, and so on) [23]. Some years later, it was realized that the appreciation of a dish by an individual is a question of art, not of technique; thus, cooking involves an artistic activity of fundamental importance. At the same time, it was understood that social context is also very important: a dish is not "good" if it is thrown in the face of the guests. All these facts led to the proposal of new program for molecular gastronomy:

1. to scientifically explore 'culinary definitions';
2. to collect and test technical added information;
3. to scientifically explore the art aspect of cooking;
4. to scientifically explore the social aspect of cooking [24].

Application in the kitchen

At the time that molecular gastronomy was introduced, we and others wanted to modernize culinary practices using what was done in scientific disciplines such as chemistry, physics or biology [25]. The idea to modernize techniques has come up many times in the history of cuisine. In 1969, Kurti [26] mentions the application of physical techniques and, since the beginning of the 1980s, I proposed the use of chemical tools [27]. The name molecular cuisine (or molecular cooking) was given in 1999, at the start of a FP5 European program called INICON. The definition of molecular cuisine is 'producing food using "new" tools, ingredients, methods' [28].

In this definition, the word new stands for what was not available in kitchens of the western countries in 1980. New tools could include siphons, used to make foams; ultrasonic probes, used to make emulsions; controlled heaters or circulators, used for cooking at temperatures lower than 100°C; liquid nitrogen, to make sorbets and many other innovative preparations; rotary evaporators and distillators, used to recover extracts; and many other types of laboratory equipment that can have useful applications in the kitchen. Concerning ingredients, many additives were no found in western kitchens of the 80's, but proved to have useful culinary applications: sodium alginate to make objects with a gellified skin and a liquid core, or spaghetti made of vegetables, and so on; other gelling agents, such as agar-agar or carrageenans; various colors; odorant compounds; and so on. Of course, not all of these items are completely new, other gelling agents from algae have been used in Asia for thousands of years, and many of these tools are used daily in chemistry laboratories, but they were not used by western chefs, and the goal was to modernize the technical component of cooking.

With regards new methods, a wealth of innovative preparations were introduced (and frequently given the name of scientists from the past), such as chocolate chantilly, beaumés, gibbs, nollet, liebig, gay-lussac, braconnot or vauquelins [21].

The term molecular cuisine was sometimes criticized, but the reasons for using it were that innovative cuisine had to be distinguished from science, and in particular from molecular gastronomy. The arguments over the name are unlikely to matter as the term molecular cuisine is likely to die out with the adoption of new techniques. A new idea is now being introduced with the name note by note cuisine [29].

The next culinary trend: note by note cuisine!
Note by note cuisine was first proposed in 1994 (in the magazine *Scientific American* [25]) at a time when I started using compounds in drinks and dishes, such as paraethylphenol in wines and whiskeys; 1-octen-3-ol in sauces for meat; limonene; tartaric acid; and ascorbic acid among others. The initial proposal was to improve food, but surely an obvious next step was to make dishes entirely from compounds.

To put it differently, note by note cuisine does not use meat, fish, vegetable or fruits to make dishes, but instead uses compounds, either pure compounds or mixtures. An analogy would be in the way that electronic music is not made using trumpets or violins, but using pure waves that are mixed in to sounds and music. For the various parts of the dish in note by note cuisine, the cook has to design the shapes, the colors, the tastes, the odors, the temperatures, the trigeminal stimulation, the textures, the nutritional aspects and more [30].

The feasibility of this new cuisine has already been shown. On 24 April 2009, the French chef Pierre Gagnaire (who has restaurants in a dozen cities of the world: Paris, London, Las Vegas, Tokyo, Dubai, Hong Kong...) showed the first note by note dish to the international press in Hong Kong. Then, in May 2010, two note by note dishes were shown by the Alsatian chefs Hubert Maetz and Aline Kuentz at the French-Japanese Scientific Meeting in Strasbourg [31]. However the first note by note meal was not served until October 2010, by chefs of the Cordon Bleu School in Paris, to the participants of the 2010 courses at the Institute for Advanced Studies in Gastronomy [32]. On 26 January 2011, at a banquet before the launching event of the International Year of Chemistry at the United Nations Educational, Scientific and Cultural Organization, Paris, a whole note by note meal for about 150 people was served by Potel et Chabot Catering Company [33]. This meal was again served in April 2011 to about 500 chefs receiving Michelin stars in Paris. And since the number of note by note initiatives is becoming too big to be tracked.

Issues
Many people are worried by note by note cuisine, asking questions about nutrition, toxicology, feasibility, economics and politics. What about nutriments, oligo-elements, vitamins? Are the compounds dangerous? Will food be liquid? Will agriculture become extinct through such a new way of cooking? All kinds of arguments are used to justify why 'traditional food', the cassoulets, stews, choucroute, should be kept. Indeed the question is important, and note by note cuisine will succeed only if we tackle the 'food neophobia' of the human species [34]: this reflex, also experienced by other nonhuman primates, leads individuals to assume that the food they learned to eat when young is 'good', and to fear new food. Our human brain, instead of making us reject novel food as nonhuman primates would do, leads us to negate new dishes and to legitimate old ones. This occurs even when the 'virtues' of the traditional food stuffs are not demonstrated [35], the worst justification being that these food types must be safe because they are old. This is a poor argument; compare this with smoked products, a traditional cooking method, that epidemiologists now clearly see the danger of through the high incidence of cancers of the digestive tract in populations in the north of Europe, who consume a lot of smoked products [36].

Food neophobia is not a good reason to discount the interest of note by note cuisine. Why should we drop traditional cuisine, and adopt note by note cuisine?

Indeed the alternative is not compulsory; as for molecular cuisine, we could keep traditional cuisine and add note by note cuisine. Or produce hybrids. . .

The technical issue

The feasibility of note by note cuisine no longer needs to be demonstrated because meals have already been produced using this techniques, but we still have to discuss the nature of the compounds used. The culinary world already uses very pure compounds, such as water, sodium chloride, sucrose and gelatine. The lay person often ignores the fact that these compounds were prepared by industry through various extraction processes, purifications and technological modifications (for example, the anti-aggregation compounds added to sucrose) [15].

Many other compounds could be prepared in the same way, such as saccharides, amino acids and glycerides, and indeed the food industry already uses some of them. The food additives industry produces pigments, vitamins, preservatives, gelling or thickening agents and so on. Additives are not currently regulated like food ingredients, but could they not be in the future? Or should the regulation of additives be suppressed, and another very different regulation be introduced?

It is difficult to make dishes from pure compounds, and so, to go back to our music analogy, another way is to make dishes in the same way electronic music is composed [37,38]. That is, to enlarge the list of usable compounds by adding simple mixtures such as those that the industry already makes by fractionation of milk or grain. Gelatine, for example, is not pure, in the sense that it is not made of molecules of only one kind: the extraction method used to make gelatine results in large variation in the molecular weight of the polypeptidic chains [39]. Also starch is not pure, as it is made of two main compounds, amyloses and amylopectins. In passing, let us not forget that, because starch is a simple fraction of grain, most traditional pastry techniques can be kept for making note by note cuisine.

Let us come back to the question of 'breaking down' plant or animal tissues to prepare fractions. The industry already extracts polysaccharides, proteins, amino acids, surfactants and other compounds from grain [39]. From milk, the industry recovers amino acids, peptides, proteins and glycerides. Could we not do the same from plant (carrots, apples, turnips. . .) or animal tissues? Could we not, using the same kind of processes (such as direct or reverse osmosis, cryoconcentration or vacuum distillation), prepare fractions that can be used later for note by note cuisine?

Many technology groups study these questions, and technologists at the Montpellier Institut National de la Recherche Agronomique Centre, for example, have devised techniques based on membrane filtration to recover the total phenolics fraction from grape juice [40]. This fraction is very different depending on the raw material, for example whether the juice is from the Syrah variety, or from Grenache, or Pinot: the diversity of the initial products is not erased by the fractionation process, so that cooks can still play with the 'terroir'.

Now we have discussed the issue of ingredients, we have to consider assembling them into dishes. We should not forget that today's food items are material systems of a colloidal nature [41-43], often with a large proportion of water in them. Many organic compounds are poorly soluble in water, and emulsification is obviously a very important process in note by note cuisine. However it is not the only process; all dispersion techniques will be useful.

During the assembly, the various biological properties of food have to be considered. Of course, the nutritional content is important [44] but it would be a mistake to forget that food has to stimulate the various sensory receptors involved in vision, odor, taste, trigeminal system and temperature [45], for instance: this creates many questions. For example, even if the individual absorption spectrum of some pigments are known, the 'color' of a mixture of such pigments is difficult to predict theoretically [46]. Also, when one mixes odorant compounds in proportions near the detection threshold, unpredictable odors are obtained. Worst still, we do not know what will happen when you mix only two odorant compounds: do they make a 'chord' or a fusion [47]?

For taste, the question is even more difficult to answer, because taste receptors and their substrates are not known [48]; it was discovered only recently (less than ten years ago) that the tongue has receptors for fatty acids with long unsaturated chains [49]. This means that other important discoveries could still be made! In the meantime, one can use citric, malic, tartaric, acetic, ascorbic or lactic acids, or saccharides such as glucose, fructose or lactose, as well as the traditional sucrose but experimental tests will be needed to appreciate the result.

For trigeminal effects, some fresh or pungent compounds are known, such as eugenol (from cloves), menthol (one of its enantiomers only), capsaicin (from chilli), piperin (from pepper), ethanol, sodium bicarbonate and many others [48]. But again the knowledge of receptors could lead to new products.

From the texture point of view, technological work can be done, because more studies are needed on the manufacture of colloidal materials. Making simple emulsions is sometimes considered difficult, but more generally one should not assume that the texturization of formulated products is fully solved, even if we now have surimi and analogous systems. Who will succeed in making the consistency of a green apple? Or a pear? Or a strawberry? Not only is there still the question of laboratory prototypes but also of mass production.

As a whole, much remains to be done and many aspects of note by note cuisine remain to be studied by science and by technology. Let us finish this paragraph with an important observation: it would be uninteresting to reproduce already existing food ingredients. As synthesizers can reproduce the sounds of a piano or a violin, note by note cuisine could reproduce wines, carrots or meats ... but why? Except for astronauts who have to travel for long periods, there is probably no value in making what already exists, and it is much more exciting to investigate flavors and dishes that were never envisioned using traditional food ingredients [50].

A simple calculation shows the immensity of what could be discovered. If we assume that the number of traditional food ingredients is about 1,000 and if we assume that a traditional recipe uses 10 ingredients, the number of possibilities is 1,000 to the power of 10 (or 10^{30}). However, if we assume that the number of compounds present in the ingredients is about 1,000, and that the number of compounds that will be used in note by note cuisine is of the order of 100, then the number of possibilities is about 10^{3000}. And, in this calculation we have not considered that the concentration of each compound can be adapted, which means that a whole new continent of flavor can be discovered.

Nutritional questions

Here we should begin by saying that traditional food is not a guaranteed to be healthy: bear in mind that the world faces a pandemia of obesity [51]! Of course, some will criticize the modern diet, but it would be rather more appropriate to observe that the new food environment is not suitable for human beings in their modern way of living. Indeed, the human species has had to face alternating times of plenty and starvation [52], and the science of nutrigenomics is now discovering mechanisms through which the human body can face these conditions [53]. For example, too much to eat does not lead to increased excretion, as we could hope, but instead increased storage in fat tissues.

Let us now consider why note by note cuisine could be interesting from a nutritional standpoint. This question relates to making 'light products'. Does the use of sweeteners lead to overconsumption? Previous studies on this could guide the study of the long-term effects of note by note cuisine.

It is certain that the science of nutrition still has questions to answers regarding the use of vitamins, oligoelements and minor nutriments. It would be a mistake to consider that we know everything regarding these elements in food; as an example, a European study of supplementation with vitamin E (a group of hydrophobic compounds with specific antioxidant properties) had to be stopped because of a higher incidence of death in the group of participants who smoke and were receiving the supplement [54].

Toxicology

This leads us to now consider the question of toxicology. Here again more studies are needed in particular when low doses of compounds are consumed for a long time. In this field, the scientific potential is huge, as beneficial effects are frequently discovered, such as cytochrome P450 polymorphism or, more recently, gene transfers between bacteria that are hosted by algae and bacteria of the human gut when algae are consumed [55].

A strange such case is estragole, which makes up to more than 50% of the total composition of the essential oils of tarragon and basil [56]. The hydroxyl derivative of this compound seems to be toxic [57], but the reason why is not understood, and there is no particular incidence of liver cancer in populations consuming a lot of such plants [58].

From a toxicity point of view, note by note cuisine will be no different from traditional cuisine, in which animal and plant tissues were never tested. It is a paradox of modern diet that novel foods are studied more than traditional food, and it is possible that some traditional foods would not be allowed if they were introduced today.

Note by note cuisine can avoid toxicity by simply not using the toxic compounds. For example, we can leave out benzo[a]pyrenes and avoid the toxic myristicin (6-allyl-4-methoxy-1,3-benzodioxole) from nutmeg; estragole; glycoalkaloids from potatoes and tomatoes; some glucosinolates from cabbages; some phenolics from plant tissues and so on [59]. The public, however, can continue to do what they want, such as continue to consume barbecued products full of benzo[a]pyrens!

The issues with the regulation of food products will then be analogous to the question of selling liquid nitrogen, ultrasonic probes and rotary evaporators to 'molecular cooks'. The evolution of practices will demand new regulations, as was the case when gas and electricity were introduced into homes. And we accept that there will almost certainly be accidents, not because note by note cuisine is more dangerous than knives or gas, but because the culinary world, as in any community, has its proportion of incautious people, such as a German man who put liquid nitrogen in a closed bottle [60]!

Primarily, what I propose to retain from this discussion is that the scientific and technological questions are considered very differently. It is time to learn about the effect of these compounds on the body.

Art, first!

The concept of art is complex but, to keep it simple, I propose that culinary art, as well as painting, music, sculpture, literature and other arts, is aimed at creating emotions [61]. Artists never stopped introducing new ideas into their works, and gourmands are longing for new flavors and new sensations. Note by note cuisine can make them happy, because it can produce a wealth of new possibilities.

Is note by note cuisine difficult? Of course cooks will have to become more familiar with the repertoire of ingredients available to them but as new recipes are introduced it will become more and more easy. Each time we have held a note by note event, cooks had to use compounds that they did not know, and they learned to use these products to make remarkable pieces, with new flavors. Of course, one can hardly describe the flavor of these dishes: how would you describe the color blue to someone who cannot see? Also what to name the dishes was difficult, but perfumery solved the issue: *N°5 from Channel, Shalimar,* and others.

For all those who are afraid of losing their traditional stew, cassoulet or choucroute, let us say that modern art does not replace old art, but simply adds to it, giving more freedom and more choice. Debussy did not make Mozart or Bach disappear; Picasso or Buffet did not prevent us from admiring Rembrandt or Brueghel. And molecular cuisine did not kill nouvelle cuisine or traditional cuisine. Note by note cuisine will be an artistic addition.

Economy

What will be the cost of note by note cuisine? Will it be more expensive than current cooking? Here, the issue of energy has to be considered because the next cost increase of energy will perhaps be key to the success of note by note cuisine. Today, to "reduce" wine or bouillon when making a sauce, cooks evaporate primarily water (but lose many odorant compounds as well, because of steam evaporation). If we assume a reduction should be by two thirds, as a professional cook would, a simple calculation shows that the energy consumed is 0.417 kWh [4], which means 0.05 euros per sauce.

The question of energy cost had not been considered in traditional cuisine, where meats are heated to greater than 200°C to produce compounds that could be immediately achieved in note by note cuisine, where mass-produced compounds could be made at a much lower cost.

In addition, it is not necessary to synthesize all the compounds used by cooks, and frequently they can be extracted from plant material, much as chlorophylls are today. Chemists know that hundreds of chemist-years were necessary to synthesize vitamin B12 [62], so agriculture and extraction remain the most efficient ways in the absence of an efficient chemical method. Note by note can use either synthetized or extracted products, regardless of where they came from.

Political and social questions

The first tests of note by note cuisine unavoidably created fear, because of the crazy idea that we would be eating 'chemicals'. Here, as for genetically modified organisms for example, political ideas confusingly mixed with other questions in the discussion. Note by note cuisine can be successful only when it is well explained, and if the 'authority argument' is used, as Augustin Parmentier well understood when he served potatoes to the king of France, at a time when the hungry country refused this ingredient [63]. But should we not be afraid that, as for genetically modified organisms, note by note cuisine will have disadvantages for human communities? How would farmers survive when - although unlikely - all food is made using note by note? These questions are more than chemists can answer, but they call for the following: as some people make money by producing wine instead of selling grapes, farmers could become richer than they are today when producing plant fractions, instead of selling the raw material.

Finally, when appreciating the value of note by note cuisine, we should not forget that humankind is facing an energy crisis: it is not definite that traditional cuisine is sustainable (it is not!) [59]; the new will always beat the old; breaking down products from agriculture and farming is already normal for milk and wheat [39]; why not carrots and apples? The objections being made to note by note cuisine today are the same made half a century ago against electronic music, and guess what you hear on the radio today? In other words, are we not now in a similar situation to the music industry in 1947, when musicians such as Varèse and others were investigating electronic music[64]?

Competing interests
The author declares that he has no competing interests.

Acknowledgements
The author thanks INRA and AgroParisTech for support, Rachel Edwards-Stuart, Pierre Gagnaire, Pierre Combris, Gérard Pascal, Jacques Risse, Jean-Marie Bourre, Annick Faurion, Jean-Louis Escudier and Robert Anton for fruitful discussions.

References
1. This H: **La gastronomie moléculaire.** *Sciences des aliments* 2003, 23(2):187–198.
2. This H: **La gastronomie moléculaire et physique.** In *PhD thesis*. Paris: VI University; 1995.
3. Trésor de la langue française ; 2006. electronic version [http://atilf.atilf.fr/tlf.htm], access 01/10/2006.
4. This H: **Solutions are solutions, and gels are almost solutions.** *Pure Appl Chem.*, ASAP article. http://dx.doi.org/10.1351/PAC-CON-12-01-01, Published online 2012-09-10.

5. Zawistowski J, Biliarderis CG, Eskin NAM: **Polyphenoloxidase.** In *Oxidative Enzymes in Food.* Edited by Robinson DS, Eskin NAM. London: Elsevier Applied Science; 1991:217–273.

6. Cazor A, This H: **Sucrose, glucose and fructose extraction in aqueous carrot root extracts prepared at different temperatures by means of direct NMR measurements.** *J Agric Food Chem* 2006, **54**:4681–4686.

7. See for example this strange "Molecular gastronomy network homepage", that some merchants are doing, using the name of a trendy scientific discipline for their business. http://www.moleculargastronomynetwork.com/recipes.html.

8. De Lavoisier AL: *Œuvres Complètes.* Paris: Imprimerie Nationale; 1862–1893:t. III, 563–578.

9. Pasteur L: **Note sur l'enseignement professionnel, adressée à Victor Duruy, 10 nov. 1863.** In *Œuvres Completes.* Paris: Masson; 1924:187. t.7.

10. This H: *Cours de gastronomie moléculaire N°1: Science, technologie, technique (culinaires): quelles relations ?* Paris: Quae/Belin; 2009.

11. This H: **Molecular gastronomy in France.** *Journal of Culinary Science & Technology* 2011, **9**(3):140.

12. **Daniel Facen, the scientific chef.** http://www.finedininglovers.com/stories/molecular-cuisine-science-kitchen.

13. This H: **Une petite histoire de la gastronomie moléculaire.** *Papilles (Roanne)* 1997, **13**:5–14.

14. Geoffroy Le Cadet M: *Mémoires de l'Académie Royale, Histoire de l'Académie Royale des Sciences, Année MDCCXXX, 312.* Amsterdam: Pierre Mortier; 1733.

15. Belitz HD, Grosch W: *Food Chemistry.* Heidelberg: Springer; 1999.

16. Perram CM, Nicolau C, Perram JW: **Interparticle forces in multiphase colloid systems: the resurrection of coagulated sauce béarnaise.** *Nature* 1977, **270**:572.

17. This H: **Froid, magnétisme et cuisine: Nicholas Kurti (1908-1998, membre d'honneur de la SFP).** *Bulletin de la Société fFrançaise de Pphysique* 1999, **119**(5):24–25.

18. Baltimore D: *Nobel Lectures in Molecular Biology 1933-1975.* New York: Elsevier; 1977:viii.

19. Weaver WT: **Molecular biology, origins of the term.** *Science* 1970, **170**:591–592.

20. Brillat-Savarin JA: **Meditation III. De la gastronomie.** In *Molecular Gastronomy,* This H. New York: Columbia University Press; 2006.

21. **Avec Hervé This: dictons, savoir et gourmandise.** http://www.pierre-gagnaire.com/#/pg/pierre_et_herve.

22. This H: **La gastronomie moléculaire et l'avancement de l'art culinaire.** *Sciences, Publication de l'Association Française pour l'Avancement des Sciences (AFAS)* 1998, **98**(7):3.

23. This H: *Cours de gastronomie moléculaire N°2: les précisions culinaires.* Paris: Belin Litterature et Revues; 2010.

24. This H: **Molecular gastronomy: a scientific look to cooking.** In *Life Sciences in Transition, Special Issue of the Journal of Molecular Biology.* Edited by Halldor S. Cambridge: Elsevier Science Ltd; 2004:150.

25. This H, Kurti N: **Physics and chemistry in the kitchen.** *Sci Am* 1994, **270**(4):44–50.

26. Kurti N: **Friday evening discourse at the Royal Institution: the physicist in the kitchen.** *Proceedings of the Royal Institution 1969,* **42/199**:451–467.

27. This H, Kurti N: **The cooking chemist.** *The Chemical Intelligencer* 1995, **1**:65.

28. **Introduction of innovative technologies in modern gastronomy for modernization of cooking.** http://www.ist-world.org/ProjectDetails.aspx?ProjectId=2df59ae3a55d4c14a67d63e25d73748d.

29. This H: *La Gastronomie Moléculaire et ses Applications.:* Keynote lecture of the Doctoriales de l'Université de Haute Alsace; 1999.

30. **Menu constructivisme culinaire pour Pierre Gagnaire.** http://www.lhotellerie-restauration.fr/journal/restauration/2009-05/Menu-constructivisme-culinaire-pour-Pierre-Gagnaire.htm.

31. **Molecular gastronomy and Note by Note cooking.** http://www.canalc2.tv/video.asp?idvideo=9432.

32. **Menu « note à note » pour le programme de l'institut des Hautes Études du Goût à l'école Le Cordon Bleu Paris.** http://www.lcbparis.com/news/noteanote102011/fr.

33. **Centre de Resources Nationales Hôtellerie Restauration: Ça vient d'avoir**

34. Pliner P, Hobden K: **Development of a scale to measure the trait of food neophobia in humans.** *Appetite* 1992, **19**(2):105–120.

35. EFSA Scientific Cooperation (ESCO) Report: **EFSA compendium of botanicals that have been reported to contain toxic, addictive, psychotropic or other substances of concern.** *EFSA Journal* 2009, **7**(9):281.

36. Ohshima H, Friesen M, Malaveille C, Brouet I, Hautefeuille A, Bartsch H: **Formation of direct-acting genotoxic substances in nitrosated smoked fish and meat products: identification of simple phenolic precursors and phenyldiazonium ions as reactive products.** *Food Chem Toxicol* 1989, **27**(3):193–203.

37. **Institute de Recherche et Coordination Acoustique/Musique.** http://www.ircam.fr/recherche.html.

38. **Gelatin Manufacturers Institute of America homepage.** http://www.gelatin-gmia.com.

39. Lorient D, Linden G: *Biochimie agro-industrielle.* Paris: Masson; 1994.

40. Escudier JL, Bes M, Mikolajczak M, Bouissou D, Samson A: **The future of the wine sector.** http://www.liendelavigne.org/HOME/ANG/DefaultANG.htm.

41. This H: **Modelling dishes and exploring culinary 'precisions': the two issues of molecular gastronomy.** *Br J Nutr* 2005, **93**(4):S139–S146.

42. This H: **Formal descriptions for formulation.** *Int J Pharmaceut* 2007, **344**(1–2):4–8.

43. This H: **Descriptions formelles, pour penser et pour la formulation.** *L'Actualité Chimique* 2008, **322**(8):11–14.

44. Roberfroid M, Coxam V, Delzenne N: *Aliments Fonctionnels.* Paris: Lavoisier Tec et Doc; 2007.

45. This H, Salesse R, Gervais R: **Gastronomie moléculaire et olfaction.** In *Odorat et goût.* Paris: Quae; 2012:439–449.

46. This H: **Apprenons enfin à cuisiner de la couleur!** In *La Couleur des Aliments, de la Théorie à la Pratique. Jacquot M.* Edited by Jacquot M, Fagot P, Voilley A. Paris: Tec et Doc Lavoisier; 2012:431–443.

47. Coureaud G, Gibaud D, Le Berre E, Schaal B, Thomas-Danguin T: **Proportion of odorants impacts the configural versus elemental perception of a binary blending mixture in newborn rabbits.** *Chem Senses* 2011, **36**(8):693–700.

48. Salesse R, Gervais R: *Odorat et goût.* Paris: Quae; 2012:439–449.

49. Laugerette F, Passilly-Degrace P, Patris B, Niot I, Montmayeur JP, Besnard P: **[CD36, a major landmark on the trail of the taste of fat].** *Med Sci* 2006, **4**(22):357–359.

50. This H: **La gastronomie moléculaire.** *Sciences des Aliments* 2003, **23**(2):187–198.

51. International Obesity TaskForce: *Waiting for a green light for health? IOTF Position Paper.* London: IOTF; 2003.

52. **Starvation.net homepage.** http://www.starvation.net/.

53. Meugnier E, Bossu C, Oliel M, Jeanne S, Michaut A, Sothier S, Brozek J, Rome S, Laville M, Vidal H: **Changes in gene expression in skeletal muscle in response to fat overfeeding in lean men.** *Obesity* 2007, **15**(11):2583–2594.

54. Yusuf S, Dagenais G, Pogue J, Bosch J, Sleight P: **Vitamin E supplementation and cardiovascular events in high-risk patients.** *The Heart Outcomes Prevention Evaluation Study Investigators, New Eng J Med* 2000, **342**(3):154–160.

55. Hehemann H, Correc G, Barbeyron T, Helbert W, Czjzek M, Michel G: **Transfer of carbohydrate-active enzymes from marine bacteria to Japanese gut microbiota.** *Nature* 2008, **464**:908–912.

56. Rietjens M, Martena MJ, Boersma MG, Alink GM, Spiegelenberg W: **Molecular mechanisms of toxicity of important food-borne phytotoxins.** *Mol Nutr Food Res* 2005, **49**(2):131–158.

57. Anthony A, Caldwell J, Hutt AJ, Smith RL: **Metabolism of estragole in rat and mouse and influence of dose size on excretion of the proximate carcinogen 1' hydroxyestragole.** *Food Chem Toxicol* 1987, **25**:799–806.

58. Nesslany F, Parent-Massin D, Marzin D: **Risk assessment of consumption of methylchavicol and tarragon: the genotoxic potential *in vivo* and *in vitro*.** *Mutat Res* 2010, **696**(1):1–9.

59. Bruneton J: *Plantes Toxiques.* Paris: Lavoisier Tec et Doc; 2001.

60. **German Heston Blumenthal blows off both hands in liquid nitrogen kitchen accident.** http://www.telegraph.co.uk/news/worldnews/europe/germany/5821433/German-Heston-Blumenthal-blows-off-both-hands-in-liquid-nitrogen-kitchen-accident.html.

lieu!. http://www.hotellerie-restauration.ac-versailles.fr/spip.php?article1615.

61. This H, Gagnaire P: *Cooking: The Quintenssential Art*. Los Angeles: University of California Press; 2008.
62. Woodward RB: **The total synthesis of vitamin B12.** *Pure Appl Chem* 1973, **33**(1):145–178.
63. **Histoire de Montdidier.** http://santerre.baillet.org/communes/montdidier/v2b/v2b4c02b54.php, last access 6 July 2012.
64. This H: *La cuisine note à note en 12 questions souriantes*. Paris: Belin; 2012.

24

Taste as a social sense: rethinking taste as a cultural activity

Susanne Højlund

Abstract

This article outlines what it means to see taste as a social sense, that means as an activity related to socio-cultural context, rather than as an individual matter of internal reflection. Though culture in the science of taste is recognized as an influential parameter, it is often mentioned as the black box, leaving it open to determine exactly how culture impacts taste, and vice versa, and often representing the taster as a passive recipient of multiple factors related to the local cuisine and culinary traditions. By moving the attention from taste as a physiological stimulus–response of individuals to tasting as a shared cultural activity, it is possible to recognize the taster as a reflexive actor that communicates, performs, manipulates, senses, changes and embodies taste—rather than passively perceives a certain experience of food. The paper unfolds this anthropological approach to taste and outlines some of its methodological implications: to map different strategies of sharing the experience of eating, and to pay attention to the context of these tasting practices. It is proposed that different taste activities can be analysed through the same theoretical lens, namely as sharing practices that generates and maintains a cultural understanding of the meaning of taste.

Keywords: Taste, Tasting, Culture, Practice, Sharing, Context

Taste as a social sense

We eat together. Although there is a constant worry that the sociality of the meal is disappearing, this is rather a myth than reality [1]. Commensality is still highly valued across cultures, even though this value is distributed differently [2]. But stating that eating is a social activity does not in itself explain how taste becomes social or culture becomes taste. As it is not the actual substance of the food that you are sharing, it is still individual what you put into your mouth, what you chew, ingest and perceive. This could lead to the argument that to analyse taste as a cultural phenomenon means, primary, to explore how individuals interpret symbolic meanings of food, e.g. the aesthetic judgement of quality in the Kantian way [3], or how the eater interpret food taboos, definitions and cultural schemes of food rules related to different cultures [4]. But there is still a missing link in explaining how this symbolism becomes a habit or a certain taste preference. The French sociologist Pierre Bourdieu has an influential contribution to this with his concept of *lifestyle* [5] stressing the need to focus not only on ideas and discursive

models but also on practice. He explains taste preferences (both the aesthetic judgements and the food choice/other types of consumption) as linked to the distribution of cultural, social and economic capital, and the learning of these preferences as a consequence of social practice [6]. This practice generates a *habitus*, he argues, that guides our choices more or less unconscious. But it leaves us with a rather passive actor [7] and do not enable us to study how one can change taste preferences [8]. Nevertheless it encourages us to see taste as a social sense, as a shared judgement, learned by actively doing taste rather that passively inheriting it from 'culture' [9].

Sharing taste

In order to move the attention from the privacy of the mouth and the subjective, internal reflections to the public space of sharing the experience of eating, it is necessary to develop methodological approaches and models of analyses that can shed light on the social processes of tasting. Many food anthropologists and sociologists are engaged in this kind of analyses focusing on food practices in relation to cultural context e.g. [10,11]. But it is seldom with an explicit focus on processes of tasting. Tasting is part of eating and drinking,

Correspondence: etnosh@cas.au.dk
Department of Anthropology, Aarhus University, Moesgaard Alle 2, Højbjerg, DK-8270 Aarhus, Denmark

but not similar hereto. With a focus on tasting rather than on eating, we stress the use of the senses and the judgement of food quality as one dimension of eating. What can be gained from seeing tasting as a practice is a way to understand how ideas of food quality and preferences for certain food stuffs are brought into the social and thereby being object for others and possible to share. I propose, thus, that outlining this field of research would include a mapping of sharing practices, seeing taste not only as something that goes into your body but also the opposite way [8].

From this analytical approach follows that different sharing practices could be studied under the same theoretical umbrella. Such different activities could be: using bodily techniques as e.g. eating with your fingers [12]; intentionally manipulating taste through cooking [13]; talking about taste [14]—from the everyday dialogues in the family to the professional chef's talks on TV [15]; to the writing of food blogs and cook books [16]; using digital media as e.g. sharing food photos at Instagram; arranging food festivals and wine tasting [17]; providing taste education in schools [18] etc. I propose that such different activities can be seen as part of the same social activity: the activity of making taste public [11,19]. Understanding how taste is externalised through culture will make it possible to also analyse how cultural taste preferences are internalised [20]. I am thus pointing to a research field of taste that focuses on the mediated space between food and eater, and an overall research question that asks how this space is constantly created and recreated through cultural mediation [21].

The cultural activity of tasting

I have in this short Opinion outlined what it could mean to see taste as a social sense and stressed that it includes an analysis of tasting as a cultural activity, rather than having an isolated focus on what a product is doing to your taste buds [7]. This is also a shift from taste to tasting. But tasting is not just tasting; tasting means different things in different contexts: The taste of mussels is not the same for a restaurant guest at a gourmet restaurant as for a fisherman in a poor part of the country side [22]. The taste of mussels will be shared in very different ways in these two situations. Mapping strategies of sharing would, thus, not be more than a long list of different types of practices if not the context is taken into account. What a certain taste sharing practice means is related to the situational as well as to the geographical, political and historical context. In order to interpret why people share certain tastes and not others, why they talk about taste in this way and not another etc. one needs to pay attention to the situations and conditions the activity of tasting take place within. This view of senses as cultural is well described, e.g. [23] what is lesser acknowledged is that the

cultural activity of tasting has this double function: being both influenced by and influencing the social.

This active use of the sense of taste is difficult to grasp perhaps because the concept of taste itself does not give us many chances to distinguish between different taste situations. It is remarkable how, e.g. the sense of vision has many related words that position the actor in relation to context and activity: you can watch, stare, scan, observe, see, notice, gaze; all these notions express how individuals are using different techniques of seeing [24] in different situations. With these concepts, one can imagine how a person puts his or her sense of vision into play in a social situation—a staring person uses this sense in another way than a person observing or gazing. With the concept of taste we do not—at the moment—have the same different possibilities to conceptualise the act of tasting in relation to context.

One of the ways forward to gain knowledge about how taste and culture influence each other is to explore how the taster is doing the act of tasting in different social situations. This will be a matter of empirical analyses studying the activities and social strategies of sharing taste: from the preparation of tastes to the moment where a taste meets a body—mediated or material—to the study of the contexts this meeting takes place within, and the analysis of which practices of taste sharing are generated under certain circumstances. Taste then becomes an experience that not only goes into the mind of the subject but also contributes to the common creation of knowledge.

Competing interests
The author declares that she has no competing interests.

Acknowledgements
I am grateful to professor emerita Carole Counihan for the inspiration, especially in relation to her and Psyche Williams-Forson's book *Taking Food Public*.

References
1. Murcott A: *Myth and Realities: A New Series of Public Debates 9th.* 2010.
2. Fischler C: **Commensality, society and culture.** *Soc Sci Inf* 2011, **50**(3–4):528–548.
3. Kant I: *Kritik af dømmekraften.* Det lille Forlag: Frederiksberg; 2007.
4. Douglas M: *Purity and Danger. An Analysis of Concepts of Pollution and Taboo.* London: Routledge and Paul Keegan; 1966.
5. Bourdieu P: *Distinction. A Social Critique of the Judgement of Taste.* Cambridge, MA United States of America: Harvard University Press; 1984.
6. Bourdieu P: *The Logic of Practice.* Stanford, California: Stanford University Press; 1980.
7. Teil G, Hennion A: **Discovering quality or performing taste? A sociology of the amateur.** In *Qualities of Food.* Edited by Harvey M, McMeekin A, Warde A. Manchester: Manchester University Press; 2004:19–29.
8. Korsmeyer C: *Making Sense of Taste. Food and Philosophy.* Ithaca and London: Cornell University Press; 2002.
9. Gherardi S: **Practice? It's a matter of taste!** *Manag Learn* 2009, **40**(5):535–550.
10. Abbots EJ, Lavis A: *Why We Eat. How We Eat. Contemporary Encounters between Foods and Bodies.* Surrey and Burlington: Ashgate Publishing Company; 2013.

11. Williams-Forson P, Counihan C: *Taking Food Public. Redefining Foodways in a Changing World*. New York and Oxon: Routledge; 2012.

12. Mann A, Mol AM, Satalkar P, Savirani A, Selim N, Sur M, Yates-Doerr E: **Mixing methods, tasting fingers. Notes on an ethnographic experiment.** *HAU: J Ethnographic Theor* 2011, **1**(1):221–243.

13. Stroller P: *The Taste of Ethnographic Things. The Senses in Anthropology*. Philadelphia: University of Pennsylvania; 1989.

14. Kuipers JC: **Matters of taste in Weyéwa.** *Anthropol Ling* 1993, **1**(4):538–555.

15. Leer J, Povlsen KK: *Media Food*. East Anglia: Ashgate; 2015.

16. Povlsen KK: **Karoline, koen og digitale måltider.** In *Det gode madliv: Karoline, maden og måltidet i kulturen*. Edited by Færch T, Møller M, Anne Kirstine Hougaard AK, Viby J. ; 2008:95–103.

17. Vannini P, Ahluwalia-Lopez G, Waskul D, Gottschalk S: **Performing taste at wine festivals: a somatic layered account of material culture.** *Qual Inq* 2010, **16**(5):378–396.

18. Strong J (Ed): *Educated Tastes: Food, Drink and Connoisseur Culture*. Lincoln: University of Nebraska Press; 2011.

19. Chau AY: **The sensorial production of the social.** *Ethnos J Anthropol* 2008, **73**(4):485–504.

20. Berger PL, Luckmann T: *The Social Construction of Reality: A Treatise in the Sociology of Knowledge*. New York: Anchor Books; 1968.

21. Boyer D: **From media anthropology to the anthropology of mediation.** In *SAGE Handbook of Social Anthropology*. London: SAGE; 2012.

22. Murcott A: **Interlude. Reflections on the elusiveness of eating.** In *Why We Eat, How We Eat. Contemporary Encounters between Foods and Bodies*. Edited by Abbots EJ, Lavis A. Surrey and Burlington: Ashgate Publishing Company; 2013.

23. Howes D (Ed): *Empire of the Senses: the Sensual Culture Reader*. Oxford, New York: Berg; 2006.

24. Hansen HP: *I grænsefladen mellem liv og død: en kulturanalyse af sygeplejen på en onkologisk afdeling*. København: Munksgaard; 2002.

25

Grape expectations: how the proportion of white grape in Champagne affects the ratings of experts and social drinkers in a blind tasting

Vanessa Harrar[1*], Barry Smith[2], Ophelia Deroy[2] and Charles Spence[1]

Abstract

Background: Champagnes (or sparkling wines that are made using the 'méthode champenoise') are composed of white and/or red wine grapes. Their relative proportions are thought to contribute to a sparkling wine's distinctive flavour profile, but this has not yet been tested empirically. We, therefore, conducted a blind tasting experiment in which the participants had to report the perceived proportion of white grapes in a range of seven sparkling wines (including six Champagnes).

Results: The participants, including four expert, six intermediate, and five novice Champagne tasters, were unable to accurately judge the percentage of white grapes in the wines. Instead, the perceived proportion of white grape was correlated with the dosage and alcohol content of the wines. The hedonic ratings for the Champagnes did not correlate with price. Further, the more expensive Champagnes were only appreciated by the expert tasters.

Conclusions: Dosage and alcohol content appear to be the two attributes that tasters rely on when judging the contribution that different grape types make to the distinctive flavour of a sparkling wine. In the case of Champagne, flavour perception relies on a complex combination of factors including alcohol content, dosage, price expectancy, and experience with the product. The present results have implications for marketing Champagnes; they might be better if focused on the distinctive characteristics of each cuvee, or simplicity (blends versus non-blends), since these might be easier characteristics to detect than the proportion of white versus red grapes.

Keywords: Blind tasting, Blends, Sparkling wine, Grape colour, Champagne, Expertise

Background

Is the reputation that Champagne has amongst many consumers attributable to the quality of its component base wines and the craftsmanship of their blending, or is branding perhaps the major contributor to the perceived prestige of this drink? In his book *Wine Scandal*, Fritz Hallgarten ([1], pp. 116-117) describes an occasion in which a group of wine consultants tried to identify the glass containing the Champagne among ten sparkling wines. Virtually no one succeeded in this task. Interestingly, though, the consultants thought that whichever sparkling wine tasted best to them was the Champagne (often, this turned out to be a less expensive sparkling wine from somewhere such as Israel or

Luxembourg). Many studies have questioned the correlation between liking and price when wines are tested under blind tasting conditions (see [2], for a review). Casual reports have suggest that people follow a similar pattern with sparkling wines - that is, rating them on the basis of liking when tasted blind- but when brands are revealed, other attributes such as price, brand appreciation, Champenoise origin, or composition of grapes influence ratings considerably (for example, the higher the price, the better the perceived quality; [3-6]; see [7] for a review).

One of the few peer-reviewed studies to have looked specifically at Champagne was conducted by Lange and colleagues [8] on social drinkers in France. They presented participants with five brut non-vintage Champagnes varying in price from €11 to €23, three of which were bottles from a selection of well-known Champagne houses from the former 'Syndicat des Grandes Marques'.

* Correspondence: vanessa.harrar@psy.ox.ac.uk
[1]Crossmodal Research Laboratory, Department of Experimental Psychology, University of Oxford, South Parks Road, Oxford OX1 3UD, UK
Full list of author information is available at the end of the article

One group of participants had to rate how much they liked each Champagne, while another group had to say how much they would be willing to pay for it. Both groups performed these tasks under two conditions: first in a blind tasting condition, and then again when given the names of the producers. After the reveal, the participants were offered a chance to re-taste. Lange *et al.* reported differences for both ratings in the two conditions. On average, the social drinkers who were tested in this study were willing to pay significantly more for the top-brand Champagnes (which they reported liking more when the labels were revealed), and less for the unknown brands. In addition, knowing the identity of the brand (and, therefore, the approximate commercial price for the wine) introduced a greater degree of separation between the price estimates than it did for the hedonic scores. These results therefore suggest that, at least amongst social drinkers, the presence of effective labelling and branding (giving the taster information about typicity, prestige, cultural associations, price, etcetera) constitutes an influential factor in driving consumer preference. One important caveat to be kept in mind here though is that those participants who generally bought their Champagne from a producer were less influenced by the presentation of the bottle, while those participants who showed little or no brand loyalty attached greater importance to the label, showing, perhaps, relatively little knowledge of Champagnes. We hypothesized that expert tasters, who might be more attuned to the sensory properties of the product (intrinsic cues), would be less affected by extrinsic factors. Experts may be more likely able to detect the distinctive contribution that the different grapes make to its flavour without necessarily seeing the label (where such information is often available).

Champagnes are typically divided into three groups based on their grape composition: Blanc de Blancs (100% Chardonnay, white grapes), Blanc de Noirs (100% Pinot Noir and/or Pinot Meunier, red grapes), and blends (of white and red grapes). Blends report the exact proportion of each grape on the label as an indication of the flavour to expect. According both to the experts and to widespread belief, each grape variety brings a set of distinctive sensory features to the sparkling wine. Chardonnay is often described as bringing elegance and finesse, with Pinot Noir providing red-berry characteristics and structure, and Pinot Meunier providing both fruit aromas and roundness. According to the well-respected British wine critic Jancis Robinson, 'the Pinot Noir...provides the basic structure and depth of fruit in the blend...[Chardonnay] imparts a certain austerity and elegance to young champagnes, but is long-lived and matures to a fine fruitiness. [Pinot Meunier] provides many champagnes with an early-

maturing richness and fruitiness'. The author goes on to note that Chardonnay 'has the greatest tendency to go toasty if aged after disgorgement, but can also develop finer, creamy, biscuity nuances' [9]. Pinot Noir 'does not retain its freshness for as many decades as Chardonnay but it arguably provides a more complex wine...goes biscuity rather than toasty, although toastiness is a common bottle aroma for this variety' [10]. Thus, the different grape compositions of Champagnes are said to result in different flavours, but how distinctive are the flavours? Tom Stevenson admits that 'a number of Blanc de Noirs can be so light that it is hard to imagine they do not contain some Chardonnay' [10]. With what degree of certainty, then, can experts work backwards from the flavour to determine the composition of a sparkling wine when tested blind?

The participants in this study ranged in expertise from novice to expert Champagne tasters. They were instructed to estimate the proportion of white grapes in seven sparkling wines (including six Champagnes) tasted blind. The participants were informed only that the sparkling wines could span the full range (from 0 to 100% Chardonnay grapes), and were chosen to provide a good range of values including 0, 22, 30, 45, 58, and 100% Chardonnay grapes, with the remainder of the grapes made up of Pinot Noir and Pinot Meunier. Commonly available Champagnes were used in the study, thus making it impossible for us to vary the composition of red and white grapes while keeping other factors roughly constant. Dosage (sugar added to champagne after bottle fermentation) varied from 6 to 12 g/l, while amount of time on the lees varied from 1 to 8 years. These elements were factored into the data analysis in order to determine whether they correlated with the perceived proportion of white grapes, pleasantness, fruitiness, or sweetness ratings.

Results

Some questions were not answered by all participants (43 responses out of 15 subjects x 7 champagnes x 8 dependent values, corresponding to approximately 5% of the data being missing). Missing data were not filled in since there were no repetitions, the wines varied considerably, and there were very few participants in each expertise group therefore there was no reasonable value that could be used as a substitute for the missing data. Degrees of freedom therefore fluctuate slightly between statistical tests. No data were removed or corrected; that is, no exclusion criteria were necessary. The results presented therefore represent all available data as is.

Analysis of the sweetness and fruitiness ratings

The sweetness and fruitiness ratings were analysed in order to determine whether there was a correlation between the two. There was no significant correlation between sweetness and fruitiness ratings when all of the data were tested together (across all participants, all wines, and all levels of expertise). Furthermore, none of the subsequent correlation analyses between sweetness and fruitiness, either at the level of individual participants, or at the level of specific groups of participants (experts, intermediate, novice), or looking at each wine separately, revealed any significant correlations either. Since sweetness ratings were not correlated with fruitiness ratings, these results argue against any kind of 'halo dumping' effect in the present data [11].

Halo dumping occurs when participants are provided with only one intensity scale (for example, sweetness) to rate a mixture of similar sensations (for example, sweetness and strawberry flavour). Forced to use one scale to describe both sensations, there is a danger that participants may 'dump' the second sensation onto the only available scale they have at their disposal. This effect disappears as soon as participants are provided with a scale for each sensation. The lack of a significant correlation between sweetness and fruitiness ratings speaks to the level of expertise, even of the novice group, in this experiment. A similar previous study (unpublished) with university students found significant correlations between the sweetness and fruitiness ratings for seven Champagnes (similar to those tested here) at all three self-reported levels of expertise. Here, however, while the novice and intermediate tasters did not report themselves as experienced Champagne tasters, the lack of any correlation between sweetness and fruitiness demonstrates their abilities in assessing wines, and gives credibility to their estimates of the proportion of white grape in the wines.

Estimated proportion of white grapes

Although certain participants accurately reported the proportion of white grape in some of the sparkling wines (see individual data plotted in Figure 1 - correct answers are points on the solid line), no one was correct more than two or three times. In order to determine whether the participants' responses were significantly different for the seven sparkling wines tasted, and in order to determine whether the pattern of responses differed reliably as a function of the three groups of responders (novice, intermediate, and expert), we compared the estimated proportion of white grapes for each of the sparkling wines with a mixed model analysis of variance (ANOVA). The within-participants factor was sparkling wine (seven levels) and the between-participants factor was experience (three levels). This analysis revealed no main effect of the type of sparkling wine [$F(6,54) = 1.07$, $P = .39$], no main effect of experience [$F(2,9) = 1.28$, $P = .32$], nor any interaction effect [$F(12,54) <1$, n.s.].

None of the groups (see also Figure 2) showed any consistent difference in their judgments concerning the proportion of white grapes in the various sparkling wines that were tasted. What about when the groups estimated the proportion of white grape for the champagnes at the two extreme ends of the scale (Blanc de Blanc and Blanc de Noir)?

The estimates for the sparkling wines containing 0 and 100% white grapes were compared in order to determine whether the participants reliably perceived the extreme difference in the proportion of white grapes. A t-test was performed in which we compared participants' responses to the Blanc de Noir (Mumm de Verzenay containing 0% Chardonnay grapes) with the Blanc de Blanc (containing 100% Chardonnay grapes). Once again, however, the results of this analysis revealed no significant difference ($t(13) = .296$, $P = .386$). Furthermore, the means were actually in the wrong direction; that is, more of the participants thought that the Blanc de Blanc contained a smaller proportion of Chardonnay grapes than the Blanc de Noir (see Figure 3). Looking at each of three groups separately, it can be seen that for both the intermediate and novice groups, the means are in the wrong direction ($t(5) = .262$, $P = .402$; and $t(3) = -1.15$, $P = .167$, respectively, see the black and grey symbols in Figure 3). That said, the responses from three out of four experts were in the right order (but not different enough to be significant, ($t(3) = .792$, $P = .243$, see the blue symbols in Figure 3).

If the perceived proportion of white grape did not follow from the actual proportion, then what sensory qualities of the sparkling wines were the participants using in order to make this assessment? None of the methods which the participants reported they were using to determine the white grape content (aroma, flavour, and/or oral-somatosensory textural cues) produced any reliable correlations, or gave any advantage in terms of correctly determining the proportion of white grapes in the mix.

The perceived proportion of white grapes was, however, correlated with some other qualities of the Champagne: Dosage - the liquor and sugar added during fermentation - was negatively correlated ($N = 99$, Spearman's rho = $-.255$, $P = .011$), while the alcohol content was positively correlated ($N = 99$, Spearman's rho = $.231$, $P = .021$, see Figure 4). Participants' estimates concerning the proportion of white grapes increased as the dosage decreased, and as the alcohol content increased. Note that true alcohol and dosage levels in the champagne were negatively correlated ($N = 7$, r = $-.813$, $P = .026$).

The certainty of participants' responses

Participants also had to rate the certainty with which they reported the perceived proportion of white wine (where 1

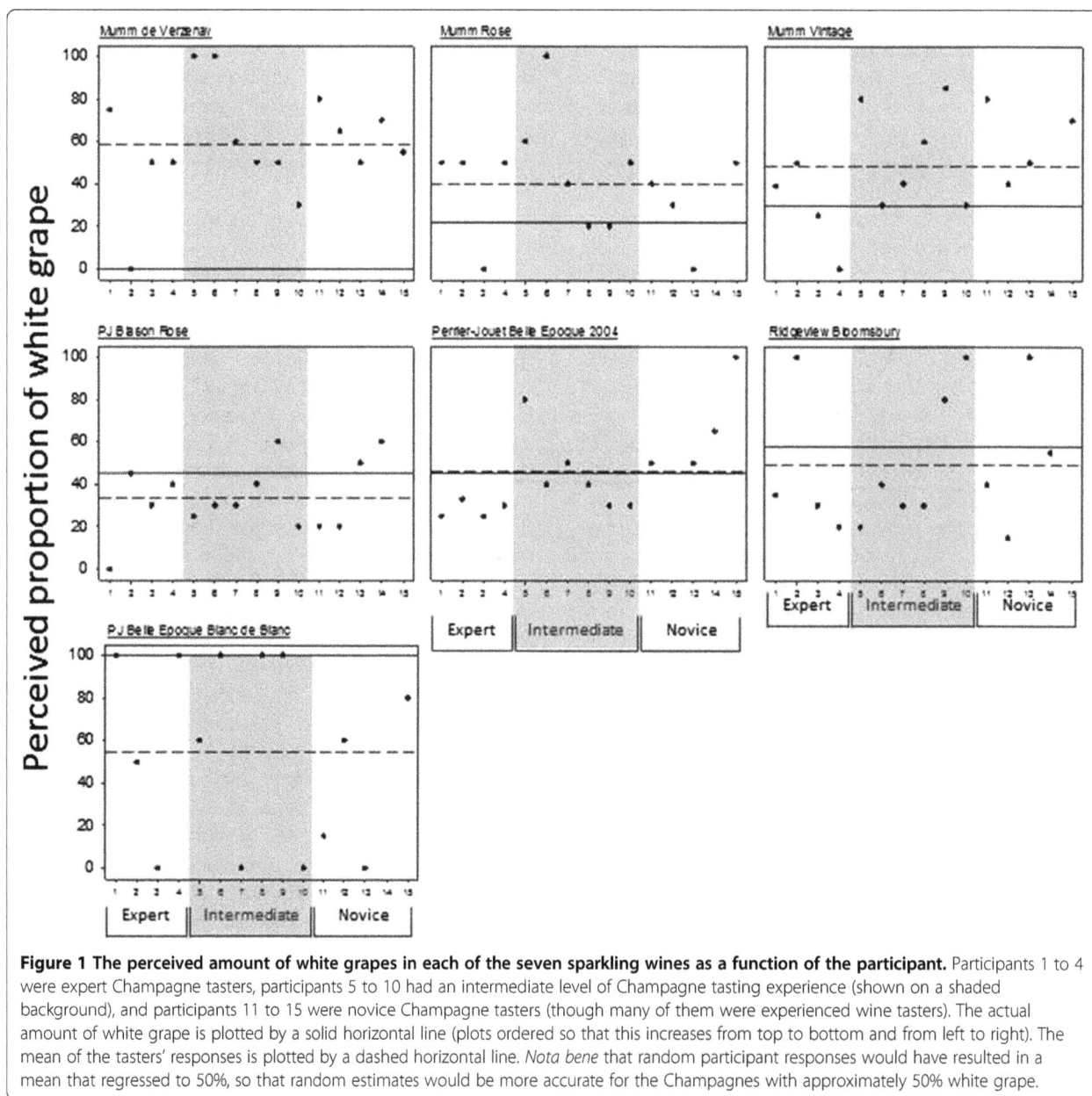

Figure 1 The perceived amount of white grapes in each of the seven sparkling wines as a function of the participant. Participants 1 to 4 were expert Champagne tasters, participants 5 to 10 had an intermediate level of Champagne tasting experience (shown on a shaded background), and participants 11 to 15 were novice Champagne tasters (though many of them were experienced wine tasters). The actual amount of white grape is plotted by a solid horizontal line (plots ordered so that this increases from top to bottom and from left to right). The mean of the tasters' responses is plotted by a dashed horizontal line. *Nota bene* that random participant responses would have resulted in a mean that regressed to 50%, so that random estimates would be more accurate for the Champagnes with approximately 50% white grape.

corresponded to 'not at all certain' and 10 corresponded to 'extremely certain'). The data concerning the certainty of participants' responses were analysed using the same mixed model ANOVA as described previously (Wine - 7 levels; Expertise - 3 levels). While expert tasters were more certain (mean certainty 5.18 ± standard error 1.19) than the intermediate (4.97 ± 1.07), or the novice tasters (3.71 ± 1.38), there was no main effect of expertise ($F_{2,9}$ = 1.051, P = .389). Further, while participants were most certain with reports concerning the Mumm Rosé (5.81 ± 0.79), and least certain of their judgments for the Mumm Vintage 2004 (4.02 ± 0.63) there was no significant main effect of sparkling wine. There was no interaction between the factors (both F <1).

Hedonic ratings

What made the participants like or dislike a particular sparkling wine? We tested correlations of hedonic ratings with the perceived proportion of white grape, perceived sweetness, fruitiness, familiarity, or vintage/NV; and known attributes of each wine (actual proportion of white grape, dosage, alcohol content, amount of time on the lees, or price). The overall score given to the sparkling wine (out of 10, where 10/10 is excellent) was correlated with fruitiness (N = 102, Spearman's rho = .355, P <.001), vintage/NV assumptions (N = 98, Spearman's rho = −.373, P <.001; negative correlation indicates that higher ratings were associated with 'vintage'), and familiarity scores (N = 95, Spearman's rho = .589, P <.001) (Figure 5).

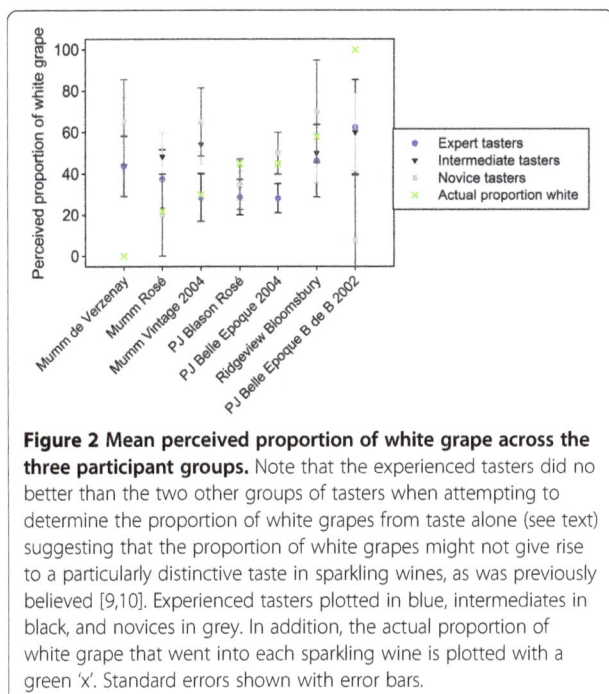

Figure 2 Mean perceived proportion of white grape across the three participant groups. Note that the experienced tasters did no better than the two other groups of tasters when attempting to determine the proportion of white grapes from taste alone (see text) suggesting that the proportion of white grapes might not give rise to a particularly distinctive taste in sparkling wines, as was previously believed [9,10]. Experienced tasters plotted in blue, intermediates in black, and novices in grey. In addition, the actual proportion of white grape that went into each sparkling wine is plotted with a green 'x'. Standard errors shown with error bars.

Most importantly, hedonic ratings did not correlate with the objective price of the bottles (Spearman's rho = .032, P = .748, see Figure 5). In fact, the price of the sparkling wines was not correlated with any of the participants' ratings or perceptions (obtained from the

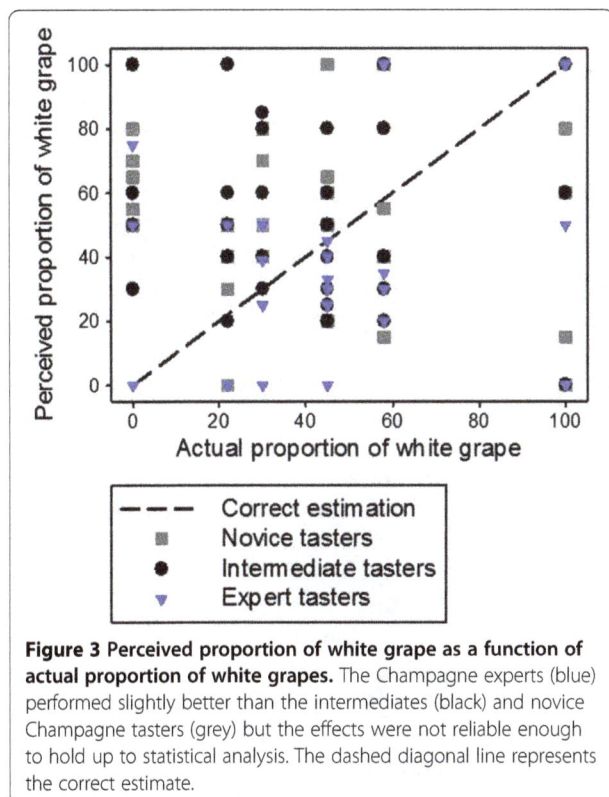

Figure 3 Perceived proportion of white grape as a function of actual proportion of white grapes. The Champagne experts (blue) performed slightly better than the intermediates (black) and novice Champagne tasters (grey) but the effects were not reliable enough to hold up to statistical analysis. The dashed diagonal line represents the correct estimate.

responses in the questionnaire); it was only correlated with the number of years that the wine had been left on its lees before being bottled (N = 7, Spearman's rho = .865, P <.012), which is a vintage characteristic. If not the most expensive, which wines were most well liked, if any?

Figure 5 shows group ratings that have been normalised based on each participant's mean rating, so that positive values indicate that the participants in that group rated the wine above their individual average. Perrier-Jouët Belle Epoque Blanc de Blanc 2002 had the lowest overall rating (rating of 5.86/10, retail price £400) Mumm Vintage 2004 and Perrier-Jouet Belle Epoque 2004 were rated highest (rating of 7.13/10, retail price £40 and £100, respectively). Although there were small variations in ratings, there were no significant differences between the ratings across the sparkling wines. Similar results were obtained in an unpublished primary and smaller scale study conducted with Oxford University students. There was also no significant interaction between the ratings on the seven sparkling wines and level of expertise.

It might be that the Perrier-Jouët Belle Epoque Blanc de Blanc 2002 (£400) was simply not showing very well; the wine professional who tasted the samples suggested that it appeared less expressive than normal on the palate and had a slight cheese or sweaty note on the nose (reminiscent of isovaleric acid, a characteristic of Brettanomyces spoilage). Alternatively, it might be the case that specialty or aged cuvées can only really be appreciated by expert champagne tasters with significant experience (that is, they may be an acquired taste). Innately unpalatable foods can often become pleasant (for example, chili and coffee) but this requires, among other things, repeated exposure (see discussion of hedonic reversals in [12], p.402).

Colour from aroma ratings

Perrier-Jouët Blason Rose was rated as having the most reddish aroma (rating of 2.44; 55% red grape) while Mumm de Verzenay had the greenest aroma (rating of 5.11; 100% red grape). Analysing the aroma-colour association (1-red, 7-green, see Figure 6), with the same mixed model ANOVA as described previously revealed no significant difference across the seven sparkling wines, no difference across expertise groups, and no interaction between the two (F <1 for each, N.S.).

Discussion

The results of the present study, conducted with a range of sparkling wines, suggests that people, no matter whether they are expert Champagne tasters, expert wine tasters, or simply social drinkers, are unable to reliably determine the proportion of white Chardonnay grapes in

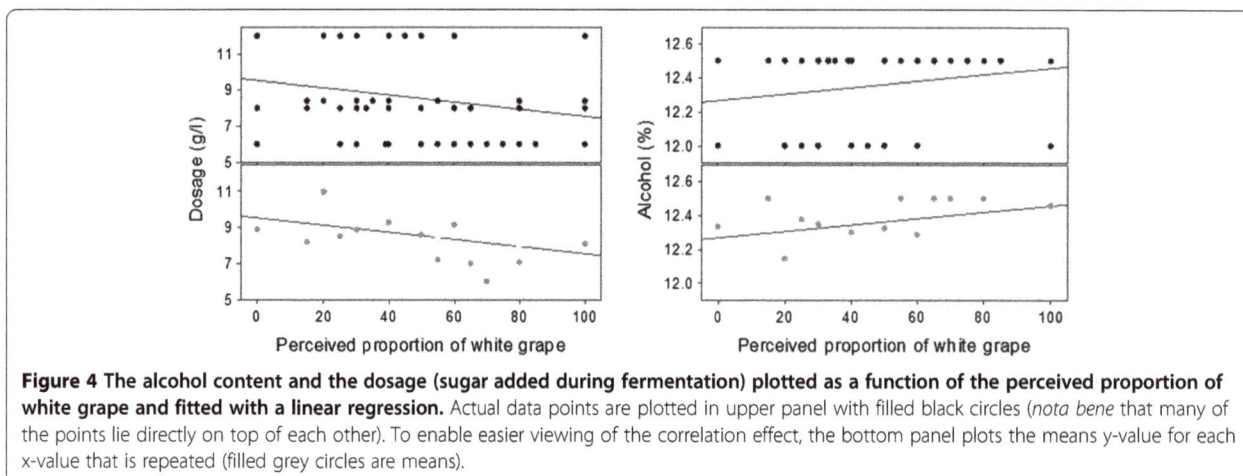

Figure 4 The alcohol content and the dosage (sugar added during fermentation) plotted as a function of the perceived proportion of white grape and fitted with a linear regression. Actual data points are plotted in upper panel with filled black circles (*nota bene* that many of the points lie directly on top of each other). To enable easier viewing of the correlation effect, the bottom panel plots the means y-value for each x-value that is repeated (filled grey circles are means).

the sparkling wines when tasted blind. We found that dosage and alcohol content are attributes that tasters actually rely on when judging the contribution each grape type makes to the distinctive flavour of a sparkling wine.

Our findings are in line with those studies of still wine that have investigated the ability of people (experts and non-experts) to sort wines in terms of the characteristics coming from the grape (that is, the 'primary aromas',) rather than properties of the wine added by vinification, bottle fermentation, lees contact (amount of time the bottle-fermented wine is left in contact with the dying yeasts before disgorgement when the yeast deposits are removed and the wine is re-corked), and elevage (the process of maturing the base [still] wine in oak barrels), or extrinsic cues from branding and labelling [13-15]. These results add to the literature on the blind tasting of sparkling wines [1,8] suggesting that this extrinsic information might contribute to what people report as being the characteristic tastes (or qualities) of certain varieties of sparkling wine.

However, the sparkling wines in this study did not vary along a single dimension. In addition to variations in the proportion of white grapes, they were also different in terms of the place the wine came from, as well as its vintage, quality, alcohol content, and dosage. These various factors introduce further complexity to the tasters' task, but at the same time make the testing/tasting situation more realistic. Still, future experiments could test an expert's ability to make finer grain comparisons of wines (for example, from the same vintage, or different cuvées or vintages from the same Champagne house, or sort white grape-only wines from different regions - see [15]). A skilful chef de cave might be able to discover, after experimenting with particular blends, the difference each component makes to the overall well-integrated flavour of the Champagne.

Although the results show that experienced Champagne drinkers are not able to judge the contribution of each grape variety to the flavour of a Champagne, tasters certainly recognised differences between the wines and rated them differently. It would therefore be wrong to infer from the fact that tasters cannot perceive the difference that the proportion of the various grape varieties makes to the blend, that it did not make a difference to the resulting flavour of the blend. Instead, we hypothesise that success in blending has made it more difficult for tasters to identify the particular contributions made by each grape variety to the overall flavour profile. This

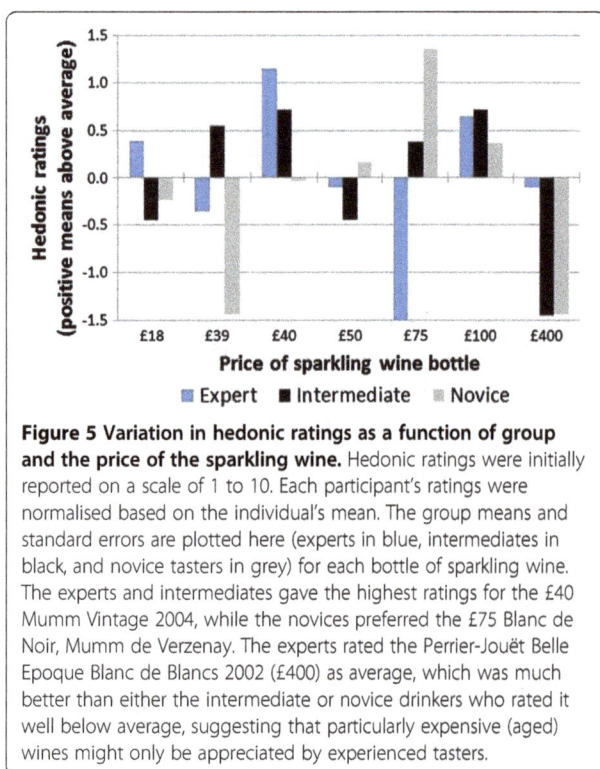

Figure 5 Variation in hedonic ratings as a function of group and the price of the sparkling wine. Hedonic ratings were initially reported on a scale of 1 to 10. Each participant's ratings were normalised based on the individual's mean. The group means and standard errors are plotted here (experts in blue, intermediates in black, and novice tasters in grey) for each bottle of sparkling wine. The experts and intermediates gave the highest ratings for the £40 Mumm Vintage 2004, while the novices preferred the £75 Blanc de Noir, Mumm de Verzenay. The experts rated the Perrier-Jouët Belle Epoque Blanc de Blancs 2002 (£400) as average, which was much better than either the intermediate or novice drinkers who rated it well below average, suggesting that particularly expensive (aged) wines might only be appreciated by experienced tasters.

Figure 6 Experimental set-up. On the left, the table is laid out with glasses and the questionnaires are shown. On the right, the first question in the response booklet is demonstrated. For this question, participants had to choose which of the seven coloured cups best matched the smell/nose of the sparkling wine. For half of the participants, the left-right ordering of the colours was reversed (that is, the deep red glass appeared on the left of the score sheet).

suggests another intriguing possibility worthy of further investigation: Whether complexity might be more easily recognised than the actual components in the wine. In support of this point, we found that Perriet Jouet Belle Epoque Blanc de Blanc was correctly identified as a non-blend by nine tasters (five were correct, while another four people guessed it was a Blanc de Noir; see Figure 1). Other samples were speculated to be non-blends by at most three tasters. It would be interesting for future studies to compare the perceptual heuristics applied to simplicity versus complexity in blended and non-blended wines.

Another interesting result to emerge from the present study is the fact that there was no correlation between the objective price of the bottles and our participants' preference ratings for the sparkling wines tasted blind. This replicates previous results [8], and extends them over a much wider price range (€11 to €23 in Lange *et al.*; £18 to £400 here). A lack of correlation between preference and price fits with the much larger body of research on still wines (see [2] for a review). We would also like to suggest, as an area for future research, that this effect might be different for experienced tasters as compared to social drinkers. Based on these results, we suggest that more expensive champagnes, with attributes related to ageing, might only be appreciated by expert tasters who are more likely to have previously experienced such flavour profiles.

Were the self-declared expert Champagne tasters in the present study a representative sample? Of course, with more expert tasters some tests would have been more powered and certain of the null results might have reached significance (for example, with about 15 experts reporting in the same way as the four that were tested here, there would be enough power to reliably conclude that experts estimate the Blanc de Blanc as having more Chardonnay than the Blanc de Noir). However, such a large sample of expert Champagne drinkers might not

be possible outside the region. Instead, as described in the Methods section, all the participants tested in this study had professional activities related to wine or spirits, including the novice and intermediate Champagne tasters. Some of those that self-assessed themselves as expert Champagne tasters were writers and journalists on the topic (note that previous verbalization of characteristics of wines tasted may have affected their taste perception [16,17]), while others indicated that they were wine merchants. The individuals we tested are likely to have had significantly more experiences and exposure to Champagne than those tested in previous research [1,8], making these results highly novel. The high baseline of tasting knowledge across the whole sample is demonstrated by the lack of correlations between the sweetness and fruitiness ratings for the wines. Using a lack of correlation between perceptually 'similar' attributes might provide an independent means of categorizing people as expert or non-expert tasters in future research.

Were the champagnes fairly presented? Participants in the present study only had limited access to cues such as the size of the bubbles, the quality of the mousse, and the amount of dissolved CO^2 since they were sampling in random order sequentially. For example, the size of the bubbles can be used to distinguish between Champagne and Cava. These cues provide quality information which could have been used by the tasters to distinguish between the wines they were sampling [18-20], but there is little evidence to suggest that they would provide information about the proportion of white versus red grapes in the wine. Colour, on the other hand, could have provided an important cue. This study therefore employed blind tasting.

Although blind tasting is a technique that is commonly used, it may not provide results that are transferrable to normal, namely sighted, taste perception. Pangborn and her colleagues [21] revealed that experts rated a rosé-coloured white wine as being sweeter than the untainted (uncoloured) white wine. By contrast, the non-experts' sweetness ratings were not influenced by the colour of the wine. As such, experts who, knowingly or unknowingly, use colour when assessing the proportion of white and red grapes in a sparkling wine might be misled [22-24]. It is also worth noting that the glass itself may also have had an influence on the flavour [25].

Conclusions

The current study balanced laboratory testing procedures (using professional black tasting glasses) with a set of realistic wines that varied in composition along multiple dimensions. The group of expert Champagne tasters was not able to reliably determine the proportion of white grapes in the seven sparkling wines tasted. One should certainly not conclude from our results that the percentage of these grapes makes no difference to the taster's experience (most of the participants distinguished between the Champagnes, as seen from the variability in their responses to each wine). Instead, it seems that the contribution to flavour that the proportion of these grapes makes cannot be detected. Indeed, the goal in blending wines is that their component parts - fruit, acidity, alcohol, lees character - are all present but with no single attribute dominating the others. That is, a 'well blended' or balanced wine may well disguise the components that have gone into producing the resulting flavour. Further experiments are therefore needed in order to assess the tentative interpretation of the results put forward here, results which could contribute significantly to the as yet ill-understood notion of simplicity and complexity in flavour perception.

Methods

Participants

Fifteen participants gave their informed consent prior to taking part in the tasting. Their median age was 30 years (ranging from 21 to over 60) and 11 were male. The studies have been approved by the Central University Research Ethics Committee of the University of Oxford and are therefore in compliance with the Helsinki Declaration. The whole sample had been involved in professional activities related to wine or spirits. To assess Champagne expertise, in particular, the participants were given the following questions: 'How would you rate your experience with Champagne? Novice, intermediate, or expert' and 'Is your professional activity related to Champagne? Yes, no, please specify'. We chose to avoid mixing kinds of expertise because perceptions of wine are known to differ among different kinds of experts [26]. For instance, experts included critics but importantly did not include any wine-makers or sommeliers. As such, there were four expert Champagne tasters who included wine merchants, well-known Champagne critics/journalists/writers specializing in champagne, and trade ambassadors for a Champagne house. There were six intermediate Champagne tasters, which included wine and spirit trade retailers, writers/journalists for food and wine columns, and sales representatives for Champagne houses. Finally, there were five self-assessed novice Champagne tasters, who included wine brand owners, people who worked for a Champagne house, wine trade, and spirit ambassadors who generally drank Champagne monthly.

Stimuli

The seven sparkling wines used in the present study consisted of six Champagnes: Mumm de Verzenay, Mumm Rosé, Mumm Vintage 2004, Perrier-Jouët Blason Rosé, Perrier-Jouët Belle Epoque 2004, Perrier-Jouët Belle Epoque Blanc de Blancs 2002, and a Ridgeview Bloomsbury Non-Vintage Sparkling wine from Sussex (UK) - the only sparkling wine not made in Champagne but made of the same grape varieties and using the same method. These wines were chosen to provide a range from 0 to 100% Chardonnay grapes (see Table 1; [see Additional file 1 for the tasting notes for each wine]).

To ensure quality control, each bottle of sparkling wine was tasted by a professional (not included as a participant in the study) before being poured. This step ensured that none of the bottles had any obvious faults. At the end of the experiment, all of the sparkling wines were re-tasted by the same professional (once they had warmed-up significantly). The expert suggested that one of the bottles of Perrier-Jouët Belle Epoque Blanc de Blancs 2002 might not have been showing all that well. It had some slight characteristics of Brettanomyces spoilage (cheese or sweaty note on the nose) and was somewhat less expressive than usual, but it was by no means definitely off.

The wines were presented to the participants in professional black ISO tasting glasses that had just been put through the dishwasher; the expert who was asked to check the glasses, and was sensitive to potential problems with residual detergent, detected no such problem. The wines were served at the same temperature (they were removed from refrigeration at the same time) and had nearly the same amount of time to breath before being tasted (the first glass was poured about five minutes before the last glass, and approximately 15 minutes before the participants began tasting).

Design

The experiment was conducted in the University of London in a quiet, well-lit air-conditioned room. Participants sat at a table with at least 1 m spacing between adjacent tasters. A line of seven opaque black glasses containing a tasting quantity of the sparkling wines (each

Table 1 The seven sparkling wines (including six Champagnes) evaluated in the present study

| Sparkling wine | Year | % White grapes | % Red grapes | | Alcohol | Dosage | Lees | Retail |
Name		Chardonnay	Pinot Noir	Pinot Meunier	(%)	(g/l)	(years)	price (£)
Mumm de Verzenay	NV	-	100	-	12.5	6	5	75
Mumm Rosé	NV	22	60	18	12.0	12	3	39
Mumm Vintage	2004	30	70	-	12.5	6	5	40
Perrier-Jouët Blason Rosé	NV	45	50	5	12.0	12	1.08	50
Perrier-Jouët Belle Epoque	2004	45	50	5	12.5	8	6	100
Ridgeview Bloomsbury	2009	58	30	12	12.5	8.4	0.67	18
Perrier-Jouët Belle Epoque Blanc de Blancs	2002	100	-	-	12.5	8	8	400

NV, non-vintage.

labelled with a random three-digit code), a questionnaire booklet, and a pen were placed in front of each participant's place prior to their arrival at the testing station (see Figure 6). The wines were randomly ordered between participants and three-digit codes were used to identify which set of answers corresponded to which wine. An opaque black spittoon was placed at each table. Each participant was also given a glass of water for rinsing as necessary.

The group of participants received a three-minute briefing at the start of the session to ensure that they all had the same background information and instructions prior to the experiment. The participants were informed that the sparkling wines could cover the full range of proportions of Chardonnay grapes (the questionnaire also said 0 to 100%). They were told that certain glasses might contain the same wine. They were informed that their neighbours might have different wines, and that they would likely be presented in a different order. They were instructed to begin tasting (and rating) with the glass on the far left and progress rightward, turning to a new page for each new wine. They were told not to go back to any of their previous glasses (or answer sheets) once they had moved on.

For each wine sample, there were a total of nine questions asked exactly as presented below (that is, there were no visual analogue scales) except that Question 1 was preceded by the image of coloured glasses displayed in Figure 6. The participants were informed that they should answer the first question before tasting (as it related to the smell/nose of the wine). Each question from the questionnaire was briefly explained and participants had the opportunity to ask for further clarification. During this process, the experimenter clarified that the overall rating (Q6) referred to hedonic preference (not the quality of wine per se), and Q9 was where tasters could hazard a guess at the brand/producer/type of champagne. There was also a blank space where tasters were invited to write down any tasting notes they might have.

Since 50% of Q9 was left blank, it was not analysed; for reference, of the 55 guesses (from 15 participants and seven Champagnes), seven were correct. The majority of the participants took no more than 40 minutes to complete the tasting.

1. Which colour best matches the smell of the Champagne? (see Figure 6)
2. What percentage of the Champagne is made from White grapes? (0 to 100)
3. How certain are you of the proportion of white grapes in the Champagne? (1-Not at all certain; 10-Extremely certain)
4. How SWEET is the Champagne? (1-Not at all sweet; 10-Extremely sweet)
5. How FRUITY is the Champagne? (1-Not at all fruity; 10-Extremely fruity)
6. Rate the Champagne (out of 10) (where a higher value indicated a higher rating)
7. Do you think the Champagne is Vintage? (Non-Vintage or, if Vintage, give a year)
8. Are you familiar with this Champagne? (1-Not at all familiar; 10-Extremely familiar)
9. What would you say this Champagne is?

Additional file

Additional file 1: The tasting notes for all sparking wines tested.

Abbreviations
ns: Not significant; nv: Non-vintage.

Competing interests
The authors declare that they have no competing interests.

Authors' contributions
VH performed the statistical analysis and drafted and revised the manuscript. VH, BS, OD, and CS designed the study, prepared the stimuli, collected the data, and revised the manuscript. All authors read and approved the final manuscript.

Authors' information

VH has a PhD in psychology from York University (Toronto, Canada) and works as a post-doc in CS's lab in the Department of Experimental Psychology at Oxford University (Oxford, UK). BS has a PhD in Cognitive Science from the University of Edinburgh and is founding director of the Centre for the Study of the Senses at the Institute of Philosophy, University of London. OD has a PhD from the Universite de Paris XII and is a Marie Curie Fellow at the Centre for the Study of the Senses at the Institute of Philosophy, University of London. CS has a PhD in Experimental Psychology from the University of Cambridge and has been a University Lecturer at Oxford University since 1997.

Acknowledgements

Vanessa Harrar holds a Mary Somerville Junior Research Fellowship from Somerville College, Oxford University, UK. Ophelia Deroy is funded by a Marie Curie IEF. Barry C Smith is funded by the AHRC for a project on Flavour Perception. We would also like to thank Matt Day for his help at the tasting event and Pernod-Ricard, who hosted the Champagne Assembly Day. Finally, we would like to thank Mumm Champagne and Perrier-Jouët for donating the Champagne.

Author details

[1]Crossmodal Research Laboratory, Department of Experimental Psychology, University of Oxford, South Parks Road, Oxford OX1 3UD, UK. [2]Centre for the Study of the Senses, University of London, Senate House, Malet Street, London WC1E 7HU, UK.

References

1. Hallgarten F: *Wine Scandal*. London: Sphere Books Ltd; 1987.
2. Spence C: **The price of everything – the value of nothing?** *World of Fine Wine* 2010, **30**:114–120.
3. Angulo AM, Gil JM, Gracia A, Sanchez M: **Hedonic prices for Spanish red quality wine.** *Br Food J* 2000, **2**:481–493.
4. Combris P, Lecocq S, Visser M: **Estimation of a hedonic price equation for Bordeaux wine: Does quality matter?** *Econ J* 1997, **107**:390–402.
5. Oczkowski E: **A hedonic price function for Australian premium table wine.** *Aust J Agric Econ* 1994, **38**:93–110.
6. Wade C: **Reputation and its effect on the price of Australian wine.** *Austr New Zealand Wine Ind J* 1999, **14**:82–84.
7. Mueller S, Osidacz P, Francis L, Lockshin L: **Combining discrete choice and informed sensory testing in a two-stage process: Can it predict wine market share?** *Food Qua Prefer* 2010, **21**:741–754.
8. Lange C, Martin C, Chabanet C, Combris P, Issanchou S: **Impact of information provided to consumers on their willingness to pay for Champagne: Comparison with hedonic scores.** *Food Qual Prefer* 2002, **13**:597–608.
9. Robinson J: *The Oxford Companion to Wine*. 3rd edition. Oxford: Oxford University Press; 2006.
10. Stevenson T: *Christie's World Encyclopaedia of Champagne Sparkling Wine*. Bath: Absolute Press; 2003.
11. Clark CC, Lawless HT: **Limiting response alternatives in time-intensity scaling: An examination of the halo dumping effect.** *Chem Senses* 1994, **19**:583–594.
12. Rozin P: **Food and eating.** In *Handbook of Cultural Psychology* Edited by Kitayama SE and Cohen DE. New York: Guilford Press; 2007:391–416.
13. Ballester J, Patris B, Symoneaux R, Valentin D: **Conceptual vs. perceptual wine spaces: Does expertise matter?** *Food Qua Prefer* 2008, **19**:267–276.
14. Parr W, Green J, White K, Sherlock R: **The distinctive flavour of New Zealand Sauvignon Blanc: Sensory characterisations by wine professionals.** *Food Qual Prefer* 2007, **18**:849–861.
15. Parr W, Valentin D, Green J, Dacremont C: **Evaluation of French and New Zealand Sauvignon wines by experienced French wine assessors.** *Food Qual Prefer* 2010, **21**:56–64.
16. Hughson AL: **Wine expertise: Current theories and findings regarding its nature and bases.** *Food Aus* 2003, **55**:193–196.
17. Melcher JM, Schooler JW: **The misremembrance of wines past: Verbal and perceptual expertise differentially mediate verbal overshadowing of taste memory.** *J Mem Lang* 1996, **35**:231–245.
18. Liger-Belair G, Beaumont F, Jeandet P, Polidori G: **Flow patterns of bubble nucleation sites (called fliers) freely floating in champagne glasses.** *Langmuir* 2007, **23**:10976–10983.
19. Liger-Belair G, Beaumont F, Vialatte M-A, Jégou S, Jeandet P, Polidori G: **Kinematics and stability of the mixing flow patterns found in champagne glasses as determined by laser tomography techniques: Likely impact on champagne tasting.** *Anal Chim Acta* 2008, **621**:30–37.
20. Liger-Belair G, Bourget M, Pron H, Polidori G, Cilindre C: **Monitoring gaseous CO_2 and ethanol above champagne glasses: Flute versus coupe, and the role of temperature.** *PLoS ONE* 2012, **7**:e30628.
21. Pangborn RM, Berg HW, Hansen B: **The influence of color on discrimination of sweetness in dry table-wine.** *Am J Psychol* 1963, **76**:492–495.
22. Morrot G, Brochet F, Dubourdieu D: **The color of odors.** *Brain Inj* 2001, **79**:309–320.
23. Spence C: **The colour of wine – Part 1.** *World of Fine Wine* 2010, **28**:122–129.
24. Spence C: **The colour of wine – Part 2.** *World of Fine Wine* 2010, **29**:112–119.
25. Spence C, Harrar V, Piqueras-Fiszman B: **Assessing the impact of the tableware and other contextual variables on multisensory flavour perception.** *Flavour* 2012, **1**:7.
26. Smith BC: **The objectivity of Tastes and Tasting.** In *Questions of Taste: The Philosophy of Wine*. USA: Oxford University Press; 2007:41–78.

Permissions

List of Contributors

René A de Wijk
Top Institute Food and Nutrition, Wageningen, The NetherlandsFood & Biobased Research, Consumer Science & Intelligent Systems, Wageningen, The Netherlands

Ilse A Polet
Top Institute Food and Nutrition, Wageningen, The Netherlands
Food & Biobased Research, Consumer Science & Intelligent Systems, Wageningen, The Netherlands

Wilbert Boek
Hospital 'De Gelderse Vallei', Ede, The Netherlands

Saskia Coenraad
Hospital 'De Gelderse Vallei', Ede, The Netherlands

Johannes HF Bult
Top Institute Food and Nutrition, Wageningen, The Netherlands
Nizo Food Research, Ede, The Netherlands

Joshua Evans
Nordic Food Lab, Department of Food Science, University of Copenhagen, Rolighedsvej 30, DK-1958 Frederiksberg C, Denmark

Roberto Flore
Nordic Food Lab, Department of Food Science, University of Copenhagen, Rolighedsvej 30, DK-1958 Frederiksberg C, Denmark

Jonas Astrup Pedersen
Nordic Food Lab, Department of Food Science, University of Copenhagen, Rolighedsvej 30, DK-1958 Frederiksberg C, Denmark

Michael Bom Frøst
Nordic Food Lab, Department of Food Science, University of Copenhagen, Rolighedsvej 30, DK-1958 Frederiksberg C, Denmark
Sensory Science Group, Department of Food Science, University of Copenhagen, Rolighedsvej 30, DK-1958 Frederiksberg C, Denmark

Georges M Halpern
The Hong Kong Polytechnic University, Hung Hom, Hong Kong SAR

Charles Spence
Crossmodal Research Laboratory, Department of Experimental Psychology, University of Oxford, South Parks Road, OX1 3UD Oxford, England

Liana Richards
London Symphony Orchestra, Barbican Center, Silk Street, EC2Y 8DS London, England

Emma Kjellin
London Symphony Orchestra, Barbican Center, Silk Street, EC2Y 8DS London, England

Anna-Maria Huhnt
London Symphony Orchestra, Barbican Center, Silk Street, EC2Y 8DS London, England

Victoria Daskal
The Antique Wine Company (AWC), London, England

Alexandra Scheybeler
The Antique Wine Company (AWC), London, England

Carlos Velasco
Crossmodal Research Laboratory, Department of Experimental Psychology, University of Oxford, South Parks Road, OX1 3UD Oxford, England

Ophelia Deroy
Center for the Study of the Senses, School of Advanced Study, University of London, London, UK

Ophelia Deroy
Centre for the Study of the Senses, University of London, London, UK

Charles Spence
Crossmodal Research Laboratory, Department of Experimental Psychology, Oxford University, Oxford, UK

Ole G Mouritsen
MEMPHYS, Center for Biomembrane Physics, Department of Physics, Chemistry and Pharmacy, University of Southern Denmark, Campusvej 55, DK-5230 Odense M, Denmark
Nordic Food Lab, 93 Strandgade, DK-1401 Copenhagen K, Denmark

Gary K Beauchamp
Monell Chemical Senses Center, 3500 Market Street, Philadelphia, PA 19104, USA

Peihua Jiang
Monell Chemical Senses Center, 3500 Market Street, Philadelphia, PA 19104, USA

Xiaoang Wan
Department of Psychology, School of Social Sciences, Tsinghua University, Qinghua Yuan, Beijing 100084, China Crossmodal Research Laboratory, Department of Experimental Psychology, University of Oxford, South Parks Road, OX1 3UD, Oxford, UK

Carlos Velasco
Crossmodal Research Laboratory, Department of Experimental Psychology, University of Oxford, South Parks Road, OX1 3UD, Oxford, UK

Charles Michel
Crossmodal Research Laboratory, Department of Experimental Psychology, University of Oxford, South Parks Road, OX1 3UD, Oxford, UK

Bingbing Mu
Department of Psychology, School of Social Sciences, Tsinghua University, Qinghua Yuan, Beijing 100084, China

Andy T Woods
Xperiment, Lausanne, Switzerland

Charles Spence
Crossmodal Research Laboratory, Department of Experimental Psychology, University of Oxford, South Parks Road, OX1 3UD, Oxford, UK

Morten L Kringelbach
Department of Psychiatry, University of Oxford, Warneford Hospital, Oxford OX3 7JX, England
Center of Functionally Integrative Neuroscience, Aarhus University, Aarhus, Denmark

Tony Conigliaro
Drink Factory, 35 Britannia Row, London N1 8QH, UK
69 Colebrooke Row, London N1 8AA, UK

Ole G Mouritsen
MEMPHYS, Center for Biomembrane Physics, Department of Physics, Chemistry, and Pharmacy, University of Southern Denmark, Campusvej 55, DK-5230 Odense M, Denmark
Nordic Food Lab, 93 Strandgade, DK-1401 Copenhagen K, Denmark

Lars Williams
Nordic Food Lab, 93 Strandgade, DK-1401 Copenhagen K, Denmark
Restaurant Noma, 93 Strandgade, DK-1401 Copenhagen K, Denmark

Rasmus Bjerregaard
Blue Food ApS, 2 Nordre Kaj, DK-8700 Horsens, Denmark

Lars Duelund
MEMPHYS, Center for Biomembrane Physics, Department of Physics, Chemistry, and Pharmacy, University of Southern Denmark, Campusvej 55, DK-5230 Odense M, Denmark

Charles Spence
Crossmodal Research Laboratory, Department of Experimental Psychology, University of Oxford, 9 South Parks Road, Oxford OX1 3UD, UK

Keri McCrickerd
School of Psychology, Pevensey Building, University of Sussex, Brighton BN1 9QH, UK

Lucy Chambers
School of Psychology, Pevensey Building, University of Sussex, Brighton BN1 9QH, UK

Jeffrey M Brunstrom
Nutrition and Behaviour Unit, School of Experimental Psychology, University of Bristol, Priory Road, Bristol BS8 1TU, UK

Martin R Yeomans
School of Psychology, Pevensey Building, University of Sussex, Brighton BN1 9QH, UK

Lilli Mauer
Department of Nutritional Sciences, University of Toronto, Toronto, Canada

Ahmed El-Sohemy
Department of Nutritional Sciences, University of Toronto, Room 310, 150 College St, Toronto, ON M5S 3E2, Canada

Charles Michel
Crossmodal Research Laboratory, Department of Experimental Psychology, University of Oxford, South Parks Road, OX1 3UD, Oxford, UK

Carlos Velasco
Cross modal Research Laboratory, Department of Experimental Psychology, University of Oxford, South Parks Road, OX1 3UD, Oxford, UK

Elia Gatti
Crossmodal Research Laboratory, Department of Experimental Psychology, University of Oxford, South Parks Road, OX1 3UD, Oxford, UK

Charles Spence
Crossmodal Research Laboratory, Department of Experimental Psychology, University of Oxford, South Parks Road, OX1 3UD, Oxford, UK

Mita Lad
Food Sciences, Sutton Bonington Campus, The University of Nottingham, Loughborough LE12 5RD, UK
Riddet Institute, Massey University, Private Bag 11222, Palmerston North, New Zealand

Louise Hewson
Food Sciences, Sutton Bonington Campus, The University of Nottingham, Loughborough LE12 5RD, UK

Bettina Wolf
Food Sciences, Sutton Bonington Campus, The University of Nottingham, Loughborough LE12 5RD, UK

Elke Scholten
Food Physics Group, Department of Agrotechnology and Food Sciences, Wageningen University, Bomenweg 26703HDWageningen, the Netherlands

Miriam Peters
Food Physics Group, Department of Agrotechnology and Food Sciences, Wageningen University, Bomenweg 26703HDWageningen, the Netherlands
FrieslandCampina Cheese & Butter, Nieuwe Kanaal 7R6709PAWageningen, the Netherlands

Nicholas Eriksson
23andMe, Inc., Mountain View, CA, 94043, USA

Shirley Wu
23andMe, Inc., Mountain View, CA, 94043, USA

Chuong B Do
23andMe, Inc., Mountain View, CA, 94043, USA

Amy K Kiefer
23andMe, Inc., Mountain View, CA, 94043, USA

Joyce Y Tung
23andMe, Inc., Mountain View, CA, 94043, USA

Joanna L Mountain
23andMe, Inc., Mountain View, CA, 94043, USA

David A Hinds
23andMe, Inc., Mountain View, CA, 94043, USA

Uta Francke
23andMe, Inc., Mountain View, CA, 94043, USA

Peter C Stewart
Psychology Program, Grenfell Campus, Memorial University of Newfoundland, University Avenue, Corner Brook, NL A2H 6P9, Canada

Erica Goss
Psychology Program, Grenfell Campus, Memorial University of Newfoundland, University Avenue, Corner Brook, NL A2H 6P9, Canada

Per Møller
Department of Food Science, University of Copenhagen, Rolighedsvej 30, 1958 Frederiksberg, Denmark

Betina Piqueras-Fiszman
Department of Experimental Psychology, University of Oxford, South Parks Road, Oxford OX1 3UD, UK

Agnes Giboreau
Institut Paul Bocuse, Château du Vivier, BP 25- 69131, Écully Cedex, France

Charles Spence
Department of Experimental Psychology, University of Oxford, South Parks Road, Oxford OX1 3UD, UK

Charles Spence
Department of Experimental Psychology, Crossmodal Research Laboratory, University of Oxford, South Parks Road, Oxford OX1 3UD, UK

Mary Kim Ngo
Department of Experimental Psychology, Crossmodal Research Laboratory, University of Oxford, South Parks Road, Oxford OX1 3UD, UK

Hervé This
INRA, UMR 1145, Group of Molecular Gastronomy, 16 Rue Claude Bernard, Paris 75005, France
AgroParisTech, UFR de Chimie Analytique, 16 Rue Claude Bernard, Paris 75005, France

Susanne Højlund
Department of Anthropology, Aarhus University, Moesgaard Alle 2, Højbjerg, DK-8270 Aarhus, Denmark

Vanessa Harrar
Crossmodal Research Laboratory, Department of Experimental Psychology, University of Oxford, South Parks Road, Oxford OX1 3UD, UK

Barry Smith
Centre for the Study of the Senses, University of London, Senate House, Malet Street, London WC1E 7HU, UK

Ophelia Deroy
Centre for the Study of the Senses, University of London, Senate House, Malet Street, London WC1E 7HU, UK

Charles Spence
Crossmodal Research Laboratory, Department of Experimental Psychology, University of Oxford, South Parks Road, Oxford OX1 3UD, UK

www.ingramcontent.com/pod-product-compliance
Lightning Source LLC
Chambersburg PA
CBHW080255230326
41458CB00097B/4997